Mars

THE MYSTERY UNFOLDS

Peter John Cattermole

OXFORD
University Press
2001

Oxford University Press

Oxford New York Athens Auckland Bangkok
Bogotá Buenos Aires Calcutta Cape Town Chennai
Dar es Salaam Delhi Florence Hong Kong Istanbul
Karachi Kuala Lumpur Madrid Melbourne
Mexico City Mumbai Nairobi Paris São Paolo
Shanghai Singapore Taipei Tokyo Toronto Warsaw

and associated companies in
Berlin Ibadan

Published in the United States of America by
Oxford University Press, Inc.
198 Madison Avenue, New York, N.Y. 10016
http://www.oup-usa.org
Oxford is a registered trademark of Oxford University Press

First published in 2001 by
Terra Publishing, PO Box 315, Harpenden, AL5 2ZD, England
E-mail: Publishing@rjpc.demon.co.uk
Website: http://www.rjpc.demon.co.uk

Library of Congress Cataloging-in-Publication Data
Cattermole, Peter John
Mars: the mystery unfolds/Peter Cattermole
p. cm.
Includes bibliographical references and index
ISBN 0-19-521726-8
1. Mars (Planet) I. Title.
QB641.C369 2001
523.43—dc21 00-047866

Printed and bound by Butler & Tanner Limited,
Frome, England

Contents

Preface

It is almost a decade since I wrote *Mars – the story of the red planet*. Although a large data archive existed – mainly derived from the Viking and Mariner 9 missions – and a fairly detailed picture of Mars' geological development could be painted, major gaps remained in our knowledge. At that time, for instance, little was known of the planet's volatile history or of the short-term changes in climate and weather. The past existence of large bodies of standing water on its surface had been considered, but evidence was disputed. Our knowledge of the planet's topography, gravity and surface composition was still rather sketchy, and the question of past life remained unanswered.

In the meantime, the mystery has definitely been unfolding. The old data archive has been turned over and over, and searched for new clues about the planet's early development. The Mars Surface and Atmosphere Through Time programme (MSATT) has greatly enhanced our understanding of Mars' volatile history and given widespread acknowledgement of the past existence of palaeolakes and oceans on the planet. New Earth-based spectroscopic and infrared data, laboratory experiments, intensive analysis of putative Martian meteorites, and observations by the amazing Hubble space telescope (HST), have all contributed to a changed perception of what makes the red planet tick.

HST has provided invaluable information about conditions on Mars and how they change. This paved the way for two highly successful spacecraft, Mars Pathfinder and Mars Global Surveyor. The former has provided an immense amount of visual, geochemical and physical data concerning the rocks, landscape and weather at the mouth of the Ares Vallis channel system. For the first time in nearly 30 years, there was new information about Martian materials that had been sampled on site. There were several surprises.

As if not to be upstaged, Mars Global Surveyor, while aerobraking on its approach to Mars in 1997, confirmed a very weak magnetic field. It then provided (and is still providing) the planetary science community with a vast new data archive that puts us in the enviable position of having more and better images of Mars than we yet have of much of Earth.

It is these recent missions, including the HST, that have thrown open the floodgates, allowing new ideas about the red planet to come pouring in. In this book I have tried to present a new perspective that, to my mind, makes Mars even more exciting than before. Its attraction to Earthlings has apparently not waned one jot since, in the early days of the twentieth century, Percival Lowell believed he had observed canals constructed by live Martians. Although we know these never existed, there is very real evidence that Mars once did have rivers of flowing water and active volcanoes of immense size, and experienced periods of extensive flooding. There may also have been shallow lakes, huge glaciers and primitive forms of life. Mars must have changed a great deal since the time when they may have been ripe for the development of primitive life, and liquid water was able to pond on the surface. May the story unfold.

Acknowledgements

No one author can have a total grasp of all aspects of Mars science, so this book is no more than my own interpretation of the research and data of many planetary scientists. I wish to acknowledge and thank all those whose work is referred to and discussed herein. Should there be any errors or misunderstandings in my interpretation of their work, they are of course my responsibility.

Figure 4.1, 9.1, 9.9, 12.10, 12.14, 15.1a, 15.2, 15.4 and 15.7, Plates 3, 10 and 11, and Tables 4.1 and 14.1 were first published in the *Journal of Geophysical Research* (Planets), copyright American Geophysical Union. Figures 3.3, 4.3, 4.6, 12.3, 12.9, 13.3, 13.4, 15.1a, 15.5 and 15.6 are copyright of the journal *Science*. Both publications kindly granted permission for reproduction. Authors' details are shown in the relevant caption and in the bibliography.

The superb imagery obtained by a variety of NASA spacecraft features prominently in this book. I wish specifically to acknowledge NASA/JPL/Malin Space Science Systems for permission to utilize the Mars Global Surveyor and Pathfinder images shown herein. The Viking and Mariner images were kindly provided by NSSDC, Greenbelt, Maryland. Dr Baerbel Lucchitta kindly supplied hard copy and granted permission to reproduce Plates 10 and 11, as did Dr Kenneth Tanaka for Plate 3; both are based at USGS, Flagstaff, Arizona. The late Paul Doherty prepared the excellent artwork for my earlier book, *Mars – the story of the red planet*, some of which is used again here.

Finally I wish to thank Roger Jones of Terra Publishing, for bearing with my idiosyncrasies and for upholding excellent standards in the production of the book.

Peter John Cattermole Sheffield February 2001

Modern water found on Mars?

In June 2000, rumours arose that an article to be published in *Science* magazine by Michael Malin and Kenneth Edgett was to present high-resolution images that pointed to the existence of modern water on Mars. To stem media fantasy, NASA quickly organized a press conference to present facts rather than fiction.

The facts are that study by Malin & Edgett (2000) of 150 high-resolution Mars Global Surveyor images returned since the beginning of January 2000 reveals 120 locations where fluid seepage and surface runoff appear to have occurred. The features lack decay or impact features, indicating that they must be very young. If water-formed, this fluid must have existed at the surface quite recently.

The specific landform grouping comprises a head alcove, associated channels and a depositional apron. On Earth, such a geomorphological package is consistent with fluid-mobilized mass movement. All occurrences are found on steep slopes, at latitudes above 30°, mainly in the southern hemisphere, where they are associated with the walls of several large impact craters, the channels Dao and Nirgal Valles, several graben and south polar pits. They are not found among ancient valley networks – formed at a time when Mars' climate and crust may have been warmer – nor in volcanic areas that could have circulated hydrothermal energy. Nor are they found in equatorial regions, but that is less surprising because protracted desiccation in these zones would have removed any volatiles at shallow depths. At first sight, their discovery appears to dispel the notion that the crust in high latitudes is deeply frozen.

However, there is another twist: just under 50 per cent are found on south-facing slopes; a mere 20 per cent face north. Since 90 per cent lie south of the equator, this means most are pole facing and receive less insolation than those facing the equator; they appear to be the least likely sites for the release of subsurface volatiles. Malin & Edgett ingeniously get around this problem by suggesting that groundwater moving towards the surface became trapped behind an ice plug that was breached only when groundwater pressure was sufficient to burst it, escape and cut the gullies.

Stephen Clifford (LPI) points out that it is currently impossible to create a near-surface aquifer on Mars, because it is simply too cold; yet the seepage appears to have emerged from layers at depths of a mere 150 m. Both Clifford and Michael Carr (USGS), suggest that if clathrates (that is ices of water and another component, e.g. CO_2) were present rather than pure water, then the freezing point would be much lower and the cold problem lessened. Other salts might also be involved and would act in a similar way.

Clifford, Victor Baker (U. of Arizona) and Kenneth Tanaka (USGS; Tanaka 2000) argue that, although it is generally held that Mars was warmer and wetter only very early in its history, this tenet now needs re-examination. For instance, it is known that precession could have increased Mars' obliquity to as much as 45° around 5 million years ago. This would have enhanced solar radiation in the polar regions and on pole-facing slopes, and caused a general warming by sending part of the ice locked inside the south polar cap into the atmosphere, increasing the greenhouse effect. Perhaps such warming melted subsurface ice?

Researchers are excited, whether water, clathrates or salts are involved. It calls into question the belief that Mars has been cold and inactive since early times.

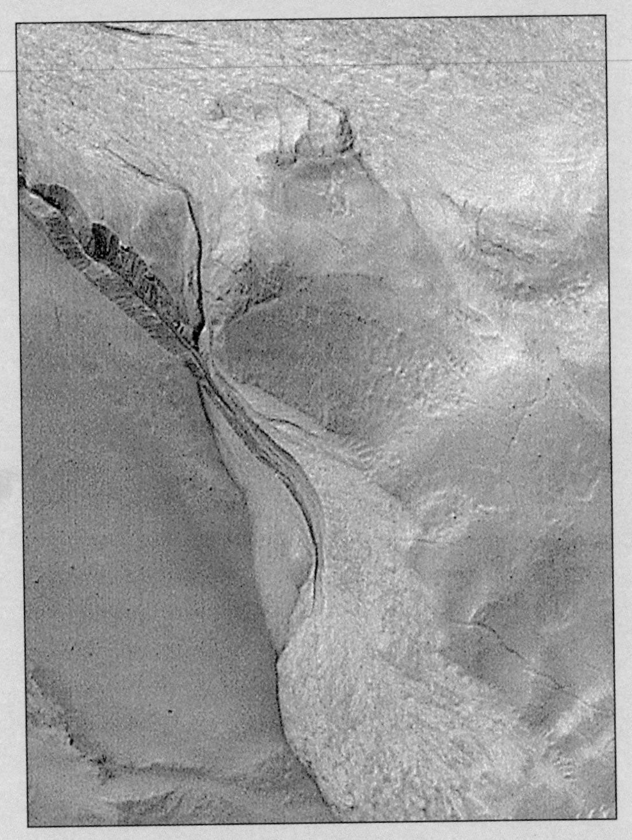

Youthful channels and aprons in East Gorgonum crater. MGS image PIA 01038; centred at 37.4°S 168.0°W.

1 Mars in the Solar System

Mars has occupied a very special place in the history of scientific thinking since the late sixteenth century. It was then that the Danish astronomer, Tycho Brahe, made careful observations of the movements of Mars, which allowed the brilliant German mathematician, Johannes Kepler, to formulate his three fundamental laws of planetary motion. The first two of these were published in 1609 after Kepler realized, having calculated the distance of Mars from the Sun at different points in its orbit, that its path was not a circle but an ellipse, as hitherto had been supposed. One focus of the ellipse was occupied by the Sun. He appreciated that this was also true of the other planets, a conclusion that led him to formulate the first of his three laws, which stated simply that planets move around the Sun in elliptical orbits, with the Sun at one focus. As it happens, Earth has an orbit that departs little from a circle (its eccentricity – a measure of its departure from circularity – is a mere 0.017, that of a perfect circle being zero), but Mars follows a much more eccentric orbit, varying from 207 million to 249 million kilometres from the Sun, and yielding an eccentricity value of 0.093.

Kepler's second law also derived from studies of Mars. He had observed that the red planet travels more quickly near perihelion than aphelion,[1] and his second law states that a straight line joining the Sun and each planet (known as the radius vector) sweeps out an equal area in equal time (Fig. 1.1). Thus, when a planet is near perihelion it travels more quickly than when near aphelion, and the planets nearer to the Sun travel more quickly than those at greater distances. For instance, Mars travels at a mean rate of about 24 km s^{-1}, compared with 30 km s^{-1} for Earth and 13 km s^{-1} for Jupiter.

The third law, published nine years later, states that the square of a planet's orbital period is proportional to the cube of its mean distance from the Sun. Taken together, laws two and three explain why Mars and other planets beyond the orbit of Earth show occasional retrograde motion against the star background. Figure 1.2 shows the apparent motion of Mars in the sky, together with the relative positions of Mars and Earth in their respective orbits. Between points 3 and 6 Earth chases Mars and eventually overtakes it, during which period Mars appears to move backwards among the constellations.

Every 780 days, Mars, Earth and Sun become aligned; Mars is then said to be at opposition (Fig. 1.3), and the planet is then well placed for observation from Earth. Because the orbits of Mars and Earth are elliptical, the distance between the two bodies is not the same at each opposition and, consequently, some oppositions are more favourable to Earth-based

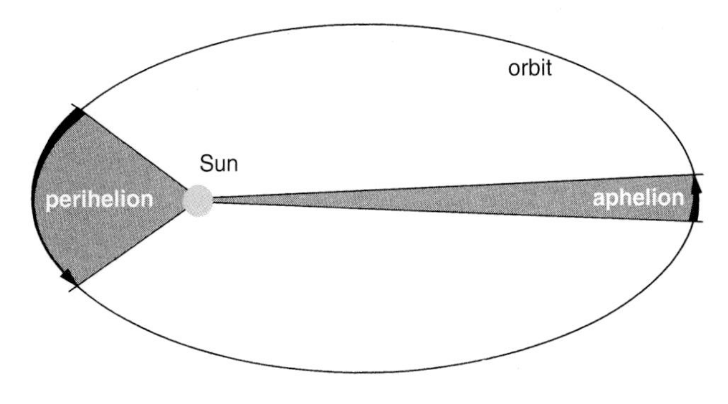

Figure 1.1 Kepler's second law. The radius vector defines an equal area whether Mars is near perihelion or aphelion; however, Mars moves quicker at perihelion than at aphelion.

1. Perihelion is the point in the orbit of a body at which it is nearest to the Sun, aphelion being the point farthest from the Sun.

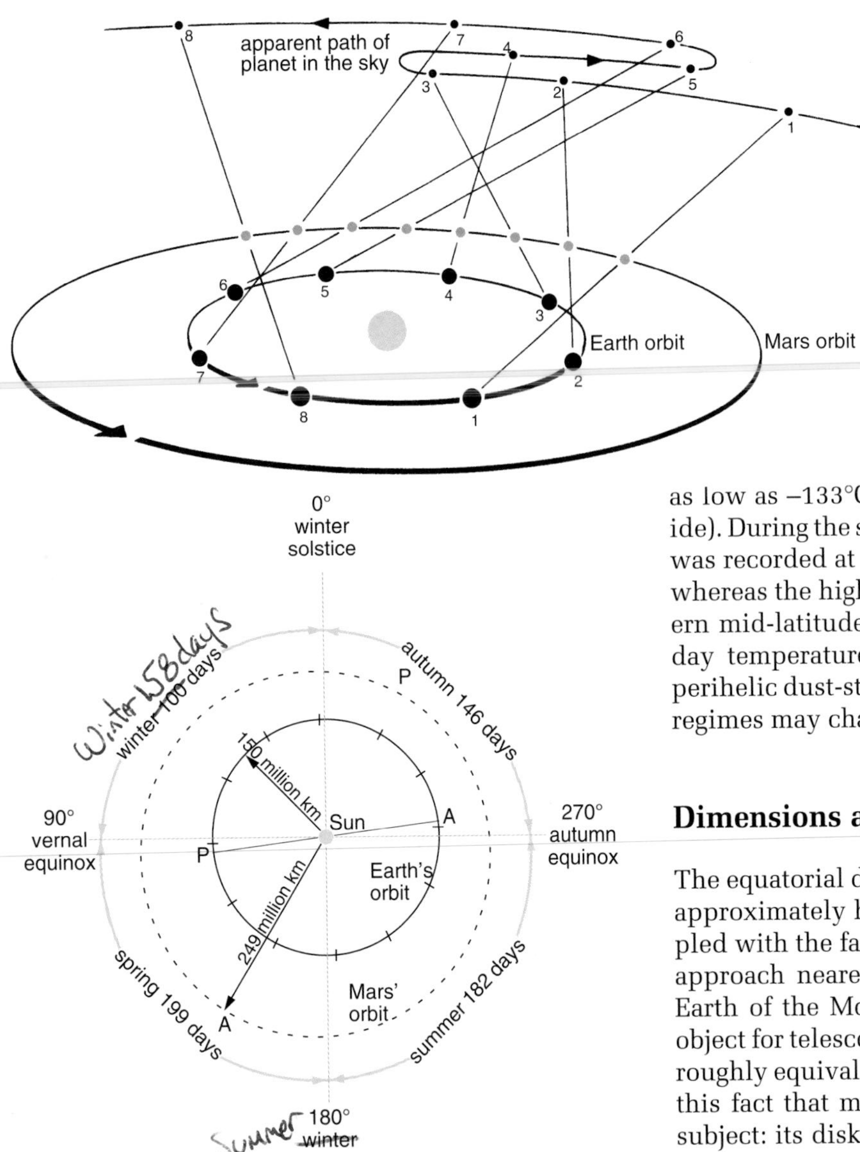

Figure 1.2 The apparent motion of Mars in the sky. When the two planets are close to opposition, Mars appears to move in a retrograde manner.

Figure 1.3 The movements of Mars and Earth around the Sun. A: aphelion; P: perihelion.

observers than others. Thus, if Mars is near perihelion, the distance may be as little as 55.7 million km, whereas at aphelion the distance increases to 101.3 million km. The interval between successive oppositions is known as the planet's synodic period, and this turns out to be longer than the Martian year, or sidereal period, which is 687 days.

Because Mars is farther from the Sun than Earth is, its mean surface temperature is much lower; furthermore, because of precession, the northern and southern hemispheres have different temperature regimes. The lowest current temperatures are at the south pole during its winter, where Viking infrared thermal mapper (IRTM) data showed temperatures may plummet to as low as –133°C (the freezing point of carbon dioxide). During the same mission, a temperature of –63°C was recorded at the north pole during their summer, whereas the highest temperatures occurred in southern mid-latitudes during their summer, where midday temperatures of +23°C were recorded. During perihelic dust-storm activity, temperature values and regimes may change significantly.

Dimensions and mass

The equatorial diameter of Mars is 6788 km, which is approximately half that of Earth. Its small size, coupled with the fact that even at its closest it can never approach nearer than 140 times the distance from Earth of the Moon, conspires to make it a difficult object for telescopic observation; it subtends an angle roughly equivalent to a large lunar impact crater. It is this fact that makes it such a tantalizing telescopic subject: its disk is just large enough for details to be seen, but never very clearly. Most of the more revealing series of Earth-based observations of the planet have been made at August or September oppositions, which is when Earth passes the perihelion of Mars' orbit; unfortunately, these are separated from one another by roughly 15 years. Even then there are difficulties for northern observers, for Mars has a far southerly declination (celestial latitude) at such times.

The mass of Mars is 644 000 billion billion kg, roughly one tenth that of Earth. It is also considerably less dense, the average density being 3906 kg m^{-3}, or about 70 per cent that of Earth. This implies that any dense metallic core that exists must be substantially smaller than Earth's. The consequently lower escape velocity of 5.1 km s^{-1} means that the atmosphere is very tenuous, exerting a surface pressure of only 8.1 mbar, a mere 1 per cent that of Earth. It is composed predominantly of carbon dioxide.

The Martian seasons

The Martian day is a trifle longer than Earth's (24 h 37 m 22 s), and the rotational axis is inclined at 25° to the plane of the orbit, just a little more than Earth's (23.5°); consequently, Mars experiences seasons much like our own, although longer. Because the eccentricity of Mars' orbit is considerably greater than Earth's, there are significant differences in the lengths of the respective seasons. As Kepler's second law predicts, the time varies for Mars to traverse any section of its orbit, time spent close to perihelion being much shorter than that near aphelion. Because the southern hemisphere is tilted towards the Sun near perihelion, spring and summer seasons in that hemisphere are respectively 52 and 25 terrestrial days shorter than autumn and winter (Table 1.1). The virtual circularity of Earth's orbit means that the eccentricity effect produces only a three-day difference.

The eccentricity also has an effect upon the maximum summer temperatures experienced by the different hemispheres; for instance, southern summers are currently shorter and hotter than northern ones, maximum temperatures in the south being about 30.1°C higher. This is because Mars is 20 per cent closer to the Sun at perihelion than at aphelion and it receives 45 per cent more incoming solar radiation.

Table 1.1 Martian and terrestrial seasons.

Hemisphere and season		Duration (days)		
Northern	Southern	Mars (Martian)	Mars (terrestrial)	Earth (terrestrial)
Spring	Autumn	194	199	92.9
Summer	Winter	178	183	93.6
Autumn	Spring	143	147	89.7
Winter	Summer	154	158	89.1
		669	687	365.3

The pattern of discovery

Galileo is held to have been the first person to observe Mars through a telescope, which he did in 1610. He could discern little on its disk but was able to confirm its phases, as he had done previously for Venus. His inadequate telescope and the small size of the planet's disk were the reasons for this. A few years later, in 1638, drawings made by the Neapolitan observer Francesco Fontana do show phase effects, but little else. It remained for Christiaan Huygens to produce the first really useful drawing of Mars, which he did on 28 November 1659; this shows a dark triangular region at the centre of the disk, which may well have represented Syrtis Major. It was christened the Hourglass Sea and, as was customary at this point in history, was believed to be an area of open water. Huygens was also able to show that the planet rotated on its axis in a period little different from Earth's day.

By 1666, the Italian, Giovanni Cassini, who subsequently became the first director of the Paris Observatory, had observed the icy polar caps; Huygens had done so by 1672. The Englishman, Robert Hooke, also made observations at this time. The periodic advance and retreat of the polar caps in response to the changing seasons was noted by several observers around this time. The next major step forwards was made by William Herschel, who in 1781 and 1784, determined, first, that the planet's obliquity was 25° and he rightly deduced that Mars must have seasons like those of Earth; secondly, by tracing the movement of features across the disk, he estimated the length of the day to be 24 h 39 min 22 s, close to the currently accepted figure. His perspicacity also led him to speculate that the icy polar caps must be relatively thin, and that the transient brightenings he observed at various points on the disk were clouds.

The particularly favourable opposition of 1830 presented the German pair of Wilhelm Beer and Johann von Mädler (using Beer's excellent Fraunhofer 950 mm refractor) with an unprecedented opportunity to draw the outlines of the light and dark regions in a way that, for the first time, strongly resembled modern telescopic charts. In 1837, these two observers noted a dark band surrounding the north polar cap, which they suggested might represent a wet region adjacent to the melting polar ice; a similar observation was made by Webb in 1856, for the southern cap. Subsequently, this "wave of darkening", as it was called by Gerard de Vaucouleurs, has been noted by many; it was, for instance, particularly noticeable at the favourable 1956 opposition. A popular view, held right up to the advent of high-resolution imagery from Mariner 9 and the Viking orbiters, was that it had something to do with the transference of moisture from the shrinking snowfields to the surrounding region. We now know that the effect has a more complex origin.

Since the mid-nineteenth century, Mars has been the most widely observed planet by amateur and professional astronomers alike. The particularly excellent opposition of 1864 spawned a fine series of observations by the Englishman, Willam Rutter Dawes. A skilful draughtsman, Dawes's drawings were superior to those of not only Beer and Mädler but also the later ones by Schiaparelli and Lowell. Important maps of the period were published by the English astronomer, Richard Proctor, and the Frenchman, Camille Flammarion; however, the most important was undoubtedly that of the Milan-based observer, Giovanni

Schiaparelli, which was published in 1878. This was influential in that it introduced a new system of nomenclature wherein permanent features were given classical or biblical names, a policy that continued until the era of spacecraft demanded a revised method. It also showed many linear markings – the famous Martian canals.

The term "canal" appears to have been first introduced during the 1860s, by Cardinal Pietro Angelo Secchi, to describe faint linear features he had observed on several occasions. Schiaparelli used the same term in 1877 for markings just visible during spells of particularly stable seeing[1] while using a 210 mm Merz refractor. During the more favourable opposition two years later, Schiaparelli was amazed to observe that several of the canals appeared to be double, an appearance that became known as gemination. Although Schiaparelli's observations attracted relatively little attention at first, as more reports of the features came in people began to take more notice. By the end of the century, the canals had become the focus of attention on Mars.

Although the authenticity of Schiaparelli's observations is not in doubt, he clearly had conceptual problems: he believed the canali to be attributable to some kind of geological process, yet could not conceive what that might be. And, whereas Percival Lowell (who in 1894 founded the famous Lowell Observatory in Flagstaff, Arizona) later was prepared to see them as the handiwork of intelligent Martians (Lowell 1906, 1909, 1910), at no point did the Italian concede this notion, although he wrote "I am very careful not to combat this suggestion, which contains nothing impossible". Lowell himself was assuredly the champion of Martian canals; he devoted much of his life to their observation, and his drawings showed about 500 of them.

Despite the immense momentum imparted by the work of Schiaparelli and Lowell to the "Canal Movement" around the turn of the century, some astronomers expressed considerable scepticism. The Greek astronomer, Eugenios Antoniadi, who incidentally spent most of his life in France, was particularly scathing in his criticism. Rightly, he pointed to the fact that, against all laws of perspective, Schiaparelli's canals were too straight to be realistic (Lowell avoided this criticism by projecting his drawings onto globes). Furthermore, because he was using much-improved telescopes, Antoniadi's observations, over a 26-year period, showed not continuous canals but lines of discontinuous blotches and streaks. He, like Evans & Maunder (1903), saw the canals as an artifice, a trick played when either poor seeing or too small an aperture caused discontinuous features to be joined up by the eye. A most stimulating account of the controversy can be found in the excellent book by Sheehan (1988).

It was to be many years before spacecraft imagery was returned to Earth and the matter of the canals was resolved once and for all. Although most of the irregular features drawn by Earth-based observers do correspond with albedo markings recorded on Mariner and Viking images, of the canals there is no sign; they have to be seen as illusory, as Antoniadi and others had unpopularly suggested. Intelligent Martians must, rather sadly, be relegated to the annals of fiction.

Albedo markings

The light and dark albedo markings that have been depicted on telescopic drawings and photographs for a century or more have maintained their general outline and position for at least this long. The fine maps by Schiaparelli and Antoniadi show these clearly and, in 1877, the former introduced a new scheme of annotation, which appended both biblical and mythical names; many of these names (or parts of them) persist to the present. In an effort to standardize and clarify nomenclature, a committee of the International Astronomical Union published a list of 128 "permanent" albedo features, together with an accompanying chart (Fig. 1.4).

Once detailed orbital mapping of the planet began in the early 1970s, it became clear that few of the albedo markings showed any correlation with topographical features. There were exceptions; for instance, one light area corresponds to the impact basin Hellas, and a prominent dark marking coincides with the equatorial canyon system, Coprates Chasma; however, there was generally little correspondence. The modern view sees the darker regions as being those relatively free of windblown dust-grade sediment, and the lighter areas as having a significant mantle of such material. On this basis it is not difficult to see how the ephemeral changes seen in the shape of areas such as Solis Lacus and Syrtis Major are related to perihelic dust-storm activity, when all or nearly all of the characteristic albedo markings may temporarily disappear. The general shape and appearance of the planet's telescopic markings has remained fairly constant over the past century, which appears to imply that the broad wind-circulation pattern must have remained constant over the period.

Although dust storms may periodically obliterate

1. A term used to describe the steadiness and/or transparency of the sky while observing; amateurs use a scale from 1 (extremely bad) to 10 (excellent).

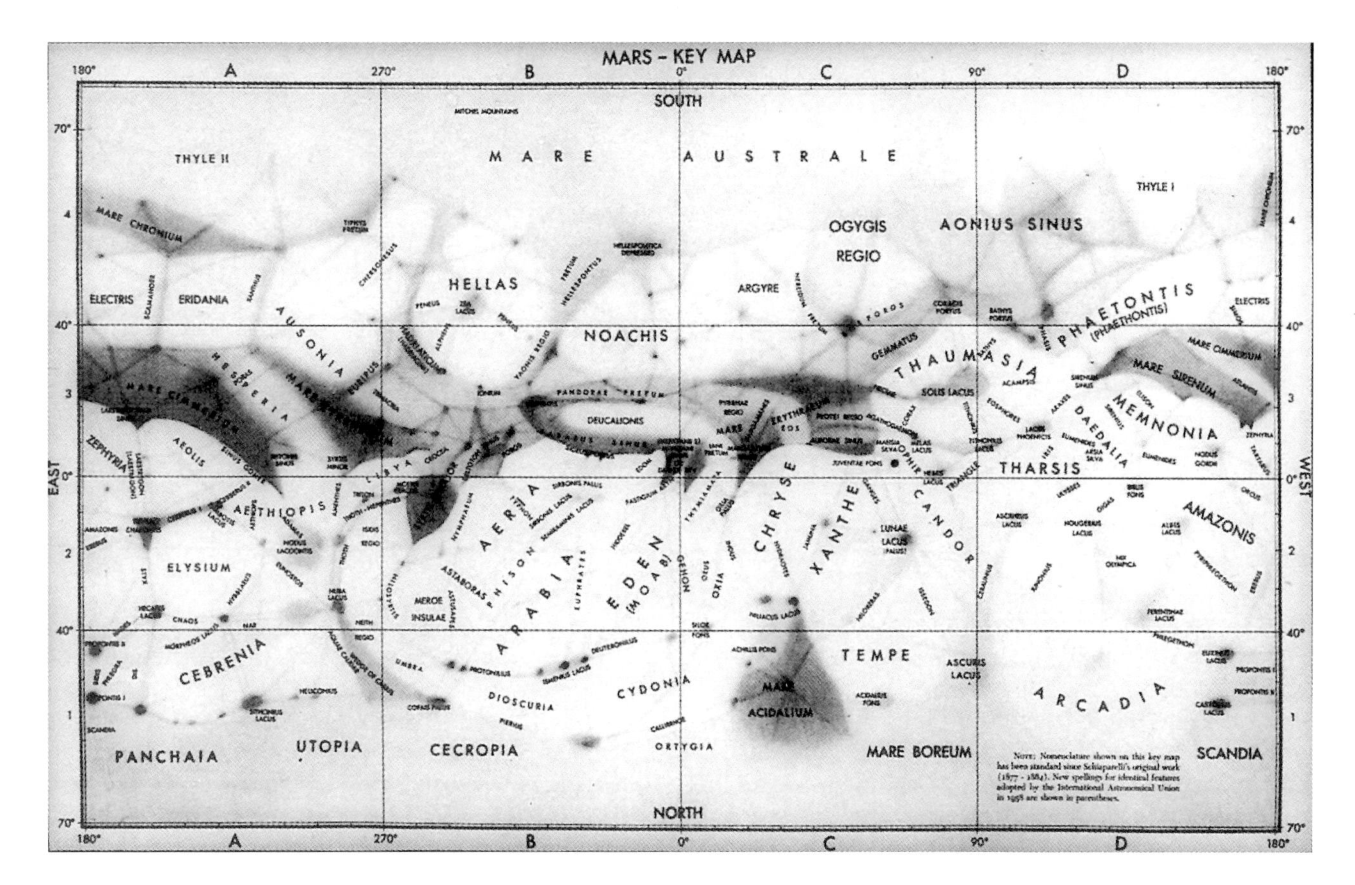

Figure 1.4 Chart of Mars, Lowell Observatory 1938, showing realized nomenclature of albedo markings.

the familiar markings, other changes are related to the formation of clouds in the polar regions and to the seasonal melting of the ice caps. A "hood" of polar clouds gradually extends over the north pole during its autumn and may eventually extend to within 40° of the equator; it remains hovering there through autumn and winter, obscuring the growing ice cap below. However, during mid-spring, the hood disappears and the northern cap may be seen shrinking at a rate of up to 20 km per day.

Other clouds are explained by condensates of carbon dioxide or water, or of dust. The former are known as white or blue clouds and they tend to focus on elevated regions such as Tharsis, Alba and Elysium; yellow clouds do not necessarily show any correlation with elevation and are generated by dust storms. More recently, exploration of Mars has proceeded apace. Since July 1965, several spacecraft have either flown past, orbited or landed on the planet. Indeed, one, Mars Pathfinder, set down a small roving vehicle that trawled across the surface taking photographs and analyzing rocks and dust. Our knowledge of atmospheric structure, weather, climate, magnetic field and geology of Mars has largely derived from this modern period. Indeed, now that the Hubble space telescope is functioning so well, we even have meteorological and climatic data for different seasons on Mars. The immense amount of data returned is still being studied by many scientists and, as is always the case with planetary exploration, the earlier missions provided the basis upon which later ones were planned. Several new spacecraft are well into the planning stage, while others are on their way to the red planet. There is, as ever, a keen interest in Mars, not least because we know that it once had liquid water at its surface, and therefore may have experienced climatic conditions appropriate for the development of carbon-based life. The archive of telescope and spacecraft images that now exists allows me to describe the fascinating history of this cold red world and to paint a fuller and appreciably different picture than I did almost a decade ago.

2 Exploration of the red planet

The very first spacecraft to reach the vicinity of Mars was the Soviet probe Mars 1, a fly-by mission that arrived on 1 November 1962 but sent back no useful data. It had been launched a mere 18 months after Yuri Gagarin's historic flight into space. Two years later, the US spacecraft Mariner 4 began its long journey to the red planet, carrying on board a variety of experiments to study the plasma and magnetic fields in the inner Solar System. During July 1965 it reached Mars, and its magnetometer recorded for the first time a weak magnetic field whose strength was only 0.03 per cent of that of Earth. On board was a radio occultation[1] experiment that allowed NASA scientists to estimate the surface atmospheric pressure of Mars to be 5 mbar. The probe also carried a single television camera which obtained 22 images along a strip of ground running from south of Thaumasia to western Amazonis. Unfortunately, the image contrast was disappointingly low. The images did reveal a high incidence of large impact craters, which gave little encouragement to those scientists who had hoped to see a greater diversity of landforms than had been found on the Moon. Surely after all this effort, Mars was not to be just another dead Moon-like world?

Almost as if NASA were willing Mars to be more forthcoming, it launched two further probes in 1969, Mariners 6 and 7. These had three times the instrument payload of their predecessor and they enjoyed a level of sophistication and computer flexibility not bestowed on Mariner 4. The carrying of both infrared and ultraviolet spectrometers was particularly significant, to take measurements of the Martian atmosphere. They confirmed that CO_2 was the predominant atmospheric gas, with nitrogen measured at less than 1 per cent. Solid CO_2 was also confirmed to be present at the poles.

Two television cameras were mounted on each instrument platform: a wide-angle camera, which had a ground resolution of about 2 km but covered an area 12 times larger than the camera aboard Mariner 4; and a narrow-angle camera that, although having a field of view only a tenth that of the wide-angle instrument, gave a tenfold improvement in resolution. The returned imagery largely reinforced the impression gained from the previous Mariner encounter: that Mars showed extensive cratered surfaces like those of the lunar highlands (Fig. 2.1). However, some images also afforded the first glimpses of blocky or "chaotic" terrain in the region of Margaritifer Sinus, of sinuous ridges, polygonal landforms and subparallel grooves in south polar latitudes, and of the very smooth floor of the Hellas basin. Contemporary reports of these findings can be found in Leighton et al. (1969) and Collins (1971).

Although Mariners 6 and 7 certainly gave hints that a greater diversity of landforms existed on Mars than had been revealed by Mariner 4, the images had neither the resolution nor the width of coverage to whip up any great excitement. However, an analysis of contemporary reports suggests that there was an underlying note of scepticism that belied considerable unease with the Mars imagery dataset at this time. How truly representative was it of Mars as a whole?

1. When one body passes in front of another, it is said to occult it. The event is an occultation. By accurately timing such events, by radio or radar, spacecraft can detect even a very tenuous atmosphere, as this will alter the predicted occultation time of two solid atmosphere-free bodies.

Figure 2.1 Mariner 6 image of Martian cratered plains, showing large impact craters and smoother intercrater areas. This image was obtained on 31 July 1969; centred at 340°W 15°S.

Mariner 9

The answer to this question was to be provided eventually by Mariner 9, one component of the next pair of US spacecraft (Mariners 8 and 9), whose objectives were the systematic mapping of about 70 per cent of Mars, the collection of information relating to the composition and structure of the atmosphere, temperature profiling and the measurement of topography. However, this was not until there had been some drama during the launch of the reconnaissance probe, Mariner 8, which on 9 May 1971 found itself abruptly and unceremoniously dumped in the Atlantic Ocean after the failure of the second-stage rocket, posing NASA scientists the unwelcome and taxing task of reassigning its companion craft within only three weeks. However, NASA rose to the occasion, and three weeks later Mariner 9 was successfully launched. It was to take 167 days to complete its journey, a distance of 400 million km. The main eyes of the probe were two television cameras: a wide-angle mapping camera of 50 mm focal length and with a potential resolution of 1 km from periapsis (lowest) altitude (the A-camera), and a narrow-angle camera of 500 mm focal length and potentially 100 m resolution (the B-camera). The mission team finally had settled on a 12-hour orbit inclined at 65° to the equator, which gave somewhat higher solar incidence angles than had originally been planned for it, but which was the best compromise after the loss of Mariner 8.

Also on board Mariner 9 was an ultraviolet spectrometer, to be used for characterizing the scattering properties of the tenuous Martian atmosphere, an infrared radiometer to measure temperature, and an infrared spectrometer to determine the composition of the atmosphere and its entrained dust. As the spacecraft went behind Mars and then reappeared after each revolution, the observed attenuation of the radio signals would be used to obtain temperature and pressure profiles and to establish surface elevations.

As the time of encounter approached, telescopic observations showed the all-too-familiar development of a perihelic dust storm of global proportions, something that had been predicted earlier by Dr Charles ("Chick") Capen of Lowell Observatory. Surely the craft had not gone all that way to have Mars clouded out? At last, on 10 November 1971, when Mariner was 80 000 km from Mars, the cameras were switched on. They revealed a virtually blank disk. The only recognizable feature was the bright south polar cap. However, it was for just such an eventuality that a high degree of flexibility was built into the probe's computer, so that last-minute changes of plan could be made. Thus, on receiving a signal from Mission Control, Mariner was able to sit out the dust storm, not wasting valuable energy in triggering camera shutters that would reveal a featureless Mars.

About two weeks after the spacecraft's arrival, the dust began to clear a little. In addition to the bright polar cap, several prominent dark spots became visible in the region known as Tharsis, one of them corresponding with the telescopic feature called Nix Olympica. As the atmosphere cleared, this was found to have a prominent complex depression at its summit. Could this be a volcanic caldera? Three similar spots lay to its south and east, lying along a roughly NE–SW line; these also had summit depressions. Since they all protruded above the dust clouds, it was clear that the spots were high mountains, and in fact we now know the group of three dark spots to be the massive shield volcanoes of Tharsis Montes, and the fourth is Olympus Mons.

Almost a month after the arrival of Mariner 9, the

decision was made to begin the systematic mapping of the planet's surface. Thereafter, each day the cameras generated two swaths of images 180° apart, each successive swath lying adjacent to the previous day's strip. By this method, planetary geologists were provided with a new terrain traverse each day. What surprises were in store . . .

During the first phase of the mapping cycle, when the region between 25° and 65°S was imaged, not only was the whole of the Hellas basin gradually revealed but also Argyre, a previously unsuspected 900 km-diameter basin situated over 2500 km to its west (as an albedo feature, Argyre was of course well known to observers). Then there were pictures of extensive fracture systems, on a scale hitherto found nowhere else within the Solar System, of plains units traversed by lunar-like wrinkle ridges, strange tributary networks and various flow-like features that appeared to have been carved by running water (Fig. 2.2). At long last Mars was beginning to reveal the secrets investigators had hoped it was hiding.

The second phase saw the mapping extended towards 25°N. This revealed not only the details of the Tharsis Montes shield volcanoes but also an immense canyon system that extended for at least 4000 km along the Martian equator and which was fittingly named Valles Marineris (Mariner Valleys) after the spacecraft that discovered it. The full extent of these equatorial canyons was revealed only as the mapping extended gradually eastwards, and it was also discovered that they ended in vast regions of blocky chaotic terrain of the type first revealed during the Mariner 6 encounter.

By the time the mapping had been accomplished, Mariner 9 had completed 698 orbits and had returned 7329 images to Earth. When the control-gas supply finally ran out on 27 October 1972, a signal was sent that switched off the probe and left the scientific community with the enormous but fascinating task of interpreting the new dataset and rewriting the history of Mars.

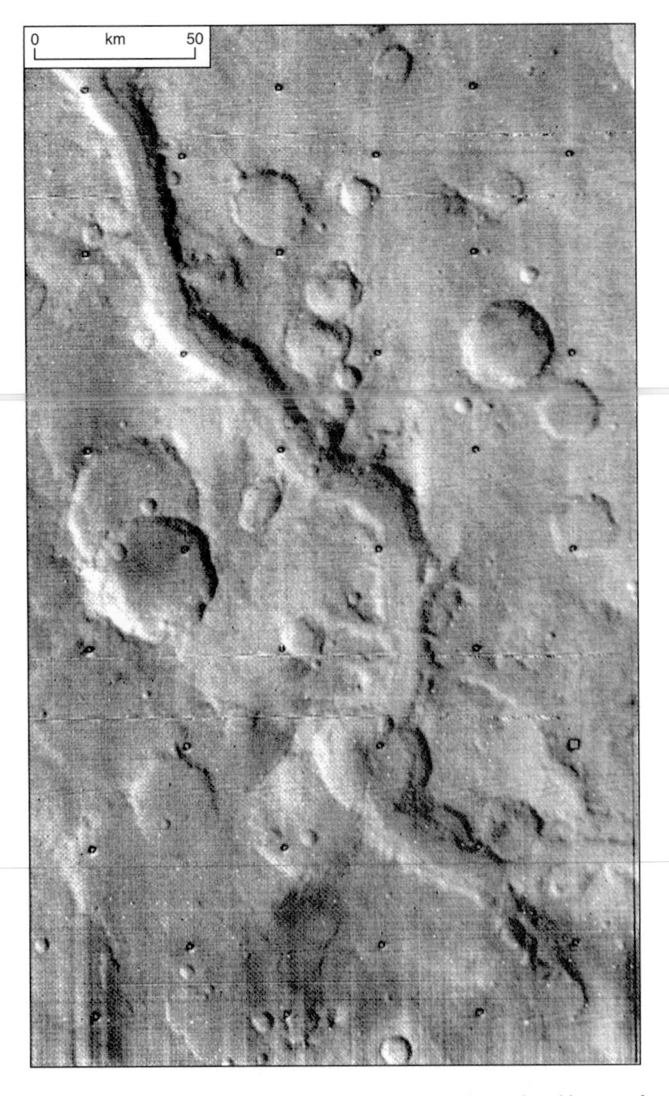

Figure 2.2 Mariner 9 image of ancient cratered terrain with part of a channel network. Centred at 183°W 22°S; Mariner 9 image 4167-18124.

Early Russian exploration of Mars

The exploration of Mars by Russia has never achieved the high degree of success typical of its Venus exploits, and bad luck has dogged them right up to the most recent launch, that of Mars 96 in 1996. Mars 2 and Mars 3 were launched on 19 May 1971 and 29 May 1971, respectively, contemporaneous with the US Mariner 9 mission. Although, as planned, both spacecraft reached Mars in late 1971 and a small lander capsule was successfully despatched onto the floor of Hellas, the latter sent back no data at all, and the two orbiters suffered much the same fate as Mariner 9 did when it first reached the planet, their cameras recording a virtually featureless disk. The difference here was that, since both probes were pre-programmed, it was not possible for them to sit out the storm while in Mars orbit, hence their singular lack of useful imagery.

However, the Russian probes did record the temperature at the Martian surface. It ranged from 13°C above zero to 93°C below, depending on latitude and time of day. At the northern polar cap the temperature dropped as low as –110°C. The Mars 3 radio telescope also recorded that the soil temperature at depths of 30–50 cm remains practically unchanged throughout the day. Presumably this is a reflection of the low thermal conductivity of the Martian soil.

Undeterred, in 1973 the Russians launched four more spacecraft: two orbiters and two landers. Of these, the two orbiters (Mars 4 and Mars 5) reached Mars on schedule, but the former failed and simply flew past Mars. However, Mars 5 was more successful and completed 20 orbits, during which it collected over 70 images similar in quality to those returned by Mariner 9. It also made measurements of atmospheric water vapour content and recorded Mars' weak magnetic field. Of the remaining probes, Mars 6 descended to the surface and is presumed to have made a soft landing, but regrettably the controllers lost contact just 0.3 s before it did so. During its descent it made useful measurements of the atmosphere. Mars 7 unfortunately missed Mars completely.

The detection of an unspecified inert gas in the Martian environment was a controversial spin-off from one of the engineering experiments associated with Mars 5. Because the only likely candidate was argon (later confirmed to be present at over 1 per cent in the Martian atmosphere), it seemed more than likely that the present atmosphere of Mars could only be a vestige of a much denser original primary atmosphere. This was a particularly interesting possibility in terms of the past climate of the planet, which, from the photogeological evidence for fluvial features, may once have been very different from that of today.

The Viking missions

Of all the missions sent to Mars, the Viking mission was undoubtedly one of the greatest successes, and the data returned by its spacecraft contributed the most to our current understanding of the Martian environment, well into the 1990s. Planning for this major project had already begun well before Mariner 9 was launched and each of Viking 1 and 2 was to consist of an orbiter and a lander. Since one of their prime objectives had been identified as the search for life on Mars, on board each lander was to be a gas-chromatograph mass spectrometer to detect any organic content in the soil, and a four-component biological experimental package. Additionally, there was to be an X-ray fluorescence analytical device, a seismometer and a set of meteorological instruments (Fig. 2.3).

The Viking orbiters were to have three main packages: a pair of high-resolution slow-scan vidicon[1] cameras, an infrared spectrometer to detect levels of atmospheric water vapour (MAWD experiment), and a series of infrared radiometers (IRTM experiment) to measure the thermal properties of the Martian atmosphere and surface. In addition, the spacecraft's radio was to be used to measure the planet's gravity field and atmospheric profiles and for topographical profiling. The cameras in particular were different from those flown on earlier missions and were arranged so that they could be shuttered alternately once every 4.48 s, thus providing a continuous swath of images, two frames wide. Each of the pair of cameras was identical, having a 475 mm focal-length telescope attached to a 37 mm diameter vidicon; individual images were subsequently constructed from around 1.25 million picture elements (pixels).

The mission controllers had arranged that each orbiter–lander pair be coupled on arrival in Mars orbit. In this way, the orbiting probe would check out the suitability of selected landing sites; if the original ones were found unsuitable, then others could be inspected before the final decision was made to send the lander craft down. This degree of flexibility was considered vital for the Viking mission to be a success. Furthermore, the orbiting probe would act as a relay for the lander below and conduct its own experiments.

Viking 1 reached Mars on 19 June 1976 and within three days the orbiter was relaying back to Earth pictures of the first of the proposed landing sites, in Chryse Planitia. The significantly better resolution of these images revealed that the site was somewhat more rugged than had been anticipated, and radar measurements suggested a degree of surface roughness that could best be explained by the presence of many boulders. Consequently, a landing was delayed while other sites were reconnoitred.

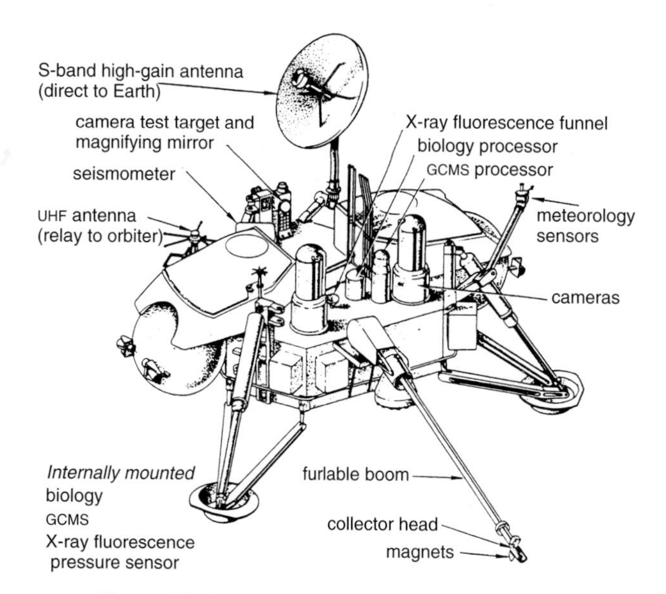

Figure 2.3 Instrumentation on board a Viking lander.

1. The vidicon camera tube has been the most widely used video imaging device for many years, taking over from the cathode-ray tube invented by Braun a hundred years ago. It uses thin-film semiconductor-type compounds as photoconductors.

It took three weeks to select the ideal site, also in Chryse Planitia, but farther west than the original, at 22.5°N 28.0°W. The Viking 1 lander made its historic descent and, landing on 20 July, provided the first ever close-up views of the surface (Fig. 2.4). On the reddish boulder-strewn surface were small dunes and, in the distance, several small impact craters. The boulders were dark and full of holes, rather like vesicular basalts; indeed, this is what they turned out to be.

Just over two weeks later, on 7 August, Viking 2 reached Mars. Eventually, on 3 September 1976, the Viking 2 lander also landed safely, this time in the region of Utopia Planitia, at 44°N 226°W. Here too, the original landing site was considered unsafe and a search had to be made for an alternative.

Both landers worked exceedingly well and pursued an intensive series of experiments, including the biological search that scientists eagerly anticipated. Disappointingly, both the gas-chromatograph and mass-spectrometer experiments showed a singular lack of organic molecules in the Martian soil, and the general feeling that emerged was that life does not currently exist on Mars, and possibly never did.

The other lander instruments worked well, too, and sent back a detailed record of changing local weather patterns, amounts of dust suspended in the air, diurnal temperature variations and the varying water-vapour content of the atmosphere. The cameras took many panoramic shots of the local region, from which details of the geology were learned. Mechanical arms allowed scoops to dig up samples of the local soil, which were analyzed on board the spacecraft by the XRF device and the results were returned to Earth.

Twenty-two samples were scooped up at the two Viking lander sites, several being large enough for XRF analysis. Naturally, the materials had to be relatively easily obtainable; consequently, the analyzed rocks were loose friable materials, known as fines. The samples from both sites contained abundant Si, together with significant amounts of Mg, Al, Ti, Ca and S (Table 2.1). However, the relative proportions of the components were unlike any known terrestrial rock, and the analyses are presumed to represent admixtures of different Martian materials. The gas chromatography experiment on board the landers detected about 1 wt per cent H_2O; this amount of water is probably bound into the samples collected.

The minerals present were largely silicates, with some oxides and carbonates. The grains themselves appear to have been cemented together with varying degrees of effectiveness, by a sulphate-rich cement, to produce what is termed a duricrust. Between 3 and 7 per cent of the material was magnetic and this suggests the presence of iron-rich minerals such as magnetite or maghemite, but some contribution from Fe–Ni meteoritic material cannot be ruled out. The consensus of opinion suggests that most of the soil is composed of iron-rich clays (such as nontronite), iron oxides, hydroxides and minor amounts of carbonate.

Table 2.1 The chemical composition of Martian surface materials

Major elements (wt%)	Viking 1	Viking 2
SiO_2	44.7	42.8
Al_2O_3	5.7	20.3
Fe_2O_3	18.2	1.0
TiO_2	0.9	5.0
MgO	8.3	<0.3
CaO	5.6	6.5
K_2O	<0.3	0.6
SO_3	7.7	
Cl	0.7	
Trace elements (ppm)		
Rb	<30	<30
Sr	60 ± 30	100 ± 40
Y	70 ± 30	50 ± 30
Zr	<30	30 ± 20

The orbiters, meanwhile, were mapping the entire surface of Mars at a resolution of 200 m; large regions were mapped at higher resolutions, even 10 m in some places. Although most images were monochrome, colour imagery was also obtained, as were several sets of stereoscopic frames. The mapping programme continued for almost two Martian years and, in addition to the visual imagery – which revealed not only the permanent geology but also the changing face of dust-storm activity and global weather patterns – the MAWD and IRTM packages collected data on the continually changing amount of water vapour and dust in the atmosphere and on thermal inertia values within the surface layer. The Martian gravity field had also been constrained by noting the effect of the gravity field upon the spacecraft's orbit. Finally, high-resolution pictures of both of Mars' satellites were also obtained.

The Viking 2 orbiter finally ran out of attitude-control gas in July 1978, and Viking 1 in August 1980; by that time over 55 000 images had been returned. Many of the images in this book have been drawn from the superb Viking dataset. For a full account of the Viking Mission, see Viking Science Team (1977).

The Russian Phobos mission

During the summer of 1988, the Russian Phobos 1 and 2 spacecraft started their ten-month journeys from Earth. Each craft had a threefold purpose: to study the activity of the Sun while en route to Mars, to study Mars itself, and, in particular, to study Mars' larger moon Phobos. This was the first time a mission had

Figure 2.4 Viking 1 lander site in Chryse, showing the blocky surface and fine-grain regolith.

the prime objective of studying one of the smaller satellites of the Solar System.

Unfortunately, early in September 1989, the ground controllers sent an incorrect signal to Phobos 1 and lost contact with it. However, Phobos 2 reached Mars on 29 January 1990, then went into a highly elliptical orbit above Mars' equator prior to entering a more nearly circular orbit just 350 km above the orbit of Phobos. While in the first, or "parking" orbit, Phobos 2 continued to observe Mars itself for three days; then, with the probe at a height of 860–1130 km, it collected the first nine television images, which it followed with two further picture-taking sessions, as the orbit was trimmed to a lower periapsis.

On 21 March, Phobos 2 entered a new orbit that brought it in as close as 120 km to the moon's surface, during which time it took further television pictures, and the mission controllers prepared to drop it as low as 35 km above Phobos on the side not facing Mars. Alas, bad luck hit again. On 27 March, with all systems apparently working perfectly, radio contact was lost and never regained. To this day it is not completely clear why this happened.

The mission, although prematurely terminated, was certainly not a failure. Important astronomical observations were made of the Sun, in particular of X-ray emissions from it, and included 140 very high-quality images of the solar corona, as well as plasma measurements. Phobos 2 also completed detailed studies of the Martian magnetosphere, which behaves rather differently from that of Earth. The very weak magnetic field surrounding Mars was found to be closely interactive with the interplanetary field, creating a natural pathway for the Sun's plasma to reach deep into the

Martian magnetosphere. Furthermore, it means that the solar wind interacts closely with ions in the Martian atmosphere, and Phobos 2 measured the flow of plasma in the solar wind and the rate at which ions of carbon dioxide and molecular and atomic oxygen were leaving Mars. Surprisingly, it was found that the atmosphere of the planet is escaping at the rate of 1–2 kg every second. This may seem insignificant, and indeed it would be for a planet such as Earth, but for Mars, which in any case has such a tenuous mantle of air, it means that it could lose its entire inventory of volatiles in much less than the lifetime of the Solar System. The extreme weakness of the magnetic field may therefore have played an important role in the gradual loss of water from Mars.

The images that Phobos 2 transmitted of Phobos were extremely interesting and added to the set made by Viking 1 and 2. Perhaps even more interesting were the infrared (IR) images that it returned of Mars itself. The Russian-made Thermoscan instrument that did this essentially consisted of a very sensitive infrared detector cooled by liquid nitrogen. Over a region of the Martian equator roughly 1500 km wide, it obtained exceedingly sharp thermal images with a resolution approaching 2 km. Their very high contrast brought out very clearly the differences in the degree of soil fragmentation from one area to the next and allowed the Russian scientists to relate this to regional geomorphological features.

Other instruments recorded the planet's spectrum at 128 different wavelengths and resolved peaks in the absorption spectrum that corresponded to different minerals. The same data should also allow scientists to establish how much volatile material is locked up in the lattices of minerals. Preliminary results suggest a relatively generous distribution of volatile-bearing rocks, presumably sedimentary in origin.

Before leaving this brief discussion, mention must made of the multi-spectral images of Phobos. The various instruments allowed for the collection of data between wavelengths of 0.32 and 3.2 μm, that is, from the ultraviolet into the thermal infrared. The measurements showed that Phobos is very dark indeed, reflecting at most 4 per cent of the incident light, the amount remaining roughly constant for the entire wavelength range. This is a property Phobos shares with certain kinds of carbonaceous chondrite, although the latter contains somewhat less water than the small moon. It is a great pity that the various analytical experiments could not be completed, since these would have provided detailed information on the chemistry of this chunk of asteroidal material and would have enhanced our knowledge of the primitive stuff from which all of the planets were made.

Mars Observer mission

The Mars Observer spacecraft was launched on 25 September 1992 from Cape Canaveral in Florida. It was to have been a global scientific mapping mission that operated over a single Martian year (687 Earth days). Regrettably, on 22 August 1993, after it had completed eleven months of its journey and was within three days of entering orbit around the planet, it fell silent. The loss of the signal coincided with the beginning of a sequence of commands to pressurize the fuel tank prior to making a 30-minute burn to place it in Mars orbit. This was the last that was heard of this unfortunate craft and it put paid to many years of preparatory work by many of planetary scientists; it was a huge disappointment.

The Mars Pathfinder mission

During the autumn of 1996, the United States and Russia launched three missions to Mars, heralding a new phase in the exploration of the red planet. The plans were to study the planet in greater detail than had ever been done before. A new series of Mars missions such as this is one of the most exciting things to burst upon the planetary science scene. The first of the three, the Russian Mars 96, was ambitious in that it involved the delivery to Mars of two small descent stations and two surface penetrators, developed to study the physical and chemical properties of the atmosphere, surface and subsurface. The second, the American Mars Global Surveyor (MGS) was a mapping mission whose orbiter carried a magnetometer, infrared spectrometer, wide- and narrow-angle cameras, and a laser altimeter. It was the first of a series of similar spacecraft to be launched once every 26 months until the year 2005 and culminating in a sample and return mission. The third mission, the American Mars Pathfinder (MPF), was a stand-alone mission that was first approved in 1992 and was the first to carry a roving vehicle with it. It consisted of a descent probe (lander) that landed on the Martian surface and deployed a small roving vehicle (rover) capable of conducting X-ray analysis of surface rocks and soils. This new wave of probes broke new ground in Mars exploration as they carried multi-spectral (MSS) imaging systems, of great value to geologists seeking to establish surface geochemical and mineralogical properties.

All three left Earth successfully. The Proton rocket carrying the first (Mars 96) rose above the Baikonur Cosmodrome at 23.49 h Moscow time on 16 November 1996; the second left Cape Canaveral at 12.00 noon eastern standard time on 7 November 1996, and the third set off on 4 December 1996, at 01.58 h from the same location. Only the two US launches were successful. Regrettably, the Russian rocket booster failed to fire and send the spacecraft out of Earth orbit, and it fell back to Earth, fortunately without disastrous consequences. Mars Pathfinder landed on Mars on 4 July 1997 and enjoyed huge success. Mars Global Surveyor is orbiting the planet, part way through completing a 687-day cycle of both global and high-resolution mapping.

Results from the second mission to be launched, but first to arrive at Mars, exceeded even the wildest dreams of mission scientists. The primary Pathfinder mission plans were for the lander to operate over 30 sols and the rover over 7 sols (one sol = 24.6 h). In the event, it operated three times longer than this, returning over 16 000 images from the lander and 550 from the rover, as well as more than 215 chemical analyses and extensive meteorological data. The probes ceased to operate in October 1997.

One of the most innovative aspects of the mission was the descent and landing manoeuvre. The probe was decelerated by a parachute as it entered the thin Martian atmosphere and, about 8 s before impacting the surface, deployed several large airbags that cushioned the lander. This then bounced across the Martian surface 16 times, covering a distance of 0.8 km, and eventually came to rest, intact, in the preselected site of Ares Vallis. The petals of the lander then opened, deploying a ramp, down which the roving vehicle eventually rolled (I say eventually, since there was an initial glitch when one of the airbags had blocked the ramp and had to be removed before the rover could reach the surface).

The Pathfinder lander weighed 360 kg on landing, measuring 2.75 m across when folded flat and had a 1.5 m-high mast with three solar panels. The masthead carried the mission's camera (known as IMP – Imager for Mars Pathfinder) with two optical paths used for stereo imaging, each being equipped with a filter wheel giving 12 colour bands in the 0.35–1.1 μm range. The camera's field-of-view was 14°, both vertically and horizontally, and it could take one frame (256×256 pixels) every two seconds (Fig. 2.5).

The ASI/MET (atmospheric structure instrument/ meteorology package) was mounted in the lander and acquired atmospheric information during descent and upon landing; it combined accelerometer and MET instruments (Seiff et al. 1997). The accelerometer contained science and engineering accelerometers that each monitored accelerations along three orthogonal axes; the MET instrument consisted of pressure, temperature and wind sensors. The pressure measurements had a maximum sensitivity a hundred times better than that available to the Viking landers. All the MET temperature and wind sensors were mounted on a 1.1 m-high mast. Atmospheric temperature was measured by four thermocouples; one designed to measure temperature during parachute descent and the other three designed for temperature measurements 25, 50, and 100 cm above the base of the mast. Wind was measured by a six-segment hotwire sensor at the mast top.

The accelerometer and MET instruments recorded data continuously while the probe descended through the atmosphere until about a minute after impact. Regular surface pressure, temperature and wind measurements began about four hours after impact at 07:00 local solar time (LST) on sol 1 (1 sol = 1 Martian day = 24.6 hours), and the MET mast was deployed at 13:30 LST.

The Pathfinder rover, known as Sojourner, was named after the African American woman crusader, Sojourner Truth, who lived during the American Civil War. It weighed 15.5 kg at launch and 10.6 kg when on the surface. It had the capacity to travel at $1\ cm\ s^{-1}$ and was 65 cm long and 30 cm high. Sojourner was equipped with three cameras – a forward stereo system and a rear colour-imaging system – and had a $0.2\ m^2$ solar array able to power it up for several hours per day. It travelled on a unique rocker-bogie system, whose six wheels provided greater stability and obstacle-crossing capabilities than four wheels only, allowing it to overcome obstacles three times larger than with only four wheels (Fig. 2.5).

The rover's analytical instrument, known as an alpha proton X-ray spectrometer (APXS), was a derivative of the Russian Vega and Phobos missions and was mounted in a retractible "snout" that could be positioned against a rock or the soil and set to analyze its elemental chemistry. Each analysis took around ten hours to complete.

Pathfinder provided the first pictures of the surface since Viking 2 had landed 21 years earlier, and the impact of these was stunning. The chosen landing site at the mouth of the outflow channel, Ares Vallis, was an area of soil and boulders, selected to offer the widest possible variety of rocks that could reasonably be expected to be encountered at any one locality. It turned out to be a very good choice indeed. The site was named the Sagan Memorial Station, in memory of the charismatic and much respected scientist, Carl Sagan, who died while the mission was in progress (Plate 6).

The surface near the lander contained bright red drift, dark-grey rocks, and soil of intermediate colour. Dark red soil also existed around the rock named Lamb. The spectra obtained at the site were calibrated

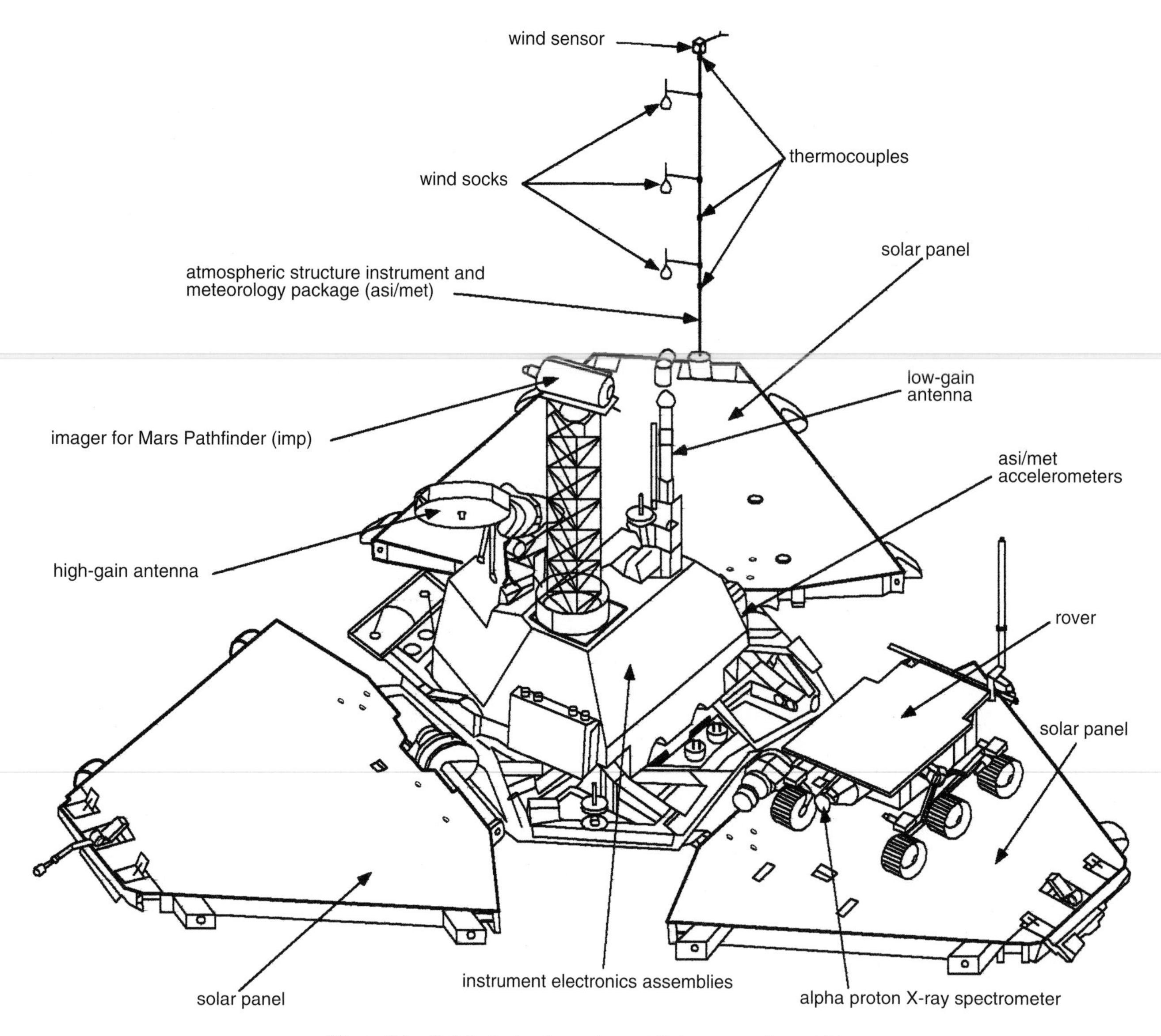

Figure 2.5 Pathfinder lander probe and Sojourner roving vehicle.

with respect to the drift in order to highlight their differences. The rocks were found to be less red and to have less ferric iron and a more unweathered composition than drift. Intermediate-colour soils appear intermediate with respect to spectral properties too. The dark soil at Lamb[1] is darker than the drift but equally red.

The shapes of the spectra of the various surface materials were successfully measured from Pathfinder's multi-spectral images (Plate 9). The left-hand image in Plate 9 shows the region southeast of the lander in true colour and, in the right-hand image, the strength of the kink in the spectrum at visible wavelengths (related to the abundance and particle size of weathered ferric iron minerals) is shown in false colour. Blue rocks are the least weathered, red soils are the most weathered, and green soils and rock faces show an intermediate state of weathering.

At least five different types of rock and soil were identified close to the lander. The typical soil was found to consist of drift and small dark-grey particles, resembling the rocks. However, two other kinds of materials are significantly different from the rocks and drift. Pinkish or whitish pebbles and crusts on some of the rocks (blue spectra) are brighter in blue light and darker in near-infrared than is the drift, and they lack the spectral characteristics closely associated with iron minerals. One of the more unusual rocks was named Ginger, located southeast of the

1. The rocks were given whimsical names by the mission science team.

lander. Parts of it had the reddest colour of any material on view, whereas its rounded lobes were grey and relatively unweathered. The very red colour of creases in the rock surface correspond to ferric material. The origin of this rock is uncertain, but it is possible that a ferric crust may have grown on it, or it may cement pebbles, in which case the sample would be classified as a conglomerate. The lovely panorama shown as Plate 6 depicts the Rover deployed against the rock Yogi, which has a weathered coating. Where the rover tracks dug into the soil, whitish material was uncovered, similar to that found at rock Scooby Doo. This may underlie much of the Ares Vallis site.

Although initial soil analyses from Ares Vallis and the Viking sites appeared very similar, further analyses showed that soils at Ares Vallis generally have higher Al and Mg, lower Fe, Cl and S. The rock named Scooby Doo turned out to be sedimentary, composed of compacted soils. When Pathfinder solid-rock analyses are compared with terrestrial rocks and meteorites inferred to have a Martian origin, they appear to plot in a gap between these fields (see Ch. 15).

Several of the rocks plotted within the chemical field of andesite, but the relatively large size of the APXS "spot" (1–2 cm) could mean that many rocks are composite, having a grain size of less than 1–2 cm. Thus, although a rock is consistent with andesite composition, it could be a breccia. To solve this problem, the rock called Barnacle Bill was targeted by the multispectral spot with all filters at maximum resolution. Under these operational conditions, if all the spectra were similar, then the rock must be homogeneous; on the other hand, if the spectra were variable, the rock might be heterogeneous, representing perhaps a melt breccia or conglomerate. In the event, spectra taken from different locations showed only two basic kinds of spectra:

- soil-like deposits
- dark rock face.

This implies that, at a spatial resolution of 1–2 cm, the rock composition is homogeneous.

There was some surprise at the discovery of andesites: such igneous rocks are usually found at convergent plate margins on Earth. Since there was a broad consensus that plate tectonics had never developed on Mars, their existence necessitated thinking of other ways in which the planet had differentiated significantly. Subsequently, the broad magnetic stripes found by Mars Global Surveyor indicated the possibility that some kind of crustal activity may have taken place. This will be discussed later in the book. The overall feeling held after the completion of this mission was that early Mars may have been much more like Earth than had been anticipated previously.

Mars Global Surveyor

Mars Global Surveyor (MGS) was the first in a program of missions aimed at sending low-cost pairs of orbiters and landers to Mars every 25 months until the year 2005. Its principal objectives were to provide a detailed global map of Mars, enabling planetary scientists to study the geology, climate and interior of the planet. Of particular interest were questions relating to the early atmosphere and the dramatic climatic changes that sent the planet into a state of deep freeze early in its history. That water had played a part in sculpting the surface of Mars was not in question, but exactly how much was once present, and where it now is, are questions that need to be addressed urgently. The general feeling is that most of it remained on Mars and has been frozen into the subcrust.

MGS carried six scientific instruments:

- A thermal emission spectrometer (TES) – to analyze infrared radiation from the surface and assist in identifying minerals in solid rocks and drift deposits. It would also provide information about short-lived clouds and dust in the atmosphere (Plate 4).
- An orbiter laser altimeter (MOLA) – to measure the height of surface features by sending a pulse of infrared light every ten seconds at a 160 m^2 area of the surface. This had the capacity to provide height accurate to within 30 m.
- A magnetometer/electron reflectometer (MAG/ER) – to search for evidence of a magnetic field and scan the surface for remnants of an ancient field.
- Radio science – by using the spacecraft's telecommunications system, high-gain antenna and time-measuring device, the probe was able to measure variations in the planet's gravity field.
- Mars orbiter camera (MOC) – the camera obtained long ribbon-like images as the spacecraft moved over the surface. Low-resolution global coverage of the planet each day was achieved via blue and red filters. Medium- and high-resolution images of selected areas were also obtained. The wide-angle camera was ideal for collecting global weather maps, imaging clouds down to a resolution of 7.5 km. The narrow-angle lens could obtain surface shots with a resolution of 2–3 m – many times better than the best achieved by Viking cameras.
- Mars relay system – a receiver/transmitter (developed for the Russian Mars 96 mission) was used to relay data from the probe to Earth.

The Global Surveyor spacecraft was captured by Mars into a highly elliptical initial orbit on 11 September 1997, with a low point 262 km above the northern hemisphere. During the next few days, each of the instruments was activated in turn, to check that all systems were functioning according to plan. Subsequently, aerobraking began, and during the next four revolutions the lowpoint of the orbit was reduced to 112 km. During this manoeuvre, on 15 September, the onboard magnetometer detected a weak magnetic field roughly 0.125 per cent as strong as that at Earth's surface. It is assumed that this is all that remains of an originally much stronger field generated by a dynamo similar to that still operating inside Earth.

Although plans to map the planet immediately had been made, because of a problem with the release mechanism of the main high-gain antenna this was postponed for almost a year, beginning eventually on 8 March 1998. Since then the probe has successfully completed a programme of global mapping and has targeted specific areas of interest for high-resolution studies. This has added significantly to our understanding of the geological structure and development of the planet. Furthermore, during April 1999, the magnetometer recorded banded patterns of magnetic fields on the surface, similar in general aspect to the magnetic stripes typical of spreading axes on Earth.

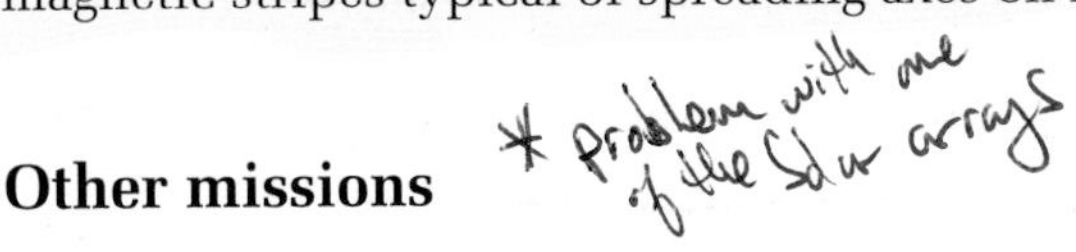

Other missions

Several spacecraft were launched during the second half of 1998 and early 1999. These included two further US probes and a Japanese mission. Together they illustrate a continuing keenness to understand the planet more completely.

Mars Climate orbiter and Mars Polar lander

Mars Climate orbiter was launched from Cape Canaveral on 11 December 1998 and was scheduled to reach Mars on 23 September 1999. Regrettably, some confusion about metric and imperial units among the technical team was responsible for the loss of this craft because of too low an entry into orbit when it reached Mars. Mars Polar lander – destined to become the first spacecraft to land near the edge of the southern polar cap – was successfully launched from the same place on 3 January 1999 and arrived in December 1999. On board the latter were two microprobes that were planned to smash into the planet's surface as a test for the technology of future descent probes. Regrettably, contact with the probe was lost just prior to landing.

Planet B mission

The Japanese Planet B spacecraft was launched from the Kagoshima Space Centre on 4 July 1998. Shortly after lift-off it was renamed Nozomi (lit. "hope"). Martian orbit insertion is scheduled to occur in 2003. The objective of this two-year mission is to learn more of the dynamics and structure of the Martian atmosphere. The craft carries a Japanese-built neutral mass spectrometer that will measure density variations of the major neutral constituents of the upper atmosphere. Also on board is the first payload to be sent into space by the Canadian Space Agency, a thermal plasma analyzer designed and built in Canada. This is a small atmospheric probe to measure low-energy particles and gases considered vital to understanding the origin and composition of the Martian atmosphere. NASA's contribution is a neutral mass spectrometer that is designed to measure the chemical composition of the upper atmosphere on a global scale.

Hubble space telescope

The Hubble space telescope (HST) has continued to generate images of Mars as the various Mars missions have performed their tasks. Images obtained in this way have given indications of the weather conditions on Mars as individual probes have reached it and have provided excellent sequences showing how dust storms develop and how clouds form and move around the planet (Fig. 2.6). This has been of particular value to atmospheric scientists and to the technicians coping with the problems bedevilling Mars landing vehicles. The overall findings from this source indicate that Mars' climate is much more chaotic than hitherto had been realized.

Hubble is particularly useful in charting the long-term changes on Mars and it operates very much like amateur observers using high-resolution devices and who, before the space age, accounted for over 95 per cent of astronomers making long-term regular observations of Mars. Already it has shown how wind-driven dust has caused changes in the appearance of long-lived albedo features such as Syrtis Major and Cerberus. It should prove invaluable to the planetary community as more and more of the planned missions are launched and it should also add to the global observations being made by Mars-orbiting spacecraft.

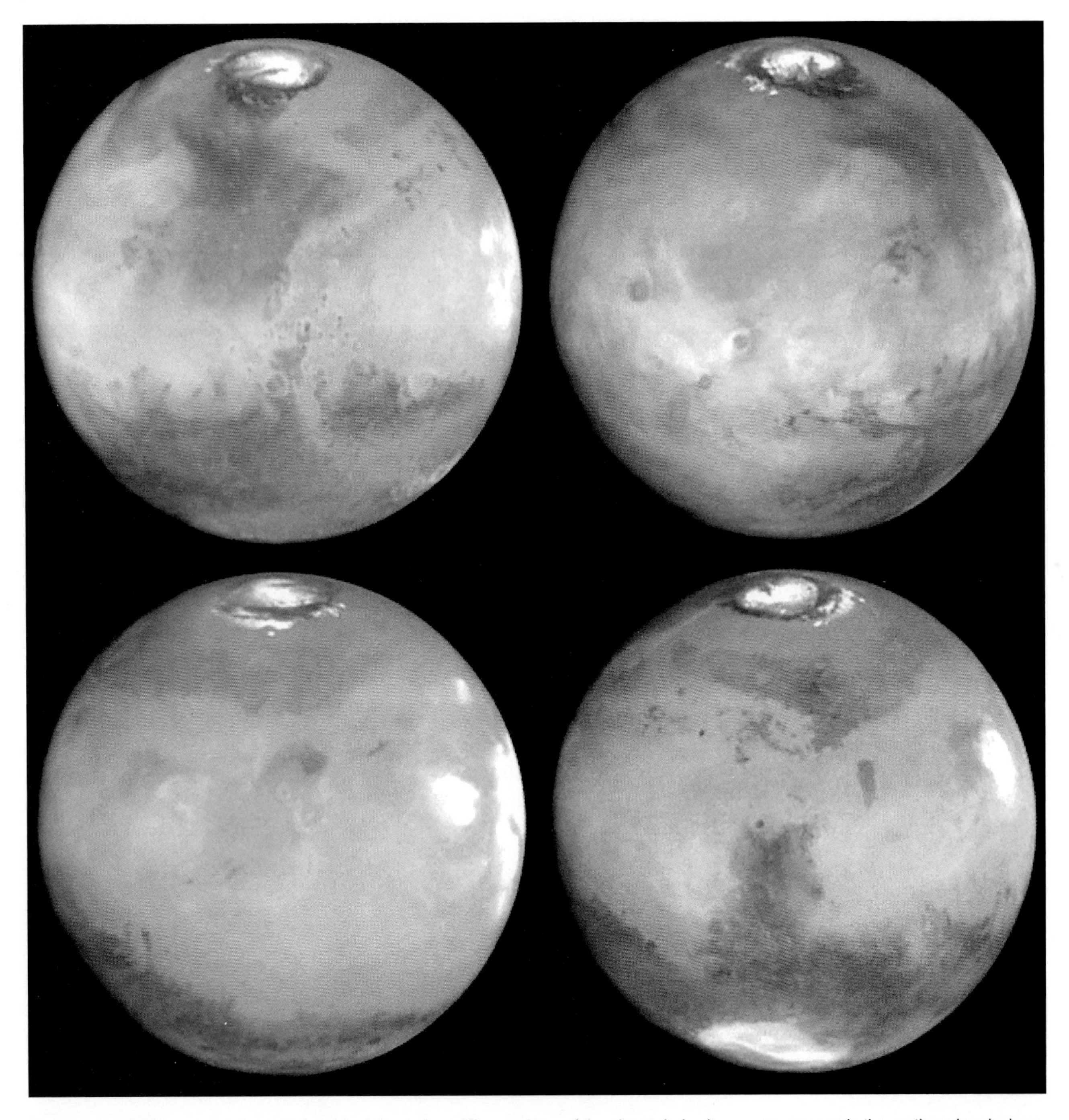

Figure 2.6 Hubble space telescope (HST) took these four different views of the planet during its summer season in the northern hemisphere during March 1997. Some of the most interesting features of the planet include the polar cap, the extinct volcanoes, its reddish colour and the atmospheric haze. Summer melting of the polar cap ice sends vapour into the atmosphere. This then causes the atmospheric hazing visible at the extreme left of the upper left-hand picture. In the second view (upper right), three of Mars' extinct volcanoes are clearly visible, the largest of which, Olympus Mons, is closest to the western hemisphere.

3 The present face of Mars

Seen through a telescope, Mars presents a reddish disk with dusky markings and white polar caps. These albedo markings bear little general resemblance to the major landforms revealed by spacecraft imagery. True, there are some correlations; for instance, the whitish spot of Nix Olympica coincides with the massive volcano, Olympus Mons, and the impact basins of Hellas and Argyre are visible as pale yellowish regions. In consequence, some of the names from old telescopic maps have survived, but the plethora of major landforms, discovered only when spacecraft and HST images were returned to Earth, has demanded a nomenclature largely different from that familiar to, say, an observer in the early twentieth century (Fig. 3.1).

The topography of Mars

Mars has an equatorial radius of 3393.4 km and a polar radius of 3375.7 km, but has a pear-shape rise in the hemisphere containing Tharsis. Because there are no oceans, a sea-level datum is not available and, therefore, before topographical maps of the planet could be prepared, a reference datum had to be selected. The datum chosen was derived from gravity data that was

Figure 3.1 The modern face of Mars, as compiled from recent spacecraft data.

combined with a 6.1 mbar atmospheric pressure surface (Christensen 1975). Prior to the Global Surveyor mission, topographical information – derived in the main from radio-occultation and gravity data – was calibrated against this 6.1 mbar datum to produce general topographical maps. More recently, altimetric data have been derived much more accurately by laser-ranging from the orbiting MGS spacecraft.

Figure 3.2 illustrates the marked topographical assymmetry of Mars, with the majority of the southern hemisphere lying 1–3 km above datum and most of the northern hemisphere below it. The line of dichotomy separating these two zones of elevation describes a great circle inclined at approximately 35° to the Martian equator and it is a zone of relatively steep slopes.

The Mars orbiter laser altimeter (MOLA) on board the Global Surveyor spacecraft made 27 million measurements of altitude between 1998 and 1999, each point being spaced 60 km apart on a global grid, with each elevation known to an accuracy of 13 m (and down to 2 m in the flatter northern hemisphere). From this immense amount of data, a new topographical map of Mars was prepared (Plate 7). The full range of heights spans a vertical distance of 30 km, which is 1.5 times the elevation range of Earth. Bearing in mind that Mars has only half Earth's diameter, it is clear that it is a world of considerable topographical variance.

In the relatively elevated southern hemisphere, the principal exceptions to the positive topography are the deeper parts of the two impact basins, Argyre and Hellas, and the region south of latitude 70°S. The most recent topographical map (MOLA data) shows the deepest part of Argyre descending to a depth of 4 km and Hellas to 9 km, the latter's surrounding ring of debris rising to 2 km above the surrounding plateau. Hellas is a major topographical feature, the volume of the surrounding ring of material being sufficient to cover the entire USA to a depth of 3.5 km. Clearly, it makes a major contribution to the elevation of the southern hemisphere. In the northern hemisphere the main exceptions to the subdatum elevations are the heavily cratered terrain between 30°W and 270°W, and the elevated volcanic provinces of Tharsis and Elysium.

In addition to the regional variations, there are local elevation differences substantially greater than those that characterize Earth. For instance, some sections of the vast equatorial canyon system, Valles Marineris, descend 7 km below the adjacent plains, whereas the summits of several shield volcanoes within the Tharsis province rise to over 20 km. This particular characteristic shows up clearly when the topographies of Mars and Earth are compared (Fig. 3.2). Incidentally, the highest point on Mars is at Olympus Mons, whose summit rises 27 km above datum.

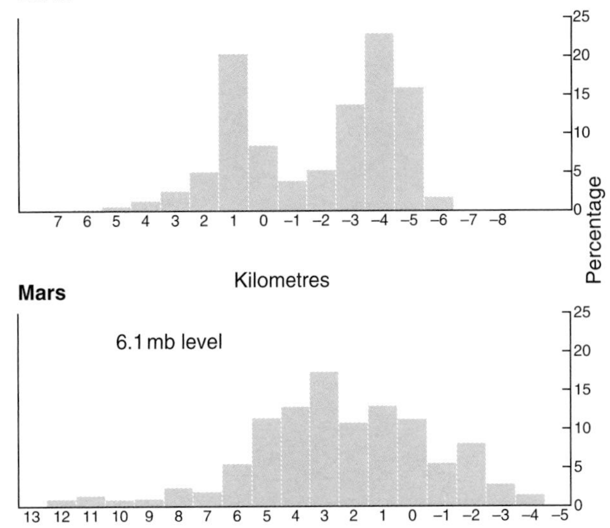

Figure 3.2 Comparison of distribution of topography for Mars and Earth.

Mention of Tharsis introduces another major topographical feature of Mars, the Tharsis Rise, a broad crustal upwarp centred at 14°S 101°W, which rises 10 km above datum and measures roughly 4000 km across. This structure or continent size straddles the boundary between the two topographical hemispheres. A similar but smaller and lower rise is centred on 28°N 212°W in the Elysium province, but the latter falls almost entirely within the lower northern hemisphere.

In essence, the northern hemisphere is a vast depression, being very smooth in the mid- to high latitudes (Smith et al. 1999a). Thus, the mean elevation north of 50°N is about –4 km, the topography varying by 3 km from this value. However, over hundreds of kilometres, the surface rises and falls by only 50 m or so. This makes it even flatter than the lunar maria and Earth's Sahara Desert. Excluding the massive Tharsis Rise, all of the topographical profiles obtained by the MOLA are either flat or slope gently upwards towards the south (Fig. 3.3). This vast flat area, which is as smooth as Earth's abyssal oceanic plains, extends across every longitude and is over 200 km wide in north–south extent. It has been described as "the flattest surface in the solar system for which we have data" (Kerr 1998a). The gradual 0.056° slope from pole to equator is a result of the separation of the planet's centre of mass from the centre of the figure along its rotational axis, and also of the 1 km increase in the mean equatorial radius attributed to the Tharsis Rise.

The steepest slopes on the planet are found to be associated with what is termed the dichotomy boundary, the region that separates the smoother and lower northern third of Mars from the rough cratered

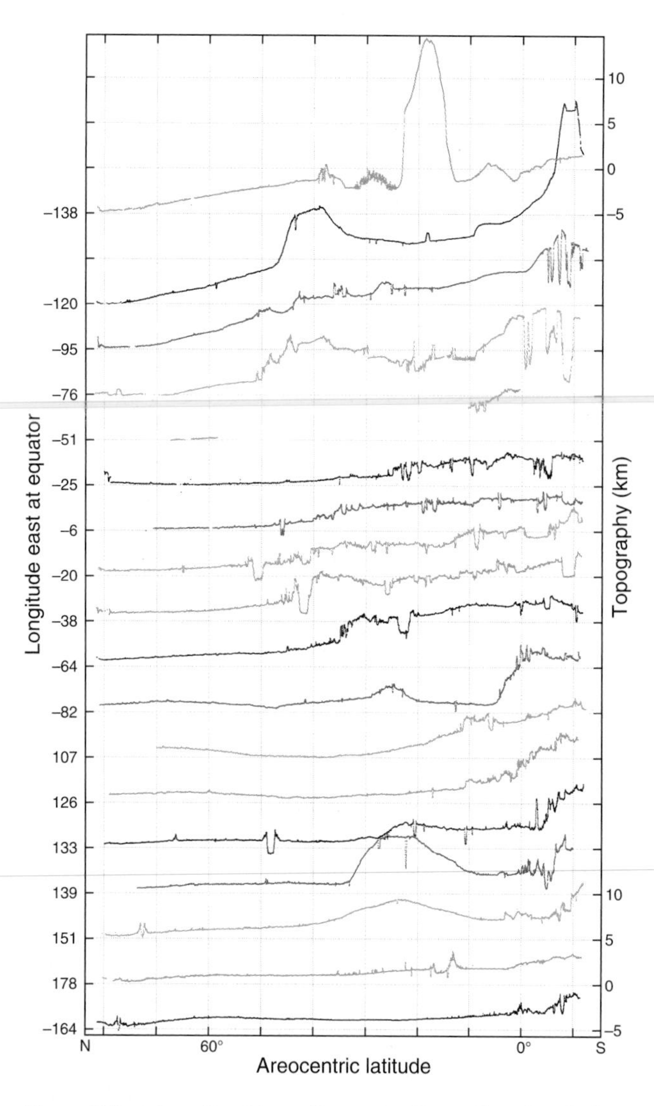

Figure 3.3 MOLA altimetric profiles across the northern hemisphere of Mars; vertical exaggeration 100:1.

highlands of the southern two thirds. Over horizontal distances of tens of kilometres or more, slopes are generally between 1° and 3°, increasing to as much as 20° at isolated spots. Such steep slopes indicate that surface modification processes have been at work, and some ancient structures may still be preserved within this zone. There is a tendency for there to be a slope from the south towards the north pole, which is generally considered to be the factor that dictated that there was a south–north flow of water on Mars in the distant past. However, there are indications that the eastern part of the equatorial canyon system slopes away from the outflow channels and also lies at least 1 km below the floors of these same channels. This being so, it is likely that, if conditions were appropriate, standing water could have ponded within parts of this vast canyon network.

Physiographic provinces

Just as the topography of Mars shows a marked asymmetry, so does the distribution of physiographic units. In fact, the physiographic features show a strong correlation with the hemispherical asymmetry in topography (Fig. 3.4). For instance, the greater part of the southern hemisphere is a landscape carved from an ancient heavily cratered surface, which at first sight is similar to the lunar highlands. However, closer inspection reveals a major difference, since the Martian cratered plateau is incised by many valley networks that had their origin in running surface water (Fig. 3.5). This heavily cratered terrain also extends northwards across the equator as a broad tongue between longitudes 300°W and 10°W, that is, north of the telescopic albedo regions known as Sinus Sabaeus and Sinus Meridiani.

Within the heavily cratered regions, the density of craters larger than 20 km in diameter is high, but smaller craters are scarcer than in the lunar highlands. Furthermore, the large impact craters are relatively shallow, having flat floors and low rims, indicating long periods of erosion and weathering. Thus, the overall aspect is less rugged than is typical of the Moon. There are also many areas of smoother and less densely cratered plains between the large impact craters.

As noted earlier, many of the intercrater surfaces are cut by short channel networks; others are ridged. Sometimes ridges may be seen crossing the intercrater plains and continuing along the same strike across the interiors of large impact craters. Spudis & Greeley (1978) estimated that the ancient cratered plains cover an area of about 29 million km^2, of which about 36 per cent is ridged. Quite extensive remnants of such ridged plains are found in Noachis Terra and the area to its east, in Memnonia and also southern Sirenum Terra (Fig. 3.6). Cratering studies suggest that these are some of the oldest plains found on Mars (Scott & Tanaka 1986). The ridges themselves, in appearance very like lunar wrinkle ridges, have been invoked by many workers to support the case for a volcanic origin (see Cattermole 1996).

Other ridged plains outcrop in the western hemisphere, within a broad zone about 1000 km wide on the east flank of the Tharsis Rise where they extend over an area of about 4 million km^2. The plains extend from Tempe Terra, through Lunae Planum, then across Valles Marineris into parts of Sinai and Solis Planae. Similar plains occur in the eastern hemisphere, particularly in Hesperia Planum, northeast of Hellas. Upland depressions such as Syrtis Major Planum and the floors of the large impact basins also exhibit similar plains.

Mars also has many large impact basins, the two

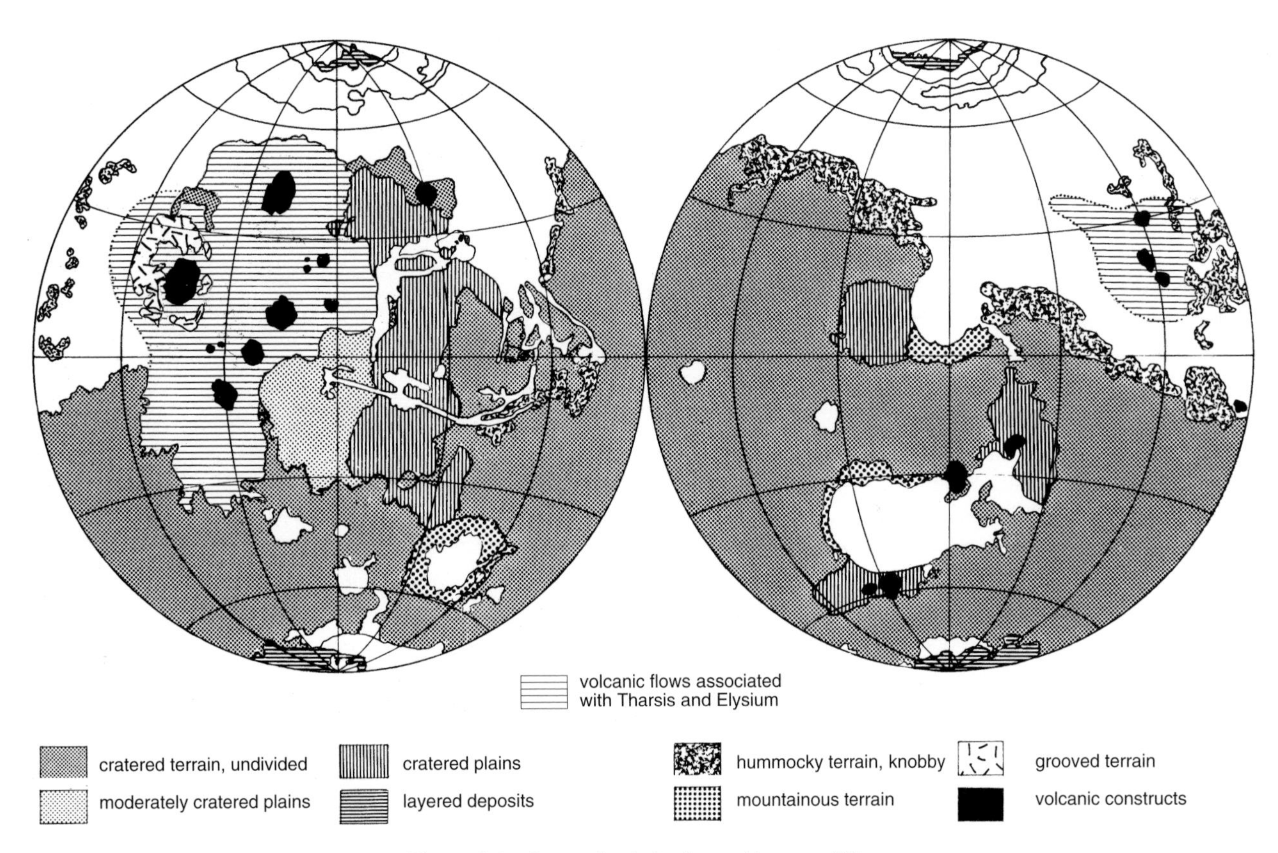

Figure 3.4 Generalized physiographic map of Mars.

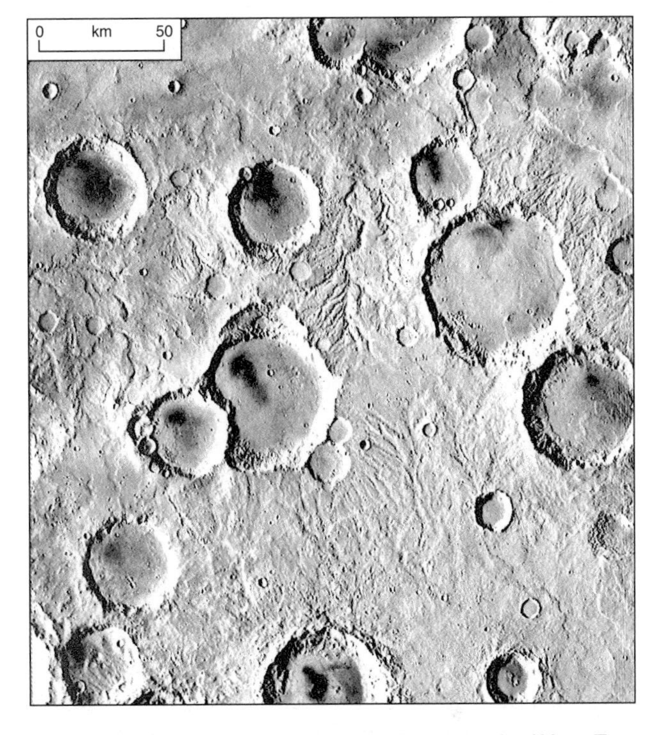

Figure 3.5 Channel networks in the cratered uplands of Mars, Terra Cimmeria. Viking orbiter image 370s54; centred at 36.95°S 212.99°W.

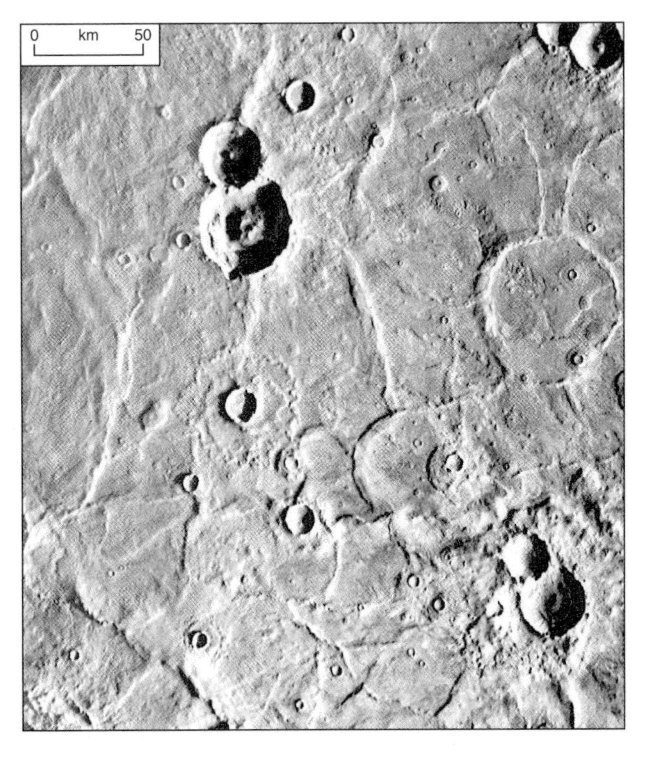

Figure 3.6 Ridged plains southwest of the Hellas impact basin. Viking orbiter image 361s13, centred at 59.25°S 313.23°W.

most obvious being Hellas (diameter 1800 km) and Argyre (diameter 800 km). The Isidis basin measures 1900 km across but has been substantially modified and is therefore visually less striking. The larger basins are multi-ringed and of dimensions similar to those on the Moon. In a non-exhaustive search for other impact basins, Wood (1980) lists ten further structures with diameters greater than 200 km and which qualify for the name basin, the largest of these being the south polar basin, and a further 40 or so have inner rings greater than 50 km across. See also work by Schulz et al. (1982), Frey & Schultz (1988) and Schultz & Frey (1990).

The polar regions, whose seasonal changes are well known from Earth-based observations, are markedly different from the rest of the planet. Around both poles and extending to latitudes of about 80°, unique laminated deposits are found; these bear witness to prolonged periods of erosion and deposition. The layered deposits are exposed in the walls of extensive scarps that girdle the polar ice caps. Because of the very low superimposed impact crater density, these must be among the youngest deposits on Mars. Around the north pole the laminated units overlie plains and are girdled by an extensive sand sea in which there is widespread development of dunes. The latter are absent from the south pole, where the laminated materials unconformably overlie old cratered terrain and a large polar impact basin with a diameter of about 850 km. Dune development here is confined to the region of Noachis and within the large impact craters, Kaiser, Proctor and Rabe. The ice caps themselves, naturally, change their extent and shape as Mars' seasonal cycle progresses. These changes have been recorded beautifully by the Hubble space telescope.

The northern hemisphere of Mars is largely given over to plains that are less heavily cratered than the older terrain to the south. Similar plains extend into the southern hemisphere around the Tharsis Rise and to the south of the equatorial canyons; they are also found both within and around the Hellas impact basin. The impact crater densities characteristic of these plains are within a factor of two or three of those of the Moon's mare plains, their surfaces exhibiting landforms that bear witness to volcanic, aeolian and fluvial activity. There is also much geomorphological evidence for the presence of subsurface ice.

The extensive volcanic flow plains that encircle the Tharsis and Elysium volcanoes are less heavily cratered. These younger deposits show widespread volcanic features, such as flow scarps, lobes, lava channels and wrinkle ridges; many of the individual units are radial about major volcanic centres (Fig. 3.7). The volcanoes themselves are extremely large and visually impressive, the most prominent group being the Tharsis Montes, a group of shields spaced at about 700 km apart and aligned NE–SW on the northwest flank of the Tharsis Rise (Plate 7). The even larger structure, Olympus Mons (Nix Olympica on telescopic charts), is situated 1600 km to the northwest and resides at the edge of the Rise. It is the youngest Martian shield volcano and also the highest (27 km). However, all of the volcanoes are immense: each of the three Tharsis Montes is over 400 km across, and Olympus Mons measures over 550 km. Other smaller volcanoes also lie close to the same tectonic line; but most are older than the main group. Lastly, on the northern flank of the rise is the lower-profile volcano, Alba Patera; this immense structure has a diameter of over 2500 km and has associated with it a prominent circumferential fracture belt.

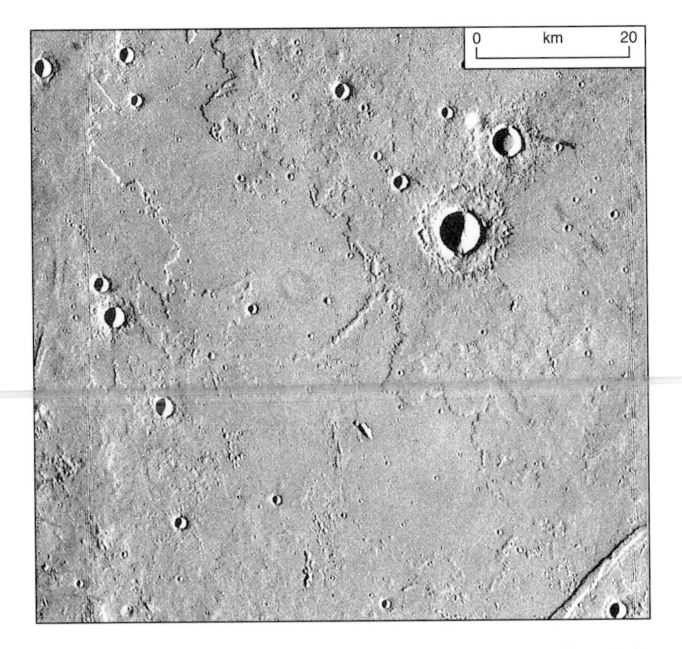

Figure 3.7 Hesperian flow plains flooding ancient cratered terrain in southern Tharsis. Viking orbiter image 056a14, centred at 31.53°S 130.73°N.

The three volcanoes of the Elysium region have a morphology somewhat different from that of their Tharsis counterparts. Both effusive and explosive volcanism appear to have played a part in their development. Lava plains associated with them extend over an area of about 3 million km^2. To the southeast of Elysium there is the rather isolated major volcano, Apollinaris Patera.

Volcanoes are also associated with the Hellas basin; these are significantly older than either of the Tharsis or Elysium structures. This group comprises what have been termed highland paterae, which are the manifestation of a much earlier phase of volcanism that took place at least 3.1 billion years ago, well before any of the Tharsis volcanoes existed. Isolated volcanic structures also occur within the ancient plains of Syrtis Major Planum and in Tempe Terra.

The character of the plains at latitudes higher than 40°N differs significantly from those at lower latitudes. Little or no direct evidence for a volcanic origin is found here; instead there is a vast blanket of sedimentary debris that gives rise to a smoothed, softened landscape. Impact craters in this region often show the effects of deep burial, sometimes their interiors being completely infilled with debris and often partially filled with linear dunes. Typically, their ejecta blankets usually stand out as pedestals, rising slightly higher than the surrounding plains, as if the ejecta had somehow resisted the erosional processes that had lowered the surrounding regions.

Equally as impressive as the giant Tharsis shield volcanoes is the amazing equatorial canyon system that extends eastwards from the crest of the Tharsis Rise, at 5°S, 100°W, running almost parallel to the equator for a distance of approximately 4000 km. This great landform, known as Valles Marineris, begins in the west as a series of interconnecting box canyons and depressions known as Noctis Labyrinthus, changes into a series of subparallel and approximately parallel-sided canyons along its central section and finally merges at its eastern end with vast areas of blocky chaotic terrain at around 15°S, 40°W. The length of this series of deep canyons is almost one quarter of Mars' circumference (Plate 7). Many smaller canyons lie to the north of the main system.

The equatorial canyon network itself is only one part of a much more widespread family of extensional fractures that splay out from the Tharsis Rise and extend over almost an entire Martian hemisphere. Fractures are particularly strongly developed to the north and northeast of Tharsis (Tempe and Mareotis Fossae), and also to the south (Claritas Fossae). Families of long curving graben also extend into the telescopic regions of Mare Sirenum and Memnonia (Sirenum and Memnonia Fossae). Less well developed fracturing is also associated with the Elysium province.

The regions of chaotic terrain, into which the Valles Marineris run at their eastern termination, outcrop approximately within the region defined by latitudes 5°N and 15°S, and longitudes 15°W and 40°W. Emerging from the northernmost areas of chaos are a series of major channels that run northwards towards Chryse Planitia, converging there with similar ones that emerge from north of the central section of Valles Marineris and enter Chryse from the west. These landforms exhibit various indicative characteristics of an origin in fluvial processes, such as scoured floors and teardrop islands. Their dimensions are typically impressive. Individual channels incised into the Chryse plains are at least 25 km wide, and the major channel system (Kasei Vallis), which emanates from valleys that run along the western side of Lunae Planum, entering Chryse from the west, is at least 2000 km long (Fig. 3.8). Other large channels are located in Amazonis, Memnonia, Hellas and Elysium.

A further, peculiarly Martian, landscape type – fretted terrain – is developed along sections of the boundary between the upland and lowland hemispheres, termed the line of dichotomy. Its classic development is found between latitudes 30°N and 45°N, and longitudes 280°W and 350°W, where it consists of extensive remnants and isolated mesas of the original high plateau, between which are lower plains units that take the form of isolated depressions and sinuous flat-floored channels; the latter extend deeply into the cratered plateau. Channels sometimes start from large impact craters; elsewhere they begin, abruptly, without any obvious connection with specific landscape features. The channel floors everywhere contain glacier-like sheets of sedimentary debris, the characteristic surface striae they exhibit indicating laminar flow in clastic material that is actively being worn from the cratered hemisphere and slowly transported onto the lower and less heavily cratered plains of the north. This zone is one of active erosion, transportation and deposition.

Thermal inertia mapping measurements

The infrared radiometers aboard each of the Viking orbiters made a series of measurements of the temperatures at the Martian surface during orbit. A wide range in temperature was recorded. Temperature varies as a function of several factors: latitude, season, time of day and properties of the surface materials where the measurements are being made. Temperature values were found to be at their lowest just prior to dawn, rising quickly during the morning and peaking just after noon; thereafter, they fell rapidly during the afternoon, this decline slowing down during the night and reaching the pre-dawn minimum.

Pre-dawn temperatures are a sensitive indicator of thermal inertia, which along with albedo is one of the two most important properties of the surface materials. Thermal inertia can be represented by $(Kpc)^{1/2}$, where K is the thermal conductivity, p is the density, and c the specific heat. It is a measure of the responsiveness of materials to changes in temperature; thus, if a material has a high thermal inertia, it will respond slowly to temperature changes; if it has a low thermal inertia, the response will be much quicker.

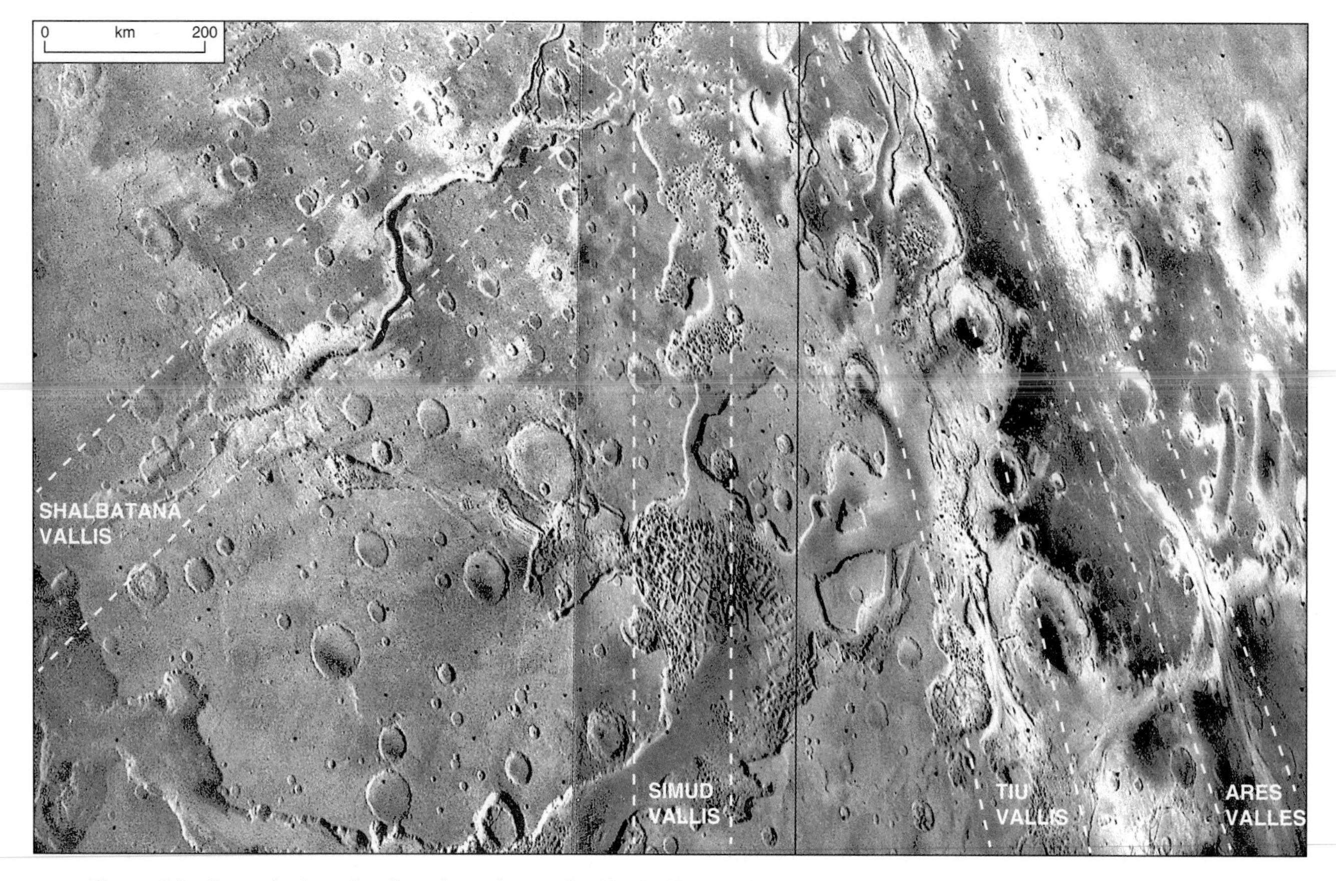

Figure 3.8 Synoptic view of outflow channels crossing Xanthe Terra. Viking orbiter image 005a08; centred at 5.72°N 42.86°W.

The dominant cause of variations in thermal inertia values on Mars is almost certainly the grain size of the surface layer. It is easy to see why grain size has this effect: as it diminishes, the number of interfaces between the constituent grains increases; consequently, because the interfaces conduct heat less efficiently than the grains themselves, the thermal conductivity diminishes. The upshot is that thermal inertia measurements can tell us where the Martian surface layer is of relatively coarse grain and where it is finer.

Kieffer et al. (1977) used the Viking IRTM experiment to produce a global map of thermal inertia values, corrected for latitude and seasonal variations. Values were found to range over one order of magnitude and show only a partial correlation with global physiography. However, the work showed how much of the Tharsis region, together with Elysium and Amazonis, has low thermal inertia, indicative of its being mantled by relatively fine-grain debris. A further region of similar values was found north of the equator, on either side of longitude 330°W. On the other hand, high values were found throughout the equatorial canyons, the regions of chaotic terrain and the major channel systems, suggesting that such areas have a relatively high incidence of blocks and boulders. Interestingly, a strong negative correlation exists between albedo and thermal inertia. Bright regions have low thermal inertia and presumably are of fine grain, whereas dark ones have high thermal inertia and so are probably of coarser grain.

This early work, and indeed other studies based on Viking data (Palluconi & Kieffer 1981), generally made the assumption that the atmosphere had the sole effect of contributing to the diminishing radiation, this being constant and equal to 2 per cent of the maximum solar radiation. This became known as the 2 per cent assumption. Subsequently, Bridges (1994) made corrections to take into account the effects of altitude and varying pressure in translating the data into meaningful particle sizes. He used the 2 per cent assumption, but assumed that this decayed with elevation with a scale height of 10 km.[1] He also made the assumption that the back radiation was 4 per cent of the maximum solar radiation at a height of –2 km, and that this and the dust opacity decayed with elevation with the same scale height. Hayashi et al. (1995) used this and the earlier work in a coupled surface–atmosphere (CSA)

1. Scale height: the vertical distance over which the ~~atmosphere vanishes, i.e. the height to which the air extends~~. Pressure changes by a factor of 1/e (i.e., decreases by 36.8% of initial value).

model (see Haberle & Jakosky 1991) to study the effects of a dusty CO_2 atmosphere on thermally derived albedo. One result of this research was the preparation of a revised map of Martian thermal inertia that takes into account atmospheric effects using a one-dimensional radiative-convective model, rather than simple scaling from maximum solar radiation. By varying the dust opacity of the atmosphere, they were able to show how particle sizes derived by using the simple 2 per cent assumption are always too large.

Particularly large discrepancies were found for some very interesting geological regions. For instance, areas with a high dust opacity (Acidalia, Chryse and Hellas) show some of the greatest differences: the CSA model thermal inertia values are about 100 times smaller than the 2 per cent model. Some low-lying regions (Valles Marineris, Vastitas Borealis, Isidis Planitia) also have large differences (–80 to –100) between the two models. Thus, particle sizes calculated for these areas by the older method would be too large. The regions are probably characterized by a fine sand-grade covering. On the other hand, certain areas of considerable altitude (Tharis Montes, Olympus Mons, Alba Patera and Elysium Mons) have CSA thermal inertias that are very similar to 2 per cent values. These bright regions appear to be covered in fine sand-grade materials. Also characterized with high albedos are both of the Viking lander sites (Chryse and Utopia), and this, together with the modelled CSA thermal inertias, suggests they are both covered in medium to coarse sand-grade material.

Although the Russian Phobos mission was not a total success, its on-board Thermoscan instrument was able to obtain both thermal and visible images of the shadow of Phobos against parts of the Martian surface (Betts et al. 1995). Thermal inertias derived for several sites (Arsia Mons, western Valles Marineris, Herschel crater) were characteristically lower than those measured by the Viking IRTM. Those for Arsia Mons are equivalent to dust-size particles (5–10 μm).

Earth-based and orbiter spectral observations

The surface materials have also been studied by radar, which allows scientists to estimate the distribution of surface slopes, wavelength-scale surface roughness and density. Broadband albedo measurements are used to observe variations in surface brightness caused by movement of dust, and radio emission observations are used to study the density of the Martian subsurface. Simple colour observations are a good guide to the composition, degree of mixing and presence of surface duricrusts.

There is a marked negative correlation between thermal inertia and albedo, but a positive correlation between thermal inertia and density, as derived by radar measurements. The latter may well be a reflection of the development of subsurface crusts, since these would increase both properties. Viking orbiter colour observations confirm the presence of three distinct kinds of surface materials:

- low-inertia bright-red dust
- high-inertia dark-grey materials, which are presumed to be an admixture of dark lithics and palagonite-like dust
- moderate-inertia dark red matter that presents rough surfaces at small scale and is assumed to be indurated material, possibly duricrust of some kind.

The suggestion that hydrated minerals are abundant on the surface of the planet is widespread in the literature and is supported by all kinds of remotely sensed data. Reflectance spectra have been obtained for Mars, beginning in the early 1960s. It is well established that there is a broad absorption band centred around 3 μm and this has been variously attributed to adsorbed, bound or frozen water in Martian surface materials (Moroz 1964, Sinton 1967, Calvin 1997). Near-infrared spectra were also collected by the imaging spectrometer carried by the Russian Phobos 2 spacecraft. This confirmed that 3 μm band depths as great as 0.62 are to be seen in the surface (Murchie et al. 1993), and variations of as much as 20 per cent in the depth of this 3 μm feature suggest lithological variability from place to place. Infrared spectrometers aboard the early Mariner 6 and 7 probes also recorded the 3 μm absorption feature and showed similar variations. Laboratory experiments conducted much more recently (Yen et al. 1998) show that the spectral measurements are consistent with the presence of up to 4 per cent water by weight in Martian surface materials, which could reduce to a mere 0.5 per cent if the hydrated phases were present as grain coatings.

The results of laboratory experiments and observations of both terrestrial and Martian spectra suggest that haematite, microcrystalline iron oxides or hydroxides, clays and palagonite glass may be present in the Martian regolith (Bishop et al. 1998). After studying both visible and infrared spectral properties of terrestrial volcanic fines produced under different kinds of weathering conditions, the above group concluded that moist environments typified by volcanic samples analyzed from Santorini (Aegean) and Tarawera (New Zealand) best fitted the kinds of environment postulated for early Mars, whereas the drier-climate samples from Haleahala (Hawaii) were more closely akin to modern Mars. As would be expected, the effects produced in infrared spectra of the same materials, subjected to different styles of weathering, are quite significant and must be taken into account in interpreting spectral information.

The overall impression gained from a variety of studies is that Mars' surface layer is relatively young and constantly re-worked; however, there appears to be little major erosion of bedrock. Global dust movement is largely from south to north, with little regional accumulation in the southern hemisphere.

The Martian surface layer

The view of the surface transmitted back to Earth by the Viking landers was of a rock-strewn desert. The Viking 1 lander site in Chryse Planitia (now called the Mutch Memorial Station, in memory of the American planetary scientist Thomas Mutch) was revealed as an undulating plain with fine-grain, often encrusted, drift deposits in the lee of large boulders. The Viking 2 site, in Utopia Planitia, was somewhat flatter, lacked the widespread drift deposits and was covered by angular, dark, pock-marked boulders, believed to have their source in the region of a 100 km-diameter impact crater located 180 km to the east. Mars Pathfinder landed on the southern side of Chryse, at the mouth of the Ares–Simud–Tiu Valles outwash valley system. It identified a rock-strewn channel floor where thin drifts of fine-grade material overlay soils that were admixed with rocks. The form and composition of the latter proved extremely interesting.

X-ray fluorescence analysis of the regolith at both Viking sites gave remarkably similar results, despite the considerable distance between the locations. This suggests that there has been rather efficient homogenization of the fines, presumably by global dust storms. Typically, the fine-grain matter has low silica content (43–45% by weight) and high iron (18–20%), implying derivation from the parent mafic igneous rocks. However, the low CaO (5.0–5.6%) and extremely low Al_2O_3 (5–6%) contents are difficult to reconcile with any terrestrial igneous rock. A fairly substantial volatile content is indicated by SO_3 (6.5–9.5%) and Cl (0.6–0.9%). Although the composition may seem strange, it should be noted that the analytical instrument was unable to detect elements of low atomic weight, such as sodium, nitrogen and carbon; furthermore, the mineralogical composition had to be inferred indirectly. One possible composition is a mixture of iron-rich smectite clay, with ferric oxides, carbonates and sulphates (Toulmin et al. 1977). If this were so, the iron-rich clay could be an alteration product of palagonite, or of basaltic glass, either of which could have been produced by interaction between hot lava and an ice-rich regolith.

The pronounced red colour of Mars has long suggested the presence of surface ferric oxides (rust). A strong absorption band below 0.5 μm and much weaker absorption features between 0.7 and 0.95 μm, detected in reflectance spectra (Singer et al. 1979) are consistent with the presence of ferric oxides. Faint bands near 1 μm suggest Fe^{+2} absorptions in mafic silicate minerals, such as pyroxene or olivine (Bibring et al. 1990). Many potential iron-bearing phases have been proposed, including goethite ($\alpha FeOOH$), limonite ($Fe_2O_3.nH_2O$), Fe^{+3}-smectite clay, weathered palagonite, and maghemite (a magnetic form of iron oxide, γFe_2O_3). More precise spectral measurements clearly indicate several distinct absorption bands around 0.68 μm and 0.85 μm, attributed to crystalline hematite (Bell et al. 1990). However, it must be stressed that the match between the Martian spectra and terrestrial iron oxides or palagonite is not exact (Bell et al. 1990) and that a close fit to the Martian data can also be obtained with a thin ferric oxide coating on weathered basalt (Singer et al. 1979). The latter interpretation is consistent with the overall mafic surface composition, vesicular basalt-like appearance of the surface rocks, and photogeological evidence for widespread basalt-like volcanism.

Spectral features of around 3–4 μm have been ascribed to condensed water (Houck et al. 1973) or hydrated minerals (Bibring et al. 1990). Combined gas-chromatograph and mass-spectrometer analyses on the Viking landers provided more direct indications for adsorbed or hydrated water on the surface (Biemann et al. 1977). Other emission and absorption features in the 5.4–10.5 μm wavelength region have been tentatively identified with sulphates and carbonates (Pollack et al. 1990).

Mars Pathfinder landed at the mouth of the Ares–Simud–Tiu Valles outflow channel system, a site chosen because rocks from a variety of geological environments and units were believed to have been deposited there by ancient floodwaters. It has since been established that these represent part of a mid-fan facies, deposition possibly having occurred in a lacustrine environment. For the first time, the successful landing of the Pathfinder probe has enabled scientists to characterize the Martian surface at the mouth of an ancient flood channel. A prominent rise to the north of the landing site appears to represent a debris tail deposited on the downstream side of two streamlined hills, named Twin Peaks. Most of the clasts found at the site probably derived from floods debouching from the Tiu Vallis system, being sourced in the cratered highlands to the south (Nelson & Greeley 1999).

Boulders as large as 7 m in diameter were identified in the vicinity of the lander, with clasts ranging from a few centimetres to boulder size. Most of the boulders and cobbles were angular, as they were at the Viking sites, but a significant proportion were well rounded, suggesting a fluvial origin. Furthermore, several tabular boulders looked rather like perched blocks, left by the late-stage flooding episode that left its mark here. Many of the rocks in what has become known as the Rock Garden (e.g. Shark, Half Dome, Moe) appear to be generally tilted in one preferred direction (imbricated), presumably by past waterflow. That such flow occurred is also indicated by the recognition of several channel features and striations on some rocks. The proportion of boulders and rocks around the site (16%) is consistent with the notion that it represents a depositional plain.

There are several explanations for the presence of rounded pebbles at this site. They could be impact melt droplets, exotic nodules brought up in volcanic eruptions, lava fountain products, or concretions. However, there is a strong feeling that they are pebbles derived from conglomerates, having been freed from their host rocks, then abraded by vigorous flood erosion, wave action or glacial activity before being dumped. Certainly, the rocks Shark and Half Dome appear to be conglomerates, their rounded 3–4 cm-diameter pebbles apparently being held in a finer-grain host of some kind, either matrix or cement (Fig. 3.9).

The form of the wheel tracks left in the fine-grain drift by the Sojourner rover indicated that the surface deposits contain substantial amounts of dust-size matter. These kinds of track are produced only in material with an average size $<40\,\mu m$, like talcum powder. In many places, the coarser soil-like deposits upon which the drifts frequently sit appeared cloddy, consisting of a mix of poorly sorted dust, clods of more coherent material, and rock clasts. These Martian "soils" (probably better termed regolith, a term that does not imply an organic component such as humus) and dust are chemically similar to those analyzed at the Viking sites. Their average density was calculated to be of the order of $1520\,kg\,m^{-3}$. The widespread recognition of sand-grade material at the Ares Vallis site lends support to the idea that fluvial activity must have played an important role in modifying the Martian surface hereabouts.

Figure 3.9 A stereo pair of images of the rock Half Dome, located at the Sagan Memorial Station. Note the rounded "clasts" set in a finer host material. Mars Pathfinder image PIA01566.

Chemical analyses obtained by the APXS mounted on board the Sojourner rover revealed that, whereas many boulders and smaller rocks yielded compositions of basaltic type (similar to those found at the Viking sites), others were more silicic, resembling terrestrial andesites. Some of these are primary igneous rocks, implying that Mars is more differentiated than previously thought. However, it is possible that some are composite rocks (e.g. volcaniclastics), with average compositions approximating that of andesite, rather than being homogeneous andesite lavas.

The Pathfinder lander also carried an experiment to measure the magnetic properties of the surface materials. Much bright red dust was attracted to the magnets mounted on it, and this had a magnetization consistent with composite particles coated with a maghemite stain or held together by cement of the same composition (Hviid et al. 1997). The most plausible interpretation consistent with the data is that iron must have been dissolved out of crustal minerals in water – suggesting that there was once an active hydrogeological cycle – and that the observed maghemite is by way of a freeze-dried precipitate.

Hubble space telescope observations

Of all the missions currently operational, the Hubble space telescope is the one that can best provide long-term observations of global and local weather, the movement of dust and changes in the polar caps. During autumn 1996, Hubble sent back two groups of images, one month apart, of a churning dust storm near the edge of the northern polar cap. This was the first time such an event had been recorded near the receding cap and it is presumed to have been caused by the large temperature differences between the polar ice and the dark regions to the south, which had been heated by the springtime sunshine.

The HST was also particularly useful to the Pathfinder mission controllers prior to the landing, since it was able to follow the progress of a major dust storm along Valles Marineris, which, had it developed into a global storm, might have affected the landing conditions at Ares Vallis. Fortunately it did not and the landing proceeded without atmospheric interference.

Comparing the appearance of the planet at the time of the 1997 Pathfinder and Global Surveyor arrival with that of Mariner 9 and the Vikings during the 1970s, the telescope showed that both spacecraft would be experiencing weather rather different from that encountered by their Viking predecessors. At the time of the earlier approach, Mars was both dusty and (relatively) warm, but now it was colder, less dusty and cloudier. It also showed that some areas of the surface had been changed quite dramatically by the movement of wind-blown dust. A classic example was the dark albedo marking known as Cerberus, which had been seen by generations of astronomers as a dark area the size of California. At the time of the Pathfinder approach, Cerberus had been reduced to three dark splotches, presumably because of the covering of most of the region by dust blown out of large impact craters by the wind.

The overall impression gleaned from Hubble observations is that the Martian climate is more chaotic than was once thought. One week the sky is pink and cloudless, the atmosphere being heavily charged with wind-blown dust; the next week, the air is dust free, the temperature lower by 40°C, and the sky dotted with brilliant water-ice clouds. Furthermore, Mars is apparently more often cloudy than dusty, which was not an impression derived from the 1970s observations. At the time of Mariner 9, Mars was at perihelion (the season during which huge dust storms sweep across the planet's surface) and it was not until the HST was able to image Mars during aphelion that it was realized that water-ice clouds are far more prevalent than dust in the Martian atmosphere. Incidentally, one major difference between Martian and terrestrial weather is that sudden changes are not merely local in effect but may change the appearance of the entire planet. This is explicable partly by Mars' more elliptical orbit, partly because of its extremely tenuous atmosphere and partly by very strong interactions between dust and water-ice clouds in the atmosphere.

Martian stratigraphy

The mapping of Mars, mainly by US Geological Survey (USGS) scientists but also in collaboration with workers from other countries and organizations researching within the NASA framework, has produced a global stratigraphy of Mars. A broad time-stratigraphic classification into three periods – Noachian, Hesperian and Amazonian – was achieved by Scott & Carr, who produced the first global geological map in 1978 (Scott & Carr 1978). Subsequently, after detailed study of Mariner 9 and Viking orbiter imagery, more detailed geological maps have been produced for the western hemisphere (Scott & Tanaka 1986), for the eastern hemisphere (Greeley & Guest 1986) and for the polar regions (Tanaka & Scott 1987). Tanaka published a detailed stratigraphy in 1986 (Tanaka 1986), subdividing the three-period time-stratigraphic units into series. This system is used throughout this book. The impact-crater density boundaries for each of the series are given in Table 3.1, and a tabulated list of the major geological units, their position within the Martian stratigraphic column, and their areal extent, are given in the Appendix. Finally, the distribution of the rocks of each series is shown in Plate 3.

Table 3.1 Impact crater-density boundaries for Martian series (after Tanaka 1986).

	Crater density (N = no. craters > (×) km diameter /10^6 km^2)			
Series	N (1)	N (2)	N (5)	N (16)
Upper Amazonian	<160	<40		
Middle Amazonian	160–600	40–150		<25
Lower Amazonian	600–1600	150–400	25–67	
Upper Hesperian	600–3000	400–750	67–125	
Lower Hesperian	3000–4800	750–1200	125–200	<25
Upper Noachian			200–400	25–100
Middle Noachian			>400	100–200
Lower Noachian				>200

4 Atmosphere, weather and climate

In many ways the dynamics of the Martian atmosphere are less complex than those of Earth. Because the atmosphere of Mars is very much more tenuous, it responds more slowly to changes in temperature and it also has a lesser heat capacity. Moreover, Mars has no oceans to influence the transport of heat or to provide extensive reservoirs of moisture with which the air can interact. However, recent HST observations have shown how very rapid changes can occur and that these may very quickly affect regions of continental size and even the entire planet, giving the climate a somewhat chaotic edge. The past climate may well have been very different, however, and there is a general consensus that the planet once had a very much denser atmosphere and that liquid water was able to survive at the surface (Pollack et al. 1987).

Mars' orbital properties are important in the context of the atmosphere. Thus, being 150 million km farther from the Sun than Earth is, the planet receives only half the amount of solar energy. Also, as its orbit is highly elliptical, it receives 40 per cent more insolation at perihelion than at aphelion.

The composition of the atmosphere

Early spectroscopic studies by Adams & Dunham, working at Mount Wilson Observatory during the 1930s, failed to find oxygen but did detect little or no water vapour in the atmosphere. Carbon dioxide certainly had not been detected at that time, and the most widely held assumption was that the main atmospheric constituent was probably nitrogen, and theory predicted that argon might also be a minor but significant constituent (see Kuiper 1952). Kuiper made the first positive identification of carbon dioxide in 1947 and it was not until the mid-1960s that the composition and atmospheric pressure became known with any degree of certainty.

The earliest accurate estimates of atmospheric pressure were those of Owen (1966) and Belton et al. (1968), who determined very low values that were confirmed later by Kliore et al. (1969) during the Mariner 6 and 7 missions. The average surface pressure has been found to be only 8 mbar – less than 1 per cent that of Earth – which means that a person standing at a level near to datum on Mars would experience a pressure close to that found at an altitude of 30 000 m above terrestrial sea level. Furthermore, there is a range in elevation of over 30 km on Mars; consequently, because pressure varies with altitude, the range in surface pressure is nearly an order of magnitude greater than on Earth. Thus, at the summit of Olympus Mons (27 km above datum) the air would be exceedingly thin and would hinder the safe landing of a Viking-type spacecraft, as the aerobraking factor would be very weak.

Because of the rarefied atmosphere and the low temperatures characteristic of the planet, liquid water becomes unstable and freezes on the surface. Another manifestation of the low atmospheric pressure is the relative ease with which particulate material can be raised into the air and transported across the Martian surface. Clouds of dust so formed may coalesce into global dust storms, which are a common feature, particularly close to perihelion.

Table 4.1 illustrates how the Martian atmosphere is composed mainly of carbon dioxide (CO_2), with nitrogen (N_2) and Argon (Ar) next in abundance; very little oxygen is present. Although water vapour is scarce, at night-time temperatures the atmosphere is close to saturation for this component almost everywhere, and

Table 4.1 Composition of the atmosphere at the surface (<120km), after (Owen et al. 1977).

Component	Proportion
Carbon dioxide	95.32%
Nitrogen	2.7%
Argon	1.6%
Oxygen	0.13%
Carbon monoxide	0.07%
Water vapour	0.03%
Neon	2.5 ppm
Krypton	0.3 ppm
Xenon	0.08 ppm
Ozone	0.03 ppm

theory predicts that water may be locked in the subsurface as ice, which may lie at quite shallow depths at latitudes higher than 40°. Dust is a component of the atmosphere not shown in the table and is believed to be present at surface level to the extent of between 5 and 30 particles per cubic centimetre.

It should be noted that the concentrations of both water vapour and dust vary widely between the seasons and, indeed, with location on the planet. Thus, the average column abundance of water vapour – measured in precipitable microns (pr-μm), which is the depth (in 10^{-6} m) of water formed if the atmospheric water vapour were precipitated on the surface as a liquid – is about 15 pr-μm and ranges between zero in the winter months at the pole to as much as 100 pr-μm at the northern pole during the early summer.

Observations made by the Viking spacecraft during 1976–8 gave the impression of a warm dust-laden atmosphere that did not permit planet-wide saturation of water vapour below an altitude of 30 km. However, we now know that the atmosphere is typically around 2°K cooler than this under "normal" conditions; indeed, Mariner 9 showed it to be even colder during the late northern spring of 1972. This being so, the average saturation height of water vapour in low to middle latitudes will vary on an annual basis, probably from <10 km at aphelion (i.e. late northern spring) to >30 km at perihelion (i.e. late northern autumn). For the period 1990–5, HST observations confirm that there are such large annual variations in water vapour saturation. Clancy & Nair (1996) noted that, because of the long photochemical lifetimes of CO and O_2, they are to be found in non-equilibrium with the much more rapidly varying (annually variable) water-vapour density.

A series of measurements were made by the HST during February of 1995, which provided an insight into the column abundances of atmospheric ozone (O_3) during the northern hemisphere spring, when the solar longitude (L_S)[1] is 62.5° (Clancy et al. 1996). In this season there was a relatively constant ozone column of 2.5–3.1 μm atm, which increased to 7 μm near the north pole. From these data it can reasonably be inferred that the "zero altitude" ozone column in lower latitudes would be around 3.1 μm atm. Although there is a measure of uncertainty surrounding the ozone abundances above 10 km altitude, the HST data are roughly twice those measured by Espenak et al. (1991) at L_S=208°, during northern hemisphere autumn.

The differences in the column abundance of ozone between the seasons are inconsistent with the predictions made in the Clancy & Nair (1996) photochemical model, which suggests that there should be a large perihelion-to-aphelion variation in Martian low- to middle-latitude ozone abundances, because of the eccentricity of the planet's orbit. The latter phenomenon results in a wide annual variation in the altitude at which Mars' atmosphere becomes saturated with water vapour.

Pressure and temperature variations

Viking lander results

The abundances of minor constituents (e.g. H_2O, O_2 and O_3) and the total surface pressure vary with the season, the former also varying with geographical location. The total pressure was measured at both Viking lander sites, where an adequate run of data was obtained to show that during southern winter there was a gradual decrease in the surface atmospheric pressure, which was then reversed as spring approached (Fig. 4.1). At the Viking 1 site the pressure ranged from 6.7 mbar during northern summer to 8.8 mbar at the beginning of northern winter (Hess et al. 1977, Ryan et al. 1978); at the Viking 2 site the equivalent data were 7.4 mbar and 10 mbar, the higher values here probably reflecting the somewhat lower elevation of this site. Similar values were recorded by Mars Pathfinder.

It seems reasonable to assume that this behaviour reflects the seasonal condensation of CO_2 on the southern polar cap, thereby reducing the atmospheric mass during southern winter. With the onset of spring, CO_2 would be released again and produce an atmospheric mass gain. Because winters in the southern hemisphere are currently both longer and colder than those in the north, the southern cap is larger than its northern counterpart and it therefore incorporates a correspondingly greater volume of atmospheric CO_2

1. Since the seasonal behaviour of the caps is discussed in the following text, the reader should be aware of the convention of identifying the Martian seasons by aerocentric longitude of the Sun (L_S) as follows: L_S= 90° is equivalent to northern summer solstice, L_S = 180° is northern autumn, L_S = 270° is northern winter solstice, and L_S = 0° is northern spring equinox.

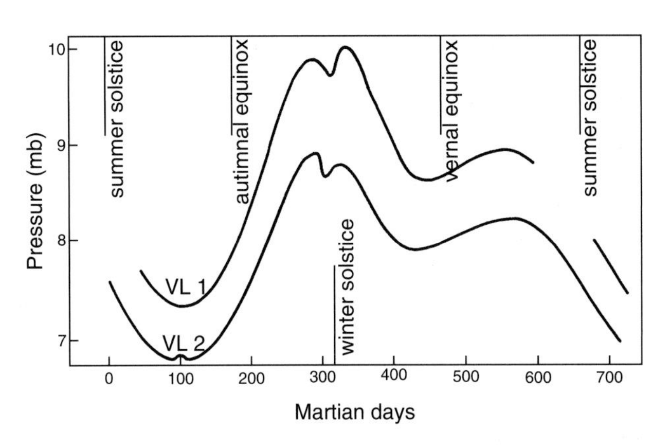

Figure 4.1 Mean diurnal pressure variation at the two Viking lander sites for one Martian year (after Hess et al. 1979).

during southern winter. Thus, it is that variations in the southern hemisphere tend to dominate the seasonal pressure cycle at the present time.

The extreme thinness of the Martian air means that it has a low heat capacity; it therefore cools and heats up much more quickly than Earth's atmosphere. Furthermore, since it is mainly composed of CO_2, which is a good infrared transmitter, large diurnal temperature ranges are experienced in the lower levels of the atmosphere. When there is little or no dust suspended in the atmosphere, Martian air absorbs little solar energy directly, its temperature profile being controlled principally by heat transfer from the ground, through either conduction or convection. The consequence of this is that, whereas there may be diurnal fluctuations in temperature of 50°C at the surface, these peter out quickly with increasing height; tidal winds are therefore weak, since only the near-surface layer is affected. However, in the presence of dust, solar radiation is absorbed and the air is heated directly. As a result, the atmosphere cools more efficiently at night, so the temperature profile becomes more nearly isothermal (Fig. 4.2). Under these conditions there will be narrower variations of temperature near the ground but wider ones at greater altitude. Furthermore, tidal winds may become stronger under dusty conditions.

Mars Pathfinder data

Data obtained by the Atmospheric Structure Investigation/Meteorology experiment (ASI/MET) during the entry, descent and landing of the Mars Pathfinder spacecraft has added much to our knowledge of the atmosphere and weather. The surface meteorology was monitored for 83 sols. During the first 30 sols, 51 equally spaced measurement sessions, 3 min long, monitored the diurnal cycle and, of course, the day-to-day variations that occurred. Then, on sol 25, a 24 h session sampled science data at 4 s intervals for one complete daily cycle. Measurements were made of atmospheric pressure and structure from 160 km all the way to the surface; but, since only the accelerometer could be utilized prior to parachute deployment, temperatures were not directly measured during the early stages of descent. Once the parachutes were opened, both temperature and pressure were measured by the MET instruments.

There were many similarities between the Viking 1 lander and Mars Pathfinder in terms of seasonal situation: Viking 1 arrived during mid-summer and Pathfinder during late summer; the sunspot count was similar on both occasions, and there was only a small difference between the two in the Mars–Sun distance (smaller during the Pathfinder landing). Furthermore, the two landing sites were at similar longitude and latitude and the amount of atmospheric dust was also comparable, but they landed at a different time of day (Viking 1 at 16.15 LST, Mars Pathfinder at 03.00 LST).

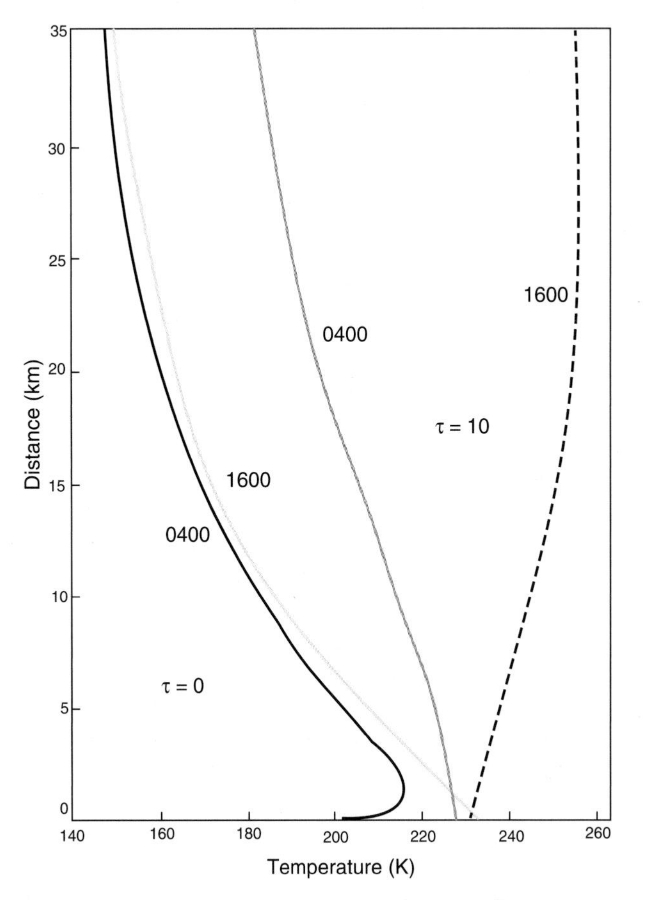

Figure 4.2 Temperature as a function of altitude. The large near-surface fluctuations die out with increasing height.

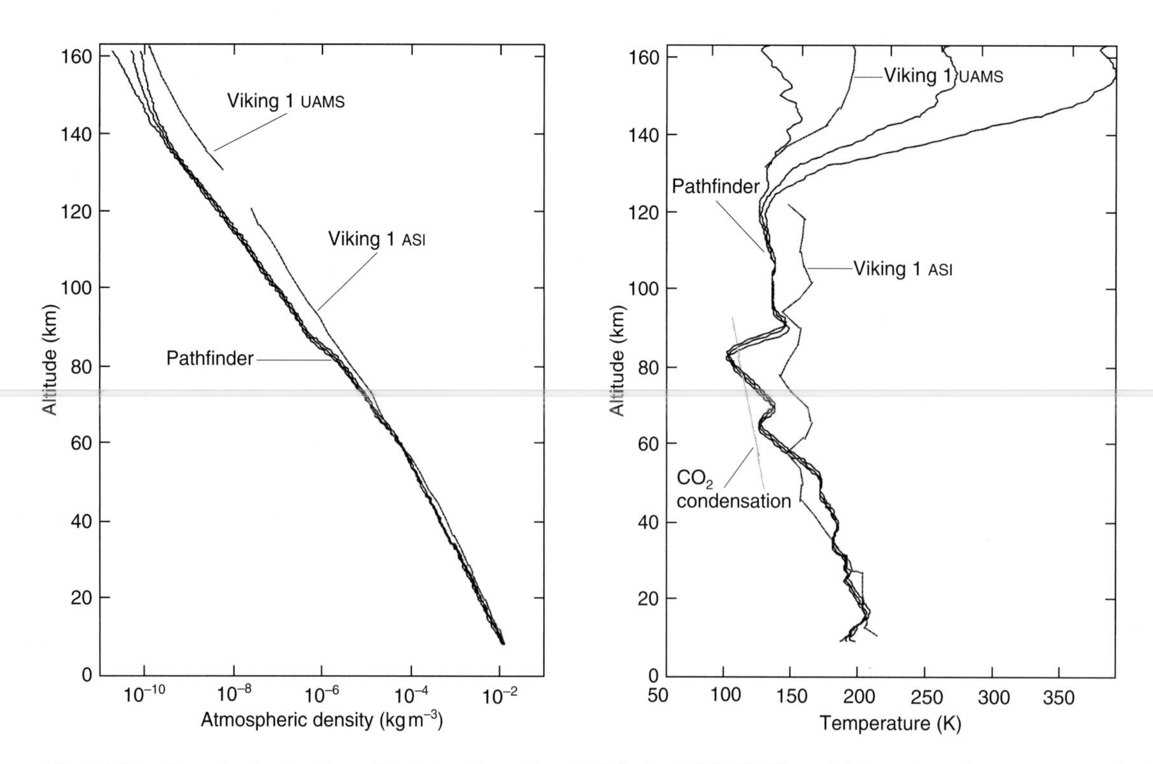

Figure 4.3 **(a)** The atmospheric density profile derived from Mars Pathfinder ASI/MET. The solid lines show the mean atmospheric density profile derived from the accelerometer data. The MET observations of surface pressure and temperature indicate atmospheric density of 176 kg m^{-3}, marked by an oval on the x-axis of the figure. Results from the Viking 1 lander atmospheric structure instrument (ASI) and the Viking 1 upper atmosphere mass spectrometer (UAMS) are also plotted for comparison. (Mars Pathfinder image PIA00926) **(b)** The atmospheric temperature profile derived from the Mars Pathfinder atmospheric density profile. In order to derive temperatures, a molecular weight versus atmospheric density model was constructed based on the results of the Viking UAMS. At lower altitudes, the Martian atmosphere is well mixed, with a constant molecular weight of 43.49. Temperature profiles from the Viking 1 lander ASI and UAMS experiments the CO_2 condensation temperature profile, and the surface temperature measured by the Pathfinder MET instrument (circle) are also shown for comparison. (From Schofield et al. 1997 and available as MPF image PIA00976.)

Atmospheric densities measured by Pathfinder on its descent varied from ~5 × 10–11 kg m^{-3} at the threshold of detection to 8000 kg m^{-3} immediately prior to parachute deployment at an altitude of 9 km (Fig. 4.3a). Between 160 km and 90 km altitude, densities range from a factor of 5 lower than Viking 1 lander values near and above 120 km to a factor of 2.5 lower near 90 km. The increase in density between 90 and 80 km, which corresponds to a deep temperature minimum (Fig. 4.3a), raises the Pathfinder densities at lower altitudes to values slightly lower than Viking 1 lander densities. The lower values of density, and therefore pressure, encountered by Pathfinder below 30 km are explicable by the lower overall mass and surface pressure of the Martian atmosphere at the time of the Pathfinder landing (Haberle et al. 1997), the decreased surface pressure being a consequence of the annual variation in atmospheric mass caused by condensation and sublimation from the polar caps (Zurek et al. 1992).

The Martian thermosphere, where temperature increases rapidly with altitude because of heating by solar ultraviolet radiation (very short wavelength), is evident above 125 km in the Pathfinder profile, and it appears that Pathfinder temperatures are close to or slightly higher than those measured by the Viking 1 lander. From 65 to 125 km, temperatures were on average 20 K lower than those observed by the Viking 1 lander (Fig. 4.3b). This contrast is responsible for the lower Pathfinder densities above 90 km. The temperature minimum of 92 K at 80 km is the lowest ever measured in the Martian atmosphere.

At an altitude of 80 km, the Pathfinder temperature profile is lower than the CO_2 condensation temperature and is possibly a response to supercooling, but it is also possible that CO_2 could be condensing at these levels to form high-altitude clouds. Below 60 km, temperatures are higher than those measured by the Viking 1 lander down to 35 km and are similar or slightly lower at lower altitudes down to 16.5 km. Below this a strong thermal inversion is present (Fig. 4.3b). At the base of the Pathfinder profile, temperature appears to increase again with decreasing altitude.

The temperature minimum in the 10 km inversion

is well below the condensation temperature of water vapour in the Martian atmosphere (Smith et al. 1997). This may mark the altitude of clouds seen in IMP images before sunrise and near the low-latitude morning terminator in HST images.

Mars Pathfinder landed in Ares Vallis, where the pressure, temperature and wind velocity were measured, allowing the variability of the Martian atmosphere to be studied during mid-summer on a range of timescales. This revealed not only the local properties of the atmosphere and its interaction with the surface but also more global information on such topics as atmospheric dust loading, air circulation and the seasonal CO_2 cycle.

During sols 1 through 30, surface pressure at the landing site underwent substantial daily variations of 0.2 to 0.3 mbar, which were associated primarily with the large thermal tides in the thin Martian atmosphere. Daily pressure cycles were characterized by two minima and two maxima per sol, together with diurnal components, although there was considerable day-to-day variability (Fig. 4.4). The large semi-diurnal tidal oscillation points to atmospheric dustiness over broad regions of Mars and over an altitude range of at least 10–20 km.

A long-term trend in daily mean pressure was also seen. Thus, the mean pressure fell slowly at the beginning of the period and rose at the end, with a minimum just under 6.7 mbar near sol 20. This time corresponds to the annual deep minimum in the seasonal pressure cycle associated with condensation and sublimation of CO_2 in the polar regions of Mars and was previously recorded by the Viking landers (Zurek et al. 1992). The instrumentation also detected pressure variations on relatively short timescales that ranged from seconds to hours and had magnitudes of 1–50 μbar. The shorter timescale variations (<10–15 min) appear to be related to wind and temperature fluctuations and tend to be greatest during the late morning and early afternoon, when the boundary layer is most turbulent. The most dramatic pressure features were minima of 10–50 μbar, usually less than a minute in duration, associated with dust devils passing over the landing site.

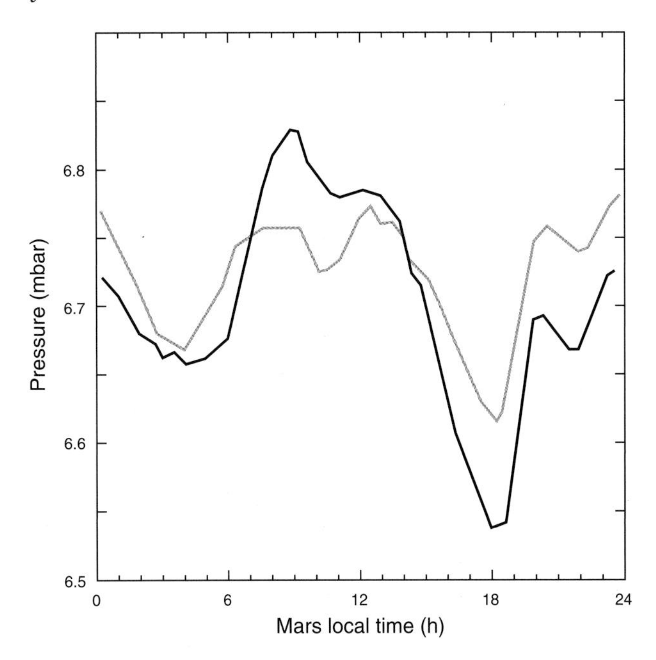

Figure 4.4 Diurnal pressure cycles for sols 9 (black curve) through 19 (grey curve), illustrating the day-to-day changes in the diurnal pressure cycle.

In common with the measurements made by Viking lander 1 at the same season, the diurnal temperature variations at the three Pathfinder levels repeat from day to day with a high degree of consistency. The diurnal cycle was sampled particularly well on sol 25 (Fig. 4.5). For the top mast thermocouple, a typical maximum temperature was 263 K at 14.15 LST and a typical minimum was 197 K shortly before sunrise at 05.15 LST. Diurnal temperature extremes at the foot of the mast exceeded those at the top because the low density of the Martian atmosphere means that near-surface atmospheric temperatures are influenced by the surface temperature cycle, which is driven by solar heating during the day and infrared cooling at night. Temperatures at the top of the mast (1.1 m) were greater by 10 K during the day and 12 K at night than those seen at 1.6 m above the surface by the Viking 1 lander, probably because of the lower albedo and greater thermal inertia of the surface at the Pathfinder landing site.

Figure 4.5 illustrates how the vertical temperature gradient also shows a diurnal variation. At sunrise the atmosphere is stratified and relatively stable, cool dense air lying near the surface; then, as the surface warms, so the air mass is heated from below until, by about 06.30 LST, all three mast temperatures are equal, indicating that the near-surface atmosphere is neutrally stable. By 07.30 LST, ground heating exceeds the ability of the atmosphere to transfer surface heat by conduction, the temperature gradient therefore reverses, the atmosphere becoming unstable, and convection begins. Later in the afternoon, the surface cools and turbulent mixing diminishes, such that by 16:45 LST the thermal profile is neutral and the observable surface winds convect heat quickly. Shortly afterwards, surface cooling causes the temperature gradient to invert and the surface boundary layer becomes stable and stratified throughout the night. Major night-time temperature fluctuations are caused by down-slope winds that disturb the surface boundary layer.

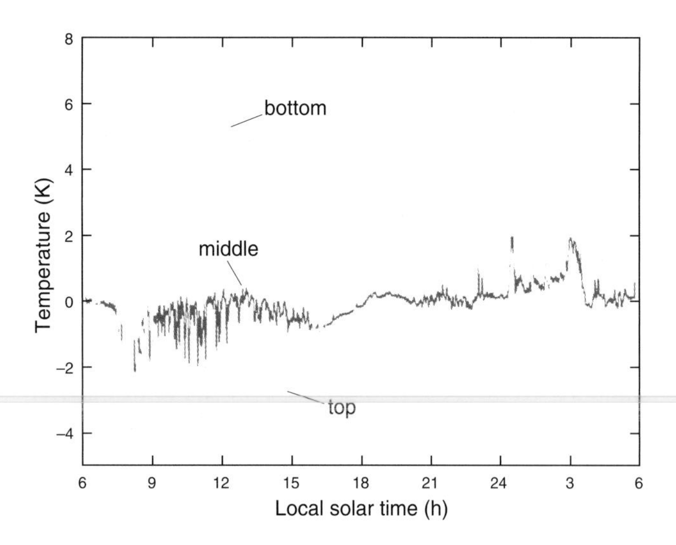

Figure 4.5 The diurnal variation of atmospheric temperature measured by the top, middle, and bottom mast thermocouples, from 06:00 LST on sol 25 to 06:00 LST on sol 26. These thermocouples are respectively 100, 50, and 2 cm above the plane of the lander solar panels. Temperatures are sampled continuously at 4 s intervals throughout this period, but the plots use 30-point (2 min) running means for clarity (this smoothing reduces the amplitude and frequency of the fluctuations that are present in the raw data).

Mars Global Surveyor thermospheric measurements

The z-axis accelerometer on board the Mars Global Surveyor obtained 200 measurements of the vertical structure of the upper atmosphere, including thermospheric density, temperature and pressure, ranging in altitude from 110 to 170 km during its aerobraking manoeuvres (Keating et al. 1998). The advantage of these data over previous Viking lander and Pathfinder measurements was that the MGS data was collected over a five-month period that spanned northern autumn, between September 1997 and February 1998.

The Martian thermosphere is the level of the upper atmosphere where the global mean temperatures increase with height above a minimum (~120 K) at 100–120 km to a maximum (200–350 K) above a height of 150 km. At the greater altitudes, the thermosphere is dominated by the absorption of ultraviolet radiation of very short wavelength and by the diffusive effects of gases such as CO_2, O, N_2 and CO. The variations measured at this level in the atmosphere are a response to diurnal, seasonal and solar-cycle variations in the amount of solar radiation received at Mars, and also by the ability of energy to propagate upwards from the lower atmospheric levels (i.e. below 100 km).

The measurements of atmospheric density within the upper thermosphere, at a height of 160 km, are shown in Figure 4.6. They are different on inbound and outbound passage and, in every case, densities to the south are greater than those to the north. During this northern autumn season, densities are observed to drop from 57°N to 33°N by 75 per cent.

Two increases in density were measured at 160 km, one maximum occurring after periapsis 50 and the other after periapsis 90. The first of these was related to a regional dust storm that developed in Noachis Terra (initially centred near 40°S, 20°E). The second may have been caused by an increase in extreme ultraviolet radiation from the Sun, something that occurs over a 27-day cycle (the rotation period of the Sun). During the former event, three days after the storm was detected the TES instrument recorded a doubling in dust opacity at mid-northern latitudes, which would have been accompanied by substantial dust-storm heating of the lower atmosphere. That the thermospheric effects were recorded up to heights of 140 km during this period gives some indication of the degree to which the lower atmosphere and thermosphere appear to be coupled.

Much of the orbit-to-orbit variation observed by the MGS accelerometer within the thermosphere appears to be explained by global-scale variation with respect to longitude at these heights. There is a clear two-wave pattern that is obvious in Figure 4.7, peaks occurring around longitudes that correspond to the highlands of Tharsis and Arabia. The clear inference is, therefore, that the considerable relief of these regions effectively forces the atmosphere upwards, propagating waves on a planetary scale.

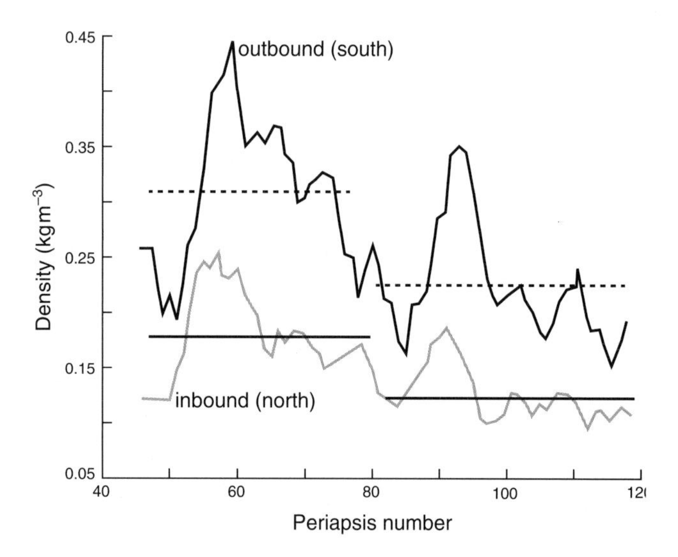

Figure 4.6 Means of thermospheric densities measured by the MGS accelerometer during five orbits, between November 1997 until January 1998, north (inbound) and south (outbound) of periapsis, separated on average by 24° of latitude. Horizontal lines show mean inbound (solid line) and outbound (dotted line) before and after periapsis 80. (After Keating et al. 1998.)

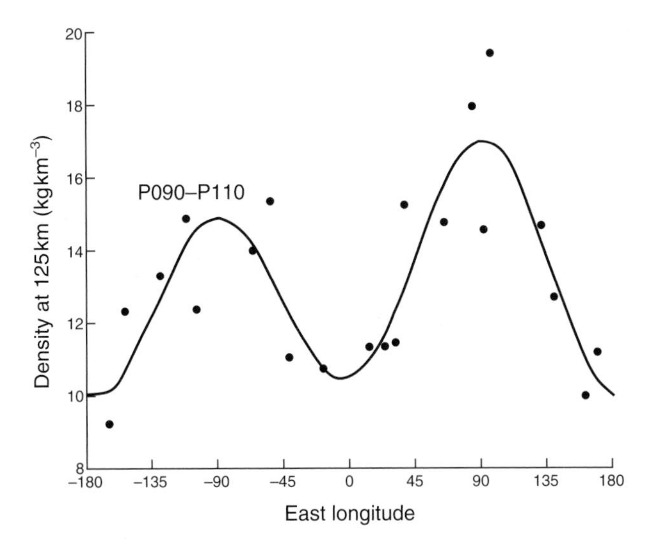

Figure 4.7 Thermospheric density, normalized to 125 km altitude, as a function of longitude, measured from P090 until P110 by the MGS accelerometer.

Winds and atmospheric circulation pattern

The absence of large expanses of open water on Mars means that the entire surface responds quickly to solar heating: solid rocks respond more rapidly than water. In this respect, therefore, the circulation of the Martian atmosphere is simpler than that of Earth. On the other hand, topographic effects are more significant and complicate the pattern to a greater degree than they do on Earth. Modelling of the Martian circulation has been undertaken by several groups, utilizing techniques first developed for Earth (Leovy & Mintz 1969, Blumsack 1971, Pollack et al. 1976, Webster 1977). More recent work has been undertaken by groups led by Haberle (Haberle et al. 1993) and Barnes (Barnes et al. 1993).

In broad terms, warm air on Mars rises over the hemisphere that is experiencing summer (summer hemisphere) and descends over that experiencing winter (winter hemisphere); the global circulation is, therefore, driven by seasonal temperature gradients. The peculiarly Martian phenomenon that sees approximately 30 per cent of the air condensing on the winter pole creates a high-latitude region of relatively low pressures. The marked pressure gradient set up by this seasonal activity induces a strong global circulation towards whichever pole is growing by the condensation of CO_2. Such condensation flow tends to dominate wind directions at all latitudes. At mid- to high latitudes in the winter hemisphere, westerly winds prevail, and both cyclonic and anticyclonic weather systems move across the planet (Plate 1). The Martian weather at these latitudes in winter is much like that on Earth. Summers in each hemisphere differ from both winter and each other. Thus, northern summers see quiet conditions and only a small east–west zonal airflow; cyclonic systems generally do not develop. In contrast, southern summers see frequent dust storms developing; these sometimes affect the entire planet and disrupt the general circulation pattern set up by the global north–south temperature gradients. Viking recorded 35 such storms during 1977, of which two were global in extent. The storms coincide with the retreat of the southern polar cap, which creates large pressure gradients between the freshly exposed surface and the shrinking cap itself. During spring and autumn, conditions are quiet in both hemispheres.

The atmospheric circulation has been extensively modelled, in particular by the NASA Ames general circulation model (GCM), a sophisticated model that allows scientists to describe the likely long-term circulation pattern and how it responds to seasonal variations and dust loading (Pollack et al. 1990). Comparisons between the derived model and data from the Mariner 9 and Viking missions allow for checks to be made, at least over the relatively short periods represented by the spacecraft data.

Meridional circulation, characterized by Hadley cells that extend to at least 20 km altitude, dominates latitudes lower than 30° (Haberle et al. 1993). The cells undergo significant seasonal changes: two roughly symmetrical cells develop at the equinoxes, sharing a common rising branch focused close to the equator (Fig. 4.8a), and at the solstices a single cell operates across the equator between 30°N and 30°S (Fig. 4.8b).

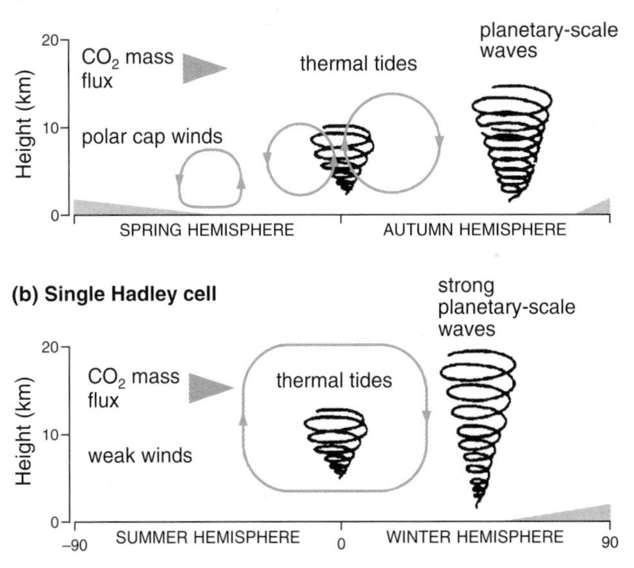

Figure 4.8 The general circulation of the lower Martian atmosphere: **(a)** at the equinoxes and **(b)** at the solstices (from Haberle in Kieffer et al. 1992).

At the solstices the zonal-mean circulation consists of a single Hadley cell that flows across the equatorial zone. When the air is modestly loaded with dust, the atmospheric flux peaks at around 10 billion $kg\,s^{-1}$ at the northern winter solstice, and roughly half this at the southern winter solstice. During both seasons, predominantly westerly winds blow in the winter hemisphere and easterlies in the summer one. The maximum velocity of the zonal winds (which occurs at around 47 km altitude) occurs at both seasons and is around $120\,m\,s^{-1}$ in the winter hemisphere and around half this in the summer hemisphere. The model predicts that westerly winds would dominate at the surface of the planet in mid- to high latitudes in the winter hemisphere, and also in the summer hemisphere close to the rising branch of the Hadley cell operating during that season.

The Hadley cell is quite a strong feature and has been described as having the form of a jet, being particularly vigorous ($>20\,m\,s^{-1}$) at the northern winter solstice. As more atmospheric dust enters the atmosphere, so the zonal mean circulation intensifies and can easily double. During the periods around the solstices, the eddies play a relatively minor role, mean meridional circulation contributing most to the maintenance of the thermal and momentum balance. Around the equinoxes, the circulation is rather like that of Earth, comprising two approximately symmetric Hadley cells characterized in mid-latitudes by westerly winds and in the tropics by easterlies. Zonal winds are predicted by the model to be roughly half the strength of those typical at the solstices. At those times the mean circulation is also relatively weak, averaging between 5 million and 100 million $kg\,s^{-1}$.

At latitudes higher than 30°, the general circulation is less regular. During the northern hemisphere winter, eastward-propagating weather systems, or "eddies", which may be either of low or high pressure, were detected by the Viking landers and showed considerable seasonal variation, virtually disappearing during the summer months in the northern hemisphere. These mobile weather systems appear to arise because of changeability in the zonal jet stream by baroclinic instability (Barnes et al. 1993), again having been modelled by the Ames GCM. In the northern part of the tropical zone in the northern hemisphere, during northern autumn, winter and spring, there is strong eddy activity. This remains strong, even with considerable atmospheric dust loadings, although it shifts somewhat northwards when dust levels are high. The longer-lived eddies (7–10 days) have a deep vertical structure, evidently generating their energy via large meridional and vertical heat fluxes.

Interestingly, during simulations with the GCM, only very weak eddy activity is predicted during southern winter. This is probably a function of the rather different topography at these southern latitudes: elevation is generally high here and the southern ice cap extensive; steep symmetrical slopes (a stabilizing influence on baroclinic instability) are absent here (but common in northern latitudes). Together these are thought to weaken transient eddy activity.

A more recent modelling exercise by Santee & Crisp (1995), of circulation during late southern hemisphere summer, indicates a two-cell circulation pattern, the air cells rising over the equator, flowing towards the poles and sinking over both polar regions. Flow returns at low altitude. The maximum poleward flow velocity was found to be $3\,m\,s^{-1}$ in tropical latitudes at an altitude of 55 km, with a maximum vertical velocity of $2.5\,cm\,s^{-1}$ downward over the north pole at a height of 60 km. Santee & Crisp calculate that if such rates are sustained over an entire Martian season, the atmosphere about the 1 mb level could be overturned in a mere 38 days.

Martian winds

Easterly winds prevail in tropical latitudes throughout the year and extend into the summer hemisphere at the solstices. In the winter hemisphere at the solstices, westerly winds are prevalent, and these extend into the middle and higher latitudes at the solstices. At higher latitudes things are less clear, the circulation being less symmetric with respect to latitude. Thus, both of the Viking landers detected eastward high- and low-pressure systems during winter in the northern hemisphere. These have been termed eddies and, during Martian summer, they virtually disappear. Nothing is known of their actual behaviour in the southern hemisphere, although modelling has been conducted (see above).

More local winds are the result of topography. This is much more pronounced on Mars because of the atmosphere's rapid response to local ground temperatures. Because temperatures are mainly a function of reflectivity and the local cycle of insulation, these may be much the same at the top of a large volcano such as Olympus Mons as at its base. The large horizontal temperature gradients so produced generate quite strong slope winds. Such winds are downslope at night and upslope during daytime.

The Mars Pathfinder ASI/MET wind sensor measured wind speed and direction 1.1 m above the base of the mast. For sols 1 to 30, wind direction generally rotated in a clockwise manner through all the points of the compass during the course of each sol (Fig. 4.9).

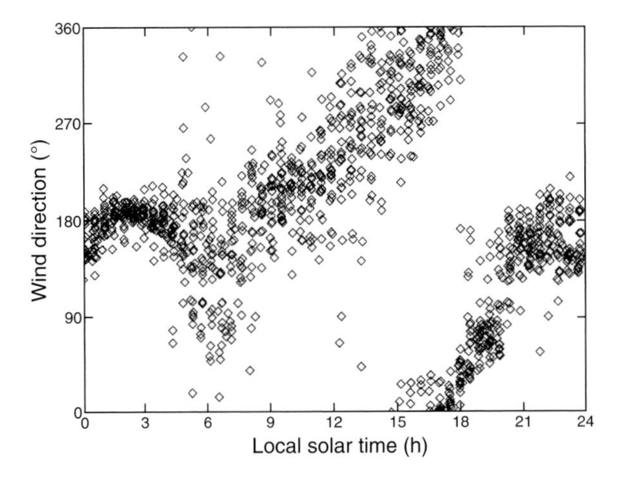

Figure 4.9 Time-averaged wind direction measured by the MET instrument over the first 30 sols of the Mars Pathfinder lander mission, plotted as a function of LST. Each point represents an average over a 3 min period. Wind direction is defined as follows: 0° and 360°, 90°, 180° and 270°.

This rotation was not uniform. Winds were consistently from the south in the late night and early morning, and then rotated steadily through west, north, and east during the day. Over the 30-day period studied, night-time wind direction was remarkably constant, but much more variation was seen during the day.

The recurrent southerly wind from late evening through morning is consistent with a drainage flow down Ares Vallis, which rises to the south of the lander, and the northerly wind seen in the afternoon is indicative of flow up the channel. Although the clockwise rotation of the wind vector agrees with that expected from the westward-migrating classical diurnal thermal tide, the time phasing of wind direction throughout the day does not. The classical tidal drive would generate a westerly maximum near 18:00 LST and a southerly maximum at about 12:00 LST. It therefore appears that local topography, or possibly non-classical tides, are controlling the wind direction at the Pathfinder location during mid-summer.

Winds measured at the Viking 1 lander site at this same season 11 Mars years earlier were generally weak (<6 m s^{-1}), exhibited a time-averaged northwesterly direction, and were approximately upslope during the afternoon and downslope during the night and early morning. Higher wind speeds accompanied the increased pressure oscillations seen during the several sols surrounding $L_S = 150$. If the winds are slope driven during this season, the differences between Viking 1 lander and Pathfinder winds are expected to reflect differences in the magnitude and direction of the slope at the two sites. Finally, preliminary estimates suggest that windspeeds were comparable with or lower than those measured by Viking 1 lander at the same time of year. Speeds were generally less than 5–10 m s^{-1}, except during the passage of dust devils, and were often less than 1 m s^{-1} in the morning hours. This may be consistent with the lower slope at the Pathfinder site.

Clouds and dust storms

Clouds

Despite the small amount of H_2O, the atmosphere is always close to saturation in water vapour; consequently, the formation of clouds and fogs is a common feature of Martian weather (Plate 1). Larger cloud masses give rise to transient brightenings over whole regions of the planet and have often been recorded by observers using telescopes.

One of the most extensive regional cloud masses is that which forms over the north polar cap during autumn. The northern autumnal "hood" is pervasive and may extend as far south as 50°N; thus far, only scattered clouds have been recorded at the southern cap during this season. However, an extensive hood does form around the southern cap during early spring. The polar hoods appear to represent thick hazes of water ice (and possibly carbon dioxide ice).

Hazes and fogs also often form in low-lying areas at dawn and dusk. These are probably composed of water ice and are produced by solar vaporization of ground frost. Early-morning fog is common along the canyon floors of Valles Marineris, particularly among the interconnecting canyons of Noctis Labyrinthus.

Clouds also form in the lee of major topographic features such as impact craters. Such wave clouds also develop around the polar caps when the hood is developed. During spring and summer, moisture-laden air is forced to rise over major features such as the Tharsis Montes and Olympus Mons, resulting in the growth of orographic clouds. These tend to develop slowly during the morning and reach a maximum during the afternoon. Convective clouds also develop in elevated regions, usually around mid-day. Areas such as Syria Planum are affected in this way. The formation of such clouds appears to be the result of atmospheric instabilities created by strong surface heating. Finally, regional or even global obscuring of the surface of Mars may be affected by dust storms; these are the yellow clouds recorded frequently by Earth-based observers.

Dust devils

Short-term variations in measured surface pressure, wind velocity and air temperature over periods of tens of seconds to several minutes suggest that small-scale

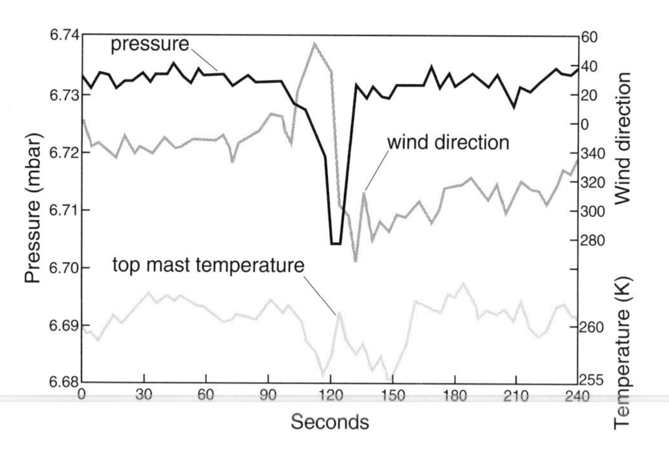

Figure 4.10 Pressure, wind, and temperature changes associated with a small-scale vortex, or dust devil, passing through the Pathfinder landing site; the measurements were taken at 4 s intervals.

convective vortices passed through the Mars Pathfinder landing site. During the passage of one of these features, pressure fell and recovered rapidly, accompanied by abrupt shifts in wind direction. These events, which generate dust devils, are probably similar in character to features noted in the Viking 1 and Viking 2 lander meteorology data (Ryan & Lucich 1983) and may have been seen by the Viking orbiter cameras. The passage of these vortices is marked by a narrow and well defined drop in pressure, but it was not seen in the Viking lander pressure record because of poor resolution and coarse temporal sampling. There are currently no Pathfinder lander images to indicate whether the apparent vortices entrain dust, but the passage of a particularly large feature on sol 62 was correlated with a short-lived reduction of about 1.5 per cent in the power generated by the Pathfinder solar panels.

Mars Global Surveyor was more successful and did capture the passage of a dust devil. This occurred at 14:15 LST on sol 25 and had the characteristics of a clockwise-rotating vortex travelling towards the south-southeast (160°), with the vortex centre passing west and south of the lander. During this event, pressure fell 0.028 mbar in 24 s to a minimum, followed by a more rapid rise of 0.027 mbar in 8 s. The pressure minimum was accompanied by a local temperature maximum recorded by the Pathfinder at all three thermocouple heights (Fig. 4.10). These temperature maxima are themselves embedded in a pronounced temperature minimum, which extended 30–40 s before and after the pressure minimum. Wind direction rotated from northwesterly, which was the mean flow direction before the vortex arrived, through north to northeasterly in the 40 s preceding the pressure minimum, before shifting abruptly through northerly to westerly at and just after the pressure minimum. The wind direction slowly returned to north-northwesterly over the minutes following the pressure minimum. This pattern is consistent with a clockwise-rotating vortex embedded in an ambient flow that has a speed less than the vortex rotational speed

Dust storms

The mantle of fine-grain dust that extends over most of the surface of Mars is occasionally lifted into the atmosphere by the winds, giving rise to dust storms of various sizes. The thinness of the air necessitates high wind strength to achieve this: *c.* 30 m sec^{-1} is the minimum required. Observations made from Earth and by orbiting spacecraft indicate that the strongest global storms develop around perihelion; a particularly large one – which left only the summits of the Tharsis Montes rising above the dust – greeting Mariner 9 as it arrived at Mars during late 1971. During the first year of the Viking mission, some medium-size (~1 million km^2) storms were observed, but only two global events. The latter, planet-encircling, storms typically occur during autumn and winter in the northern hemisphere, beginning as small local storms in southern subtropical latitudes and then expanding, first in a west–east, then in a north–south, direction.

The HST made a particularly fine observation of a regional storm during September 1996 (Fig. 4.11), images taken one month apart revealing the movement and change in area of the event. This kind of polar storm probably develops because of the wide difference in temperature between the polar ice and

Figure 4.11 Hubble space telescope image of dust storms near the north pole of Mars.

the dark regions to the south, which are heated by the spring sunshine. In November 1997, Mars Global Surveyor observed that a regional dust storm in the southern hemisphere triggered an unexpectedly large thermospheric response at mid-latitudes in the northern hemisphere (Keating et al. 1998). It has been suggested that this behaviour may be caused by topographically controlled waves propagating upwards from the lower atmosphere.

Modelling of size-dependent movement of dust-size particles within the atmosphere has been undertaken by Murphy et al. (1993) and indicates the importance of considering the full range of potential particle sizes within major dust storms. Thus, the extent of dust transport away from a prescribed source is a strong function of particle size. Enhanced latitudinal transport is achieved when dust of a range of non-spherical particle sizes (radii from submicrons to tens of microns) is entrained in the atmosphere. Compared to the amount of transport achieved for a single-size spherical fraction, a mixed-size fraction can be moved farther by virtue of the lower sedimentation speeds and lengthier suspension lifetimes of the smaller particles. This enables them to be raised more readily above the ground and transported farther horizontally, before settling towards the surface.

The HST has observed storms and is continuing to monitor them. From multi-spectral observations made of dust activity in the northern polar regions and around the Hellas basin on 8 April and 21 August 1995, the team of Wolff et al. (1997) were able to derive lower bounds for the atmospheric dust loading. A range in optical depth of from 0.5 to 0.8 for the higher latitudes was derived in this study.

Precessional effects and climatic change

As we saw in Chapter 1, the Martian seasons are similar to, if longer than, those of Earth, and are a function of the same orbital and rotational phenomena. Longer-term changes in the seasons are a manifestation of very slow variations in the various orbital and rotational parameters.

Precessional effects

Precession is a general term that describes the slow conical motion of a rotational axis, such as that of a spinning top. The rotational axis of Mars precesses over a period of 175 000 years, during which time there is a slow rotation of the line of intersection of the equator with the orbital plane. The axis of Mars' orbit also precesses and this causes a line joining aphelion and perihelion (termed the line of apsides) to rotate. This cycle is completed every 72 000 years and it causes a gradual shift in the position of perihelion in space. The combined effect of these two phenomena is to produce a 51 000- year cycle of climatic change (Leighton & Murray 1966). As we have seen, at present the southern hemisphere is tilted towards the Sun at perihelion; however, in about 25 000 years the northern hemisphere will be in a similar position and will experience short hot summers. The cycle will then repeat itself.

There are also some longer-term climatic changes that occur because of variations in orbital eccentricity. Today the eccentricity has a value is 0.093, but this ranges from 0.004 to 0.141 over a shorter period of 95 000 years and also over a longer period of about 2 million years (Ward 1974). Obliquity also changes; at present it is 25°, but over a period of 1.2 million years it varies between 14.9° and 35.5° (Ward 1974). This has an effect on the amount of solar radiation that strikes different latitudes; thus, when obliquity is high, more radiation reaches the polar regions, with the result that polar temperatures rise while equatorial temperatures fall. At low obliquities the reverse applies. This has an effect on the volume of volatiles that can be stored in the polar caps (see Ch. 12).

Precessional effects of this kind are used to explain the strongly layered deposits that characterize both polar regions. These are among the youngest deposits on the planet and they consist of a series of huge sheets, each approximately 30 m thick, that decrease in area upwards through the sequence, and which are crowned by the residual ice caps. The uniformity and continuous depositional nature of these geological units suggest that they were laid down by some kind of atmospheric sedimentation process, determined by precessional changes (Toon et al. 1980). Thus, variations in obliquity would have the effect of alternately increasing the temperature at the poles relative to lower latitudes (high obliquity) and reversing this trend at low obliquities. The result would be that, at high obliquities, the annual insolation would increase significantly at the poles and decrease at the equator. When the poles became relatively warm, CO_2 stored within the high-latitude regolith would be driven into the atmosphere, increasing the pressure, in turn allowing more dust to enter the atmosphere and thereby promoting dust sedimentation at the poles. The reverse would happen when obliquity was low, at which time the polar regions would become cooler, and permanent ice caps would form. Under these conditions of lower pressure, dust storms would be relatively rare and sedimentation rates extremely low.

An important spin-off from this changing pressure pattern is the effect it would have on water. At low obliquities, the permanent CO_2 ice caps would trap any water brought into contact with them and permanent water-ice caps would form. Then, when obliquity increased and dust storms became more frequent, the water-ice caps would become buried in dust sheets and eventually insulated from the atmosphere by ice. In this way, the oscillatory precessional effects produced layered terrain that comprises ice sheets created during periods of low obliquity and also dust layers generated when obliquity was high.

Climate history of Mars

Currently, Mars is locked in an icy grip by subzero temperatures, and geological activity is restricted largely to the polar regions, where dust storms and the seasonal cycle of cap advance and retreat are modifying the surface constantly. Elsewhere, surface modification is slow and is attributed predominantly to aeolian activity. This suggests that the planet appears to be well past the peak of its geological and climatic activity.

Naturally enough, the climatic history of Mars played an important part in the development of its surface and there is now a substantial body of evidence favouring a significantly warmer and wetter Mars in the past. For instance, there is now unequivocal evidence that fluvial activity was extensive during Noachian times, implying that water could exist in a stable form on the planet's surface at that time. Subsequently, during the Hesperian period, catastrophic floods that had their origins in extensive regions of collapsed and chaotic ground south of the dichotomy boundary scoured out huge outflow channels. Their formation is attributed to the massive release of volatiles from huge aquifers within the brecciated cratered uplands. Mars Pathfinder landed near the mouth of one such channel, Ares Vallis. It has been estimated that the volume of water required to cut the channels would have been equivalent to a global ocean at least 50 m deep.

Evidence from the landing site that supports the notion of a warmer and wetter planet includes:

- pebbly rocks, some resembling conglomerates
- soil horizons
- streamlined hills that must have been sculpted by rapidly flowing fluids
- weathering horizons that involve oxides of iron
- lag deposits
- grooved and striated rocks abraded by sand-size fines.

It was thought for many years that these floodwaters must have been extremely short-lived. However, the discovery of possible ancient shorelines, beach terraces and palaeolakes in the region of the northern plains, and of lake beds within the equatorial canyon system, challenges this old assumption. Although recent MGS imagery does not actively encourage the shoreline interpretation of landforms tentatively believed to be the result of "coastal" activity, this does not invalidate the palaeolake or ocean hypothesis. Not surprisingly, there has been and still is considerable research activity in this area, aimed at understanding whether or not such oceans and other bodies of standing water could exist and, if so, for how long, and also whether they could exist under a cover of ice. Since some of the stratified (supposed lake) deposits within the canyons are 5 km thick, this has been a very stimulating area of enquiry.

Given the notion that Mars once enjoyed warmer, conditions, the obvious question to be asked is what brought these about. Sagan & Chyba (1997) entered this argument by suggesting that enhanced amounts of reduced greenhouse gases, such as CH_4 and NH_3, would have helped to warm the planet early in its history and probably contributed to putting water at the surface. In other words, we are talking about greenhouse gases. Forget & Pierrehumbert (1997), following a similar line of argument, argued that the presence of a CO_2 and H_2O could have kept early Mars warm if the atmosphere has been regularly filled with CO_2 clouds. Such clouds tend to scatter rising infrared radiation instead of absorbing and re-radiating it; consequently, they would form a partially reflecting layer at infrared wavelengths. Since such clouds would also tend to form in the upper troposphere (as opposed to near the surface), they would have a strong warming effect.

Finally, if the planet did start out with a much more massive atmosphere, how might it have evolved into its present state? The most widely quoted method is by the conversion of CO_2 in the atmosphere into carbonates at the surface (Haberle 1998). Thus, in the presence of a water-laden atmosphere, CO_2 would have been removed from it via the weathering of silicate volcanic rocks. On Earth, these would have been recycled back into the atmosphere via plate tectonics. Since it seems unlikely that such a process ever operated for lengthy periods on Mars, the most likely way was by the re-cycling of CO_2 on Mars through volcanism, a phenomenon known to have been extremely important on early Mars.

5 Mars and volatiles

Volatiles are particularly vital to Earth, since it is their presence that has permitted life to evolve and which helps drive the process of plate tectonics. The latter not only constantly modifies the surface layer but also recycles Earth's crust and the chemical components from which it is built, some of which have been fixed in it from the atmosphere. Thus, there is a constant interchange of chemical elements, the more volatile of which are returned to the atmosphere for later recycling by the same process.

Earth initially had sufficient volatiles, and was able to retain enough of these, for a breathable atmosphere to evolve, something that played a vital part in the development of life, some 3850 million years ago. The interior is more or less continually degassing, releasing volatiles and other elements into the atmosphere and hydrosphere, there being a delicate balance between this and the fixing of various components in silicate and carbonate rocks. This is all a part of the hydrological and geochemical cycles of our planet. Other planets evolved in rather different ways and Mars is a prime case of an Earth-like world whose climate and volatile history has gone along a very different path.

Volatile distribution among the inner planets

During the early stages of Solar System evolution, the proto-Sun sent out huge numbers of charged particles during its T Tauri stage (early phase of vigorous particle activity). These constituted the solar wind, which continues to blow today but with considerably less vigour. The primordial atmospheres of both Mars and Earth were almost certainly rich in water vapour, carbon dioxide, carbon monoxide, nitrogen, hydrogen chloride and hydrogen. Most of the hydrogen quickly escaped into space, whereas some of the water vapour in their upper atmospheres was broken down by sunlight into hydrogen and oxygen, the latter escaping and combining with gases such as methane (CH_4) and carbon monoxide (CO) to form water (H_2O) and carbon dioxide (CO_2). The remaining hydrogen and helium was stripped off during the T Tauri stage and blasted out of the solar system into interstellar space. This cosmic cleansing process saw the stripping away of the primary atmospheres of all of the inner planets, so that only their solid parts remained.

When the inner planets had grown to about a tenth of their present volume, gravitational attraction became strong enough to cause the infall of large numbers of freely orbiting planetesimals (small accreting solid bodies). These struck at very high velocities, high enough to vaporize them upon impact and, when this happened, as it did on millions of separate occasions, volatiles were released, and these formed the primitive atmospheres. Non-volatile elements produced by the same process were condensed and they gradually built up an outer layer of molten silicate rock. As a result, early Mars, Venus and Earth were probably made up of a cool volatile-rich interior surrounded by a thick ocean of molten rock, perhaps 1000 km deep. The first solid particles to condense from the solar nebula were refractory materials (with high melting points) composed of metal oxides, chiefly of tungsten, aluminium and calcium. Then, as the nebular temperature slowly fell, these reacted with gases in the cloud to form silicates, chiefly of magnesium, iron and calcium.

With the passage of time, the decay of long-lived radioactive isotopes locked inside the inner planets gradually heated up their interiors, melting at least parts of them. Thus, under the influence of gravity, the denser materials sank towards the centre of mass, causing chemical redistribution, known as differentiation. Each planet developed an iron-rich core surrounded by layers of mainly silicate material. This was the position reached by Venus, Earth and Mars at this primitive stage, but much has happened since that time, the course of evolution of each of the planets taking a somewhat different path.

The present atmosphere of Mars is entirely secondary, as is Earth's, and has been derived from the interior by a process of degassing, which occurred as the mantle layer solidified. By the time that this took place, much of the metallic iron had sunk towards the core, and oxidizing conditions promoted by the relatively oxidized mantle resulted in the production of oxidized gaseous forms of nitrogen, sulphur and carbon (i.e. NO_2, SO_2 and CO_2).

Isotopic data relevant to Martian volatiles

The present loss rate of the Martian atmosphere is equivalent to a global sheet of water 3m deep. This rate was probably higher in the distant past and possibly sufficient to represent a water layer 80m deep. Not surprisingly there is considerable interest in establishing the size of the current reservoir.

Clues to the evolutionary history of the Martian atmosphere can be gleaned from the study of certain isotopes, especially those of nitrogen, argon, neon, xenon, oxygen and carbon. In particular, the noble gases are useful indicators, since, being inert, they are not removed from the atmosphere by chemical reactions. The present atmosphere of Mars may be quite different from the original one, in terms of both the relative proportions of the components and the actual amounts. For instance, much of the original CO_2 and H_2O may now be locked into the subsurface rocks and regolith, whereas exospheric processes in the upper reaches of the atmosphere may have stripped away a significant fraction of the original air.

Nitrogen, oxygen and carbon all have the potential to escape from the Martian gravity field, albeit very slowly. The potential for escape is biased towards the lighter isotopes of these elements (e.g. ^{12}C, ^{14}N and ^{16}O), which means that, should they escape from the upper levels of the air mantle, the lower levels should become relatively enriched in the heavier isotopes (e.g. ^{13}C, ^{15}N and ^{18}O). In 1976, Nier et al. reported that the naturally occurring heavy isotope of nitrogen (^{15}N) is enhanced by a factor of 1.7 over the terrestrial value. Subsequently, this was confirmed for the Viking 1 samples by Owen et al. (1977). On the basis of such an enrichment, McElroy et al. (1977) estimated that a significant portion of the nitrogen inventory of Mars must have been lost, with a corresponding loss of original water equivalent to a layer of water 130m deep, averaged over the whole planet. There is, surprisingly, no similar enrichment in ^{18}O, and McElroy and his colleagues attributed this fact to the existence of a large reservoir of oxygen in close proximity to the Martian surface, which could be in the form of H_2O or CO_2. Such a source would readily be interchangeable with oxygen in the atmosphere and, if large enough, could lead to dilution of the fractionation effects, even after 4.5 billion years of Martian evolution. On the assumption that the oxygen isotopes are fractionated by about 5 per cent, they estimated that the reservoir (possibly composed mainly of water) could be equivalent to a layer of water 13m deep over the entire surface of the planet. However, the actual degree of fractionation may be considerably less than this, with the consequence that the amount of degassed water could be far higher. Should this be so, then the estimates relating to nitrogen isotopes (a 130m layer of water) could be nearer the truth.

Owen et al. also made determinations of carbon and oxygen isotopes, finding them to be similar to terrestrial values, insofar as they could tell within the error limits of their measurement procedures (±10%). They also studied ^{36}Ar, and established that the ratio of ^{36}Ar to ^{40}Ar was only about one tenth the terrestrial value. When the Viking 2 lander was successfully deployed, they made further isotopic measurements and established that, with respect to the other xenon isotopes, ^{129}Xe was much more abundant (^{131}Xe and ^{132}Xe) in the Martian atmosphere than in that of Earth.

In an attempt to synthesize all the chemical information, Anders & Owen (1977) argued that two particular elements are important in terms of the way certain elemental groups behaved in the solar nebula during the formative stages of planetary development. These two, potassium (K) and tellurium (Tl), provide keys to the groups of elements that condense respectively at temperatures between 1200K and 600K, and <600K; tellurium (Tl) therefore provides clues to the group that contains carbon, nitrogen, chlorine, bromine and the noble gases, and naturally is of particular significance to studies of Mars. After examining the abundances of these two elements in meteorites, on the Moon and on Earth, Anders & Owen estimated that the Martian atmosphere contains 100 ppm K and 0.14 ppb Tl, their quoted figure for Tl being estimated from the measured ratio of ^{40}Ar to ^{36}Ar. Now the terrestrial value for Tl (4.9 ppb) is substantially greater than the Martian, and they argue that the high ratio for $^{40}Ar/^{36}Ar$ cannot be explained away by more effective outgassing of ^{40}Ar, because this would also release ^{36}Ar; consequently they assert that amounts of ^{36}Ar

must always have been low, and along with it, all other elements of the Tl group.

Since the present abundance of ^{36}Ar should approximate to the amount that has been outgassed (little should have escaped), it seems reasonable to suggest that Mars' outgassing efficiency could only have been about 0.27 that of Earth. With this figure in mind, Anders & Owen estimated the amount of CO_2 that had degassed and, assuming all of this had been held in the atmosphere at the same time, suggest it would have exerted a surface pressure of 140 mbar. Had Martian outgassing been as efficient as Earth's, this figure should have been as high as 520 mbar.

Since the Anders & Owen work, fresh data derived from the Pioneer Venus mission appear to have rendered some of their arguments invalid. For instance, it appears that the non-radiogenic rare gases behave independently of other volatiles during planetary accretion, and Pollack & Black (1979) argued that the abundances of the rare gases within the atmospheres of the inner planets reflect the original nebular gas pressure during accretion, whereas abundances of other volatiles provide a more accurate measure of outgassing efficiency. Their calculations indicate that the nebular gas pressure in the vicinity of Mars was between 5 and 20 per cent that of Earth, and they estimate that a volume of water equivalent to a uniform layer 60–160 m deep must have outgassed. This is comfortingly close to the figures arrived at by McElroy et al. (1977) after analysis of the oxygen and nitrogen isotopic data. The general consensus is therefore that, because Mars' atmosphere is deficient in most volatiles with respect to Earth, its degassing is significantly less complete: the present atmosphere can represent only a small fraction of the total amount of volatile elements outgassed by the planet.

Much has been inferred from the study of meteorites of Martian provenance, particularly the SNC group (McSween 1994). There have been various attempts to establish the amount of mass fractionation of the upper atmosphere by using argon isotopes (Hutchins & Jakosky 1996). The latter authors concluded that volcanic outgassing alone is insufficient to account for the present-day abundances of ^{36}Ar and ^{38}Ar in the Martian atmosphere. Similar calculations for ^{20}Ne suggest outgassed volumes of between 100 and 1800 times above that attributable to volcanism. This implies that sources of argon other than volcanism are required to produce the observed abundances.

A more recent reappraisal of the $^{36}Ar/^{38}Ar$ ratio by Bogard (1997) suggests that earlier values for this ratio (4.1 ± 0.2) may be too high. The basis for this assertion lies in the observation that the atmospheric gases that became trapped in meteorites of Martian provenance, is apparently composed of two distinct components: atmospheric and mantle. These can be recognized since they show substantial differences in their $^{36}Ar/^{38}Ar$ ratios and their Ar/Kr/Xe elemental ratios. The $^{36}Ar/^{38}Ar$ ratio of the mantle component has an unfractionated value (perhaps similar to Earth's), whereas the largest Ar/Xe and Kr/Xe ratios are found in impact-generated glass (Zagami and EETA79001), which contains relatively large amounts of noble gases that have been implanted into the Martian atmosphere by impact shock processes. The trapped $^{36}Ar/^{38}Ar$ ratios (after correction for cosmogenic argon) in the shocked-glass samples are significantly lower than in terrestrial samples or in solar values and they strongly suggest major mass fractionation effects. By making reasonable assumptions about end-member compositions of Martian atmospheric and mantle argon, the observed $^{36}Ar/^{38}Ar$ ratio can be used to estimate the mixing proportions of the two components. Bogard derives an upper limit for atmospheric $^{36}Ar/^{38}Ar$ of 3.9 and a possible lower limit of ~3.0–3.6. This range is substantially lower than the widely utilized figure of 4.1 ± 0.2 and it implies that the amount of mass fractionation during the loss of argon from the Martian upper atmosphere is greater than previously anticipated.

Deuterium–oxygen systematics of the atmosphere (i.e. the ratio of heavy to light isotopes of hydrogen) also provide useful information. Deuterium (the heavier of the two isotopes) is preferentially concentrated in a planet's air when the lighter isotopes have escaped into space; this gives an enhanced D/H ratio. Owen et al. (1988) point out that Martian water is enriched in deuterium; indeed, the D/H ratio is six times that of Earth. Although the business of making deductions from this isotopic ratio is model dependent (being dependent upon what fractionation factor (F) is chosen) they argue that if $F = 0.32$ is used, the present exchangeable reservoir volume – assuming a layer 3 m deep had been lost – would best be represented by a global layer of water 0.2 m deep. However, a recent set of Lyman-α emission results from the HST, indicate that the fractionation factor (F) actually should be nearer 0.02 (Krasnopolsky et al. 1998). This would greatly enhance the volume of the current reservoir (Table 5.1).

Martian provenance meteorites have also been

Table 5.1 Estimates of exchangeable water reservoir sizes (after Krasnopolosky et al. 1998).

	3 m lost		80 m lost	
Fractionation factor	Global layer (m)	Polar cap diameter (km)	Global layer (m)	Polar cap diameter (km)
$F = 0.02$	0.5	400	13	1100
$F = 0.32$	0.2	200	5	700

studied with respect to $\Delta^{17}O$ (Farquhar et al. 1998). Measurements of $\Delta^{17}O$, $\Delta^{17}O$ and $\Delta^{18}O$ in carbonates from meteorite sample ALH84001 support the notion that there were two oxygen reservoirs at the time the carbonates grew: the atmosphere and the silicate crust of the planet. The cause of the atmospheric oxygen isotope anomaly may be attributable to an exchange between CO_2 and ozone, mediated by the metastable atom $O(^1D)$, because of the decomposition of ozone. This being so, the planetary regolith may provide a sink for ^{17}O-depleted oxygen, to maintain the mass balance.

Surficial evidence for subsurface volatiles

Early on in the interpretation of Mariner 9 and Viking imagery, it became clear that the nature of ejecta blankets associated with many impact craters, even at quite low latitudes, indicated interaction between ejecta and subsurface volatiles (Fig. 5.1). It was argued by some that, since only craters with a diameter greater than 6 km exhibited the telltale morphology, this represented some minimum depth at which the impacting bolide encountered stored volatiles (Carr et al. 1977, Boyce 1979). In addition, many pedestal craters were found to be present, generally in latitudes higher than 40°. These raised features had evidently been produced by deflation of original, perhaps blocky, ejecta, which probably had become indurated by trapped volatiles.

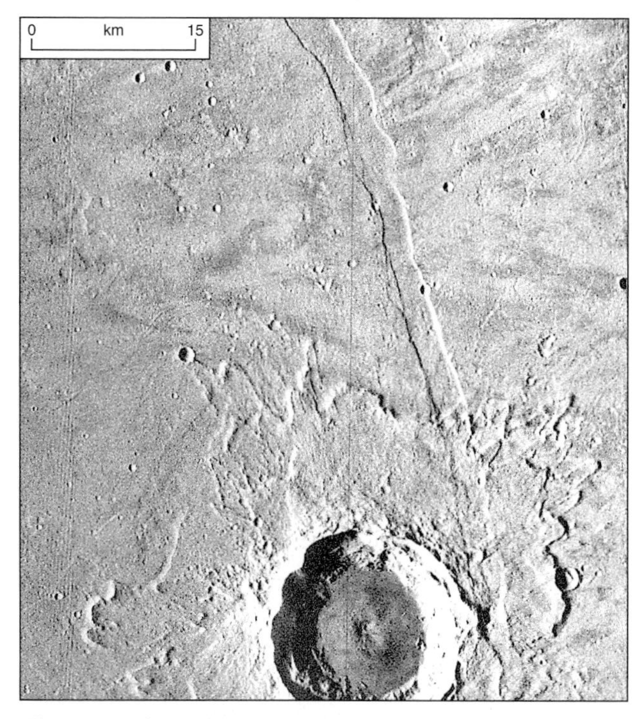

Figure 5.1 Small Martian impact crater showing overlapping lobes of fluidized ejecta deposits. Viking orbiter image 623a59; centred at 30.28°N 124.12°W.

The telltale fluidized ejecta blankets provided an early pointer to the presence of volatiles locked into the Martian subcrust. Then, as more and more images were closely studied, it became apparent that there were several different kinds of topographic features that confirmed this impression. Styles of landforms visible in the northern plains strongly implied that fluvial sedimentation was important in that region at some time in the distant past. As it is into these plains that the extensive outflow channel systems debouched, logic alone indicates that this region would be a massive sink for volatiles.

In the region of the dichotomy boundary there is a widespread development of fretted channels, whose floors show evidence for mass wasting of upland plateau material and its transfer to the low-lying ground to the north. Surface striations on the material that floors these channels bears a strong resemblance to the shear ridges found on terrestrial glaciers, implying that a mixture of sediment and ice may be involved in the transportation process. In addition, the channels do not have the dendritic branching style of fluvial networks, but are characterized by amphitheatre-like headwalls and, frequently, an origin in large impact craters. Such an occurrence strongly suggests some kind of sapping origin, probably involving the release of volatiles frozen into the subcrust.

There are also extensive areas of chaotic terrain concentrated in the equatorial regions. Their morphology indicates them to have been produced by large-scale collapse of the surface, whose origin is most logically to be found in the melting of subsurface ice, which, in turn, is presumed to have generated the massive flooding that occurred during the early history of Mars. The fact that they developed at the same general time as Tharsis volcanism peaked adds weight to the notion that volcanic heat melted ground ice locked beneath the surface in low latitudes, causing massive ground collapse and the release of huge volumes of volatiles.

Another landscape type that has widely been interpreted as reflecting the activity of subsurface volatiles is terrain softening. This occurs polewards of latitude 30°, where landforms have a smoothed appearance. A common interpretation sees this as a result of relaxation of topographic highs by ice-enhanced creep. The absence of softening in equatorial latitudes is cited in support of this notion. However, there are many rampart craters in the lower latitudes, for which the traditional origin is impact into ice-rich crustal rocks. To explain this apparent discrepancy, the assumption has been made that these were produced by impacts into not ice-laden but water-laden materials.

Clifford & Zimbelman (1988) pointed out that the latter is inconsistent with observation and also faces formidable objections in terms of thermodynamics. The first objection is based on the fact that thermal models suggest that, in equatorial latitudes, the base of the cryosphere lies at depths of 1–3 km (Fanale 1976). On this assumption, theoretical modelling confirms that, wherever groundwater exists, ground-ice will form and be replenished more rapidly than it can be removed, even in equatorial latitudes.

However, a more powerful objection derives from the rheological properties of ice on Mars, which for even the slightest increase in temperature necessitates a large increase in shear stress; this also increases if particulates are present. Thus, deformation of ice could occur only at depths below 1 km and, if so, ice should survive even near the equator and therefore there ought to be evidence of terrain softening in low latitudes. While acknowledging that objections to the traditional explanation of terrain softening do not preclude the existence of subsurface ice in high latitudes, Clifford & Zimbleman offered an alternative explanation: that the high-latitude landscapes are generated by latitude-dependent deposition of atmospheric dust, which creates a widespread mantle of aeolian material.

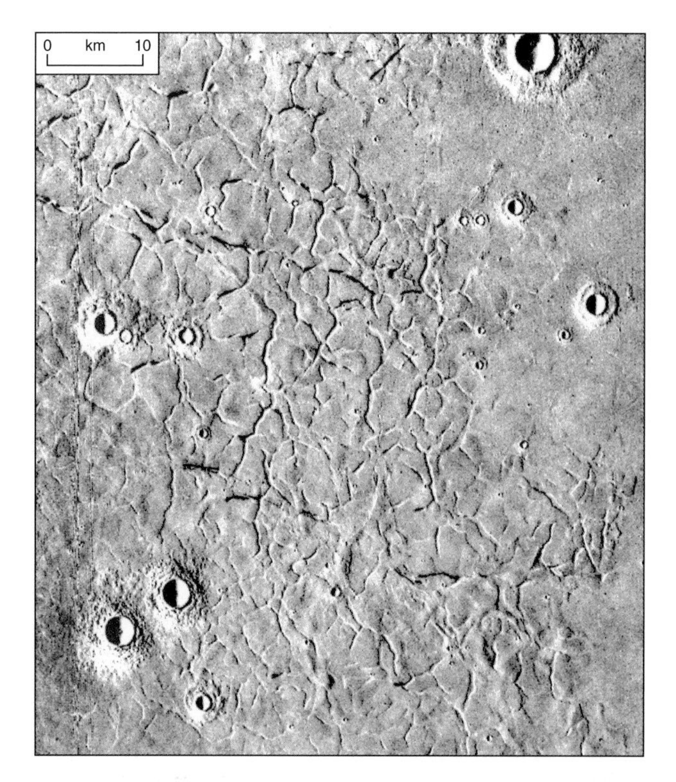

Figure 5.2 Fractured terrain in the northern lowlands. Viking orbiter image 32a18, centred at 14.79°W 43.43°N.

Palaeolakes and oceans

The northern plains of Mars have been interpreted in various ways, but throughout the history of post-Viking research there has been a strong lobby for the idea that sedimentary rocks are widely distributed north of the dichotomy boundary and that palaeolakes and even ancient shorelines can be identified in these low-lying and relatively smooth regions.

One of the more distinctive geomorphological imprints seen on the lowland plains is a polygonal fracture pattern that divides the plains into a series of giant blocks, anything from a couple of kilometres to 10 km across (Fig. 5.2). In two short papers McGill (1985a,b) proposed that such patterns had developed in relatively thin layers of sedimentary rocks, where contraction and compaction had occurred over the underlying topography. In this way, some curvilinear fracture patterns would represent outlines of impact craters buried beneath the sediment cover. A study by Lucchitta et al. (1986) confirmed this suggestion. There are several reasons why the idea seems plausible: first, polygonally fractured deposits tend to be located in low-lying areas that apparently received an influx of sediments; secondly, the fracture patterns are concentrated near embayments in the upland hemisphere where major outflow channels debouch (Fig. 5.3); and thirdly, crater ages indicate the channels and fractured plains to have similar ages.

Although the giant polygons are at least an order of magnitude larger than any terrestrial analogue, they have been interpreted by McGill (1985a,b) and by Lucchitta et al. (1986) to be giant desiccation or compaction features in frozen outflow-channel sediments. However, Pechmann (1980) considered them as tensional fracture patterns developed in the plains by regional doming. However, the pattern of cracking implies uniform tension in all directions within the plane of the surface. Closely similar patterns (in form, not scale) occur on Earth where horizontal extensional stresses develop because of cooling or desiccation.

De Hon (1988) showed that, in many locations close to the mouths of major outflow channels (e.g. Maja and Ladon Valles), there are relatively featureless plains units that appear to represent lacustrine deposits. Temporary ponding in channels would have been expected under Martian conditions, so local sedimentation probably occurred, both within channels and near their mouths. Recent discovery of stratified rocks within several large impact craters, in Terra Meridiani and inside Candor Chasma, implies that sedimentary processes were operative earlier in Mars' history than previously thought, and that extensive, possibly lacustrine, rocks are also to be found south of the line of dichotomy (Malin & Edgett 2000).

Studies first undertaken during the early 1990s identified potential strandlines that were considered

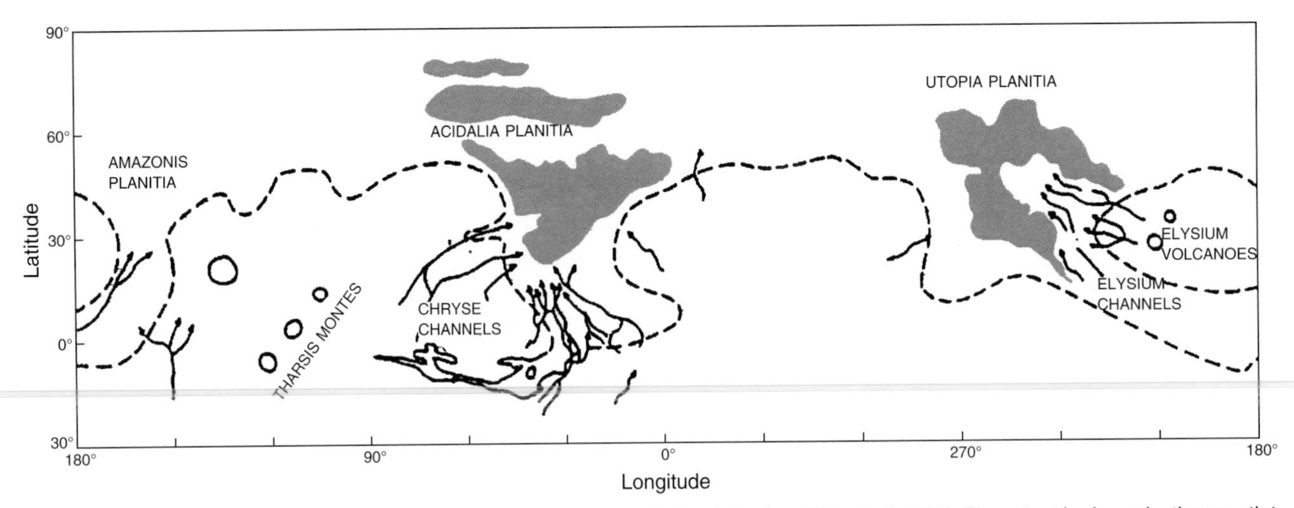

Figure 5.3 Map of polygonally fractured terrain in the northern plains of Mars (after Lucchitta et al. 1986). The extent is shown by the grey tint.

to represent coastal landforms pointing to at least two horizons of previously higher sea level (highstands) of a former northern ocean, possibly as recently as Early Amazonian times (Parker et al. 1993). Calculations indicate that the volume of water debouched from the circum-Chryse outflow channel system alone is more than enough to have produced large bodies of standing water (assuming a sufficiently temperate climate). This is but one part of the global channel system and there is fairly convincing evidence that an earlier phase of flooding contributed water to a much larger basin, generating a relatively long-lived ocean that may have existed beneath a cover of ice.

This idea received something of a setback when studies of Mars Global Surveyor images of the appropriate areas did not deliver positive support for such an interpretation (Malin & Edgett 1999). However, this fact notwithstanding, the probability that palaeolakes and even an ocean once existed is still very much under discussion as a live concept.

Ancient glaciers?

The notion that glaciation affected Mars in the distant past is not new, the effects of this process having been described by Lucchitta (1981, 1982, 1993), Hodges & Moore (1979), Chapman (1994), Lucchitta et al. (1986), Kargel & Strom (1992) and Christensen (1989), to name but a few. Although some of the landforms identified on Viking images resembling terrestrial glacial features are open to more than one interpretation, MGS imagery is throwing fresh light on this issue.

Distinctive geomorphological features occur on the plains within a few tens of kilometres of the lowland/upland boundary. Like much of the polygonally fractured ground, these are located near the mouths of the Chryse channels that debouch into Acidalia Planitia and below the highland scarp in western Utopia. One particularly interesting set of features comprise sinuous ridges 0.5–1 km wide, set in elongate depressions. Lucchitta et al. (1986) observed that they bear a striking resemblance to the ridges that are generated near the mouths of Antarctic ice streams and on ice shelves, particularly where shoals are developed. For this reason they suggested that the Martian uplands were bordered by frozen materials that had mechanical properties similar to those of modern Antarctic ice shelves. If their suggestion is accepted, then the northern lowland materials represent outflow channel sediments that were laid down in a partially frozen ocean. However, Parker et al. (1986) compared the ridges to terrestrial lacustrine or shallow marine coastal spits and barriers, which imply a predominantly liquid environment in which waves are responsible for the features seen. More recent studies of these include those by Kargel et al. (1995). The term thumbprint terrain has been attached to certain kinds of high-latitude landscape that are characterized by these sinuous ridges and isolated knobby features.

Martian volatiles and nature of reservoirs

The evidence presented above suggests the presence of subsurface volatiles on Mars. That water once flowed across the surface is indicated by the very extensive outflow channel systems and the more localized valley networks within the ancient cratered hemisphere. The structure of large areas of chaotic terrain has been taken widely to imply surface collapse of large areas of crust because of the removal of subsurface frozen volatiles. Each of these areas of evidence adds to the realization that Mars once was a watery planet.

Measurements of the hydrogen escape flux from the atmosphere, together with kinetic energy calculations, show that, over the course of its evolution, Mars must have lost the equivalent of a global layer of water 3 m deep by processes such as photodissociation and exospheric escape. This implies that most of the original volatile inventory must still reside somewhere. The potential reservoirs are the polar caps, the atmosphere and the regolith.

Recent estimates of past extent of the polar cap deposits and the volume of water and other volatiles locked into the polar regions have been forthcoming from the MGS mission. On the basis of the topography of the region of Olympia Planitia and the distribution of residual ice patches in the mantled plains that lie adjacent, Zuber et al. (1998) proposed that the units that underlie the longitudinal dunes in Olympia represent old polar-cap sediments, whereas the residual ice patches are what is left of a once more-extensive polar ice cap. On this assumption, and by interpolating a topographic surface beneath the cap and then removing the inferred basement topography from the measured topography, they estimated that the polar cap has an area of 1.2 ± 0.2 million km^3, covering an area of 1.04 million km^2 with an average thickness of 1030 m. The calculated volume of ice locked in such a body would be equivalent to a global layer of water 9 m deep. If the north polar cap melted, then its southernmost limit would be in the Chryse basin and, assuming the water filled to the –4680 m contour, the average water depth would be 270 m.

The MAWD experiment on the Viking spacecraft showed that, if all the water locked into the atmosphere condensed on the planet's surface, it would produce a layer a mere 15 μm deep. Similarly, measurements of the volumes of the perennial polar ice caps show quite clearly that they could not generate a planet-wide ocean more than a few tens of metres deep, which is also what the MGS results indicate. Thus, both of these reservoirs fall short of the volume indicated by the geomorphological evidence. The only logical conclusion, therefore, is that most of this volatile material must remain locked into the subsurface.

Could the regolith hold a massive volume of fluid? Well, the answer appears to be yes. Yes, at least, if we are right in assuming that the Martian regolith is similar to that of the Moon, which is known to be porous and brecciated to a depth of at least 20 km. By scaling to account for differences in the gravity between the two worlds, there is every likelihood that the Martian crust may be relatively porous to a depth of at least 10 km. On this basis, it could store a volume of volatiles equivalent to a global layer between 0.5 and 1 km deep. To test this theory, it was necessary to land on Mars and probe beneath the surface, from either a mobile roving vehicle or a manoeuvrable balloon. This is exactly what happened with Mars Pathfinder and what had been planned for the post-millennium missions, particularly the Mars Polar orbiter, which was to send two probes into the high-latitude plains. In addition, the huge amount of superb high-resolution and synoptic imagery and other data sent back by Mars Global Surveyor have done much to dot the i's and cross the t's in this absorbing area of research.

Recent data from the MGS thermal emission spectrometer for low-albedo surface materials suggest a 4:1 mixture of pyroxene to plagioclase, plus ~35% dust, best fits the spectra obtained. The qualitative upper limit for carbonate concentration is <10% in the limited regions observed (Christensen et al. 1998). These low-albedo regions are considered to be composed predominantly of pristine igneous rocks, and the low carbonate concentrations suggest that, if carbonates are volumetrically important, they are likely to be concentrated at specific locations that favoured either their initial deposition or their subsequent preservation. In a sense, this is a similar scenario to that for oxygen, since there must be hitherto unidentified sinks for the heavier isotopes of oxygen too, possibly in ferric oxides and hydroxides in the regolith or subcrust.

The largest modern reservoirs are the northern and southern polar caps. Currently, the southern ice cap is areally smaller than its northern counterpart, but the layered deposits that surround the southern cap are more extensive than those in the north. The distinct plateau-like regions encircling both polar ice caps correlate with the well known layered terrain, which are believed to be formed from ice-rich materials. Using the high-resolution altimetric data from the MOLA instrument aboard Mars Global Surveyor, Smith et al. (1998) estimated that the total surface volatile inventory for the polar deposits is 3.2 million to 4.7 million km^3, equivalent to a global water layer 22–33 m deep. This is significantly more than was estimated by Zuber et al. (1998) above.

Finally, if the planet did start out with a much larger atmosphere, how might it have evolved into its present state? The most widely quoted process is the conversion of CO_2 in the atmosphere into carbonates at the surface. Thus, in the presence of a water-laden atmosphere, CO_2 would have been removed from the atmosphere via the weathering of silicate volcanic rocks. On Earth, these would have been recycled back into the atmosphere via plate tectonics. Since it seems unlikely that such a process ever operated for lengthy periods on Mars, the most likely phenomenon for the recycling of CO_2 on Mars is volcanism, a process that we know was extremely important on early Mars.

6 The ancient cratered terrain

Although the heavily cratered upland plains to the south of the line of dichotomy bear some similarities to the lunar highlands, they show widespread evidence for much more modification than the ancient cratered terrain of the Moon. Certainly, early impact cratering left an indelible imprint on both worlds, and subsequent impact erosion severely degraded the more ancient craters and basins. However, on the Moon, little changed thereafter, whereas on Mars fluvial and aeolian activity, volcanism, tectonism and non-impact erosion each played an important part in modifying the ancient crust and producing its present geomorphological nature. Although the upland hemisphere looks like a lunar surface, closer appraisal reveals not only a difference in the cratering record but also in the morphology of the craters and of their associated ejecta. In lower latitudes, volatiles were entrained in the ejectamenta as they were emplaced. Within the upland terrain there are many dendritic valley networks. These are incised into the plains between the larger impact craters and they sometimes appear to have originated within them. They were evidently cut by fluids (probably water), aided and abetted perhaps by sapping processes.

The Moon's main volcanic episode saw the partial filling of the larger nearside impact basins with highly fluid basaltic lavas, the Mare basalts, an episode that occupied the period between 3.8 and around 2.5 billion years ago. Highland volcanism was largely confined to small-scale dome building and eruption of a suite of unusual lavas (KREEP,[1] etc.) that became mixed up with the highland breccias. On Mars, the situation was very different, since there are several large central volcanic structures, most of which are distributed around the great Hellas impact basin, well within the outcrop of the ancient cratered terrain. This early phase of central volcanism, perhaps largely explosive in style, is all a part of the geological development of the southern hemisphere of Mars.

Volcanism around the Hellas basin may well have been a response to the formation of this massive structure. Recent topographic data from MGS has shown that the total relief of Hellas, from the bottom of the floor to the top of the surrounding ring massifs, is at least 9 km. Earlier work noted that there was little sign of basin ejecta surrounding Hellas, but MOLA data show that there is a distinct 2 km-high annulus lying at a diameter of ~4000 km, which may comprise both uplifted crustal blocks and ejecta. The formation of this huge basin must have had shattering implications for the ancient cratered hemisphere.

The cratering record

Our knowledge of the distribution of the different kinds and sizes of interplanetary object currently circulating within the Solar System comes largely from studies undertaken in the vicinity of Earth (Dohnanyi 1972). It appears that within the Solar System today there are solid objects that range in mass from less than 1 billion (micrometeorites) to about 100 billion billion kg (asteroids). Because of the shielding effect of the Martian atmosphere, at the present time only objects with masses more than about 1 kg produce craters on its surface (Gault & Baldwin 1970). Compared with the Moon, therefore, there should be fewer smaller craters on its surface.

1. KREEP is a NASA acronym for early lavas enriched in K (potassium), rare-earth elements (REE) and phosphorous (P).

The cratering history of the Moon is fairly well constrained, because it has been possible not only to derive crater ages for its different surfaces but also to date returned lunar samples radiometrically; for Mars we do not have this luxury. It has become clear that about 4 billion years ago the cratering rate for the Moon was very high indeed and that it declined sharply around 3.9 billion years ago. In consequence the lunar highlands, which formed before the decline in impact rate, are nearly saturated with craters, whereas surfaces that were produced after this reduction in the flux (e.g. the maria) are relatively sparsely cratered. On the Moon, very few surfaces exist that exhibit intermediate crater densities (i.e. the crater density record is bimodal), so it has been assumed that the impact flux fell away very rapidly. Soderblom et al. (1974) suggested that Mars was similar, a view that has generally been accepted; however, debate continues about both the time at which such a decline occurred and the present cratering rate.

There are many statistical studies of lunar cratering (Carr 1981: 55). As a solid surface ages, the tally of superimposed impact craters increases to the point at which the number of new craters being formed is exactly balanced by those being destroyed as a result of the impact process. The surface is then said to have reached an equilibrium, termed saturation. This appears to have occurred on the lunar highlands and over much of Mercury; however, the situation on Mercury is complicated by the subsequent (and possibly coeval) development of intercrater plains, many of which have a volcanic origin (Cattermole 1989b). The situation for Mars is somewhat similar to Mercury in this respect, although even more complex and possibly less well understood.

There is little doubt that the most densely cratered parts of the Martian upland plateau are representative of the most ancient surfaces on Mars and may be comparable with the lunar highlands. However, the size–frequency distributions for lunar and Martian craters are markedly different, since there are few craters smaller than about 20 km in diameter. The resultant sharp change in the slope of the size–frequency curve (Fig. 6.1) was noticed by, among others, the astronomer Ernst Opik (1965, 1966), who attributed it to an obliterative process that he thought might have removed most of the smaller craters and certainly modified many of the larger ones.

Subsequent work by the US Mariner 6 and 7 teams also noted a bimodal distribution of craters, in terms of not only their size–frequency relationships but also their morphology. Thus, craters larger than about 5 km seemed to have relatively low ramparts and to be shallow; smaller craters were mostly bowl shape, like small lunar craters. The more definitive studies associated with the later Mariner 9 mission led to the realization that the size–frequency curves consisted of not two but three segments (Fig. 6.2): craters smaller than 5 km and those larger than 30 km yielded a slope often exceeding –2, whereas those between 5 and 30 km diameter fell on a curve of shallower slope verging on –1. The more recent Viking data has confirmed this. In seeking to explain this behaviour, Hartmann (1973) surmised that craters larger than 30 km in diameter

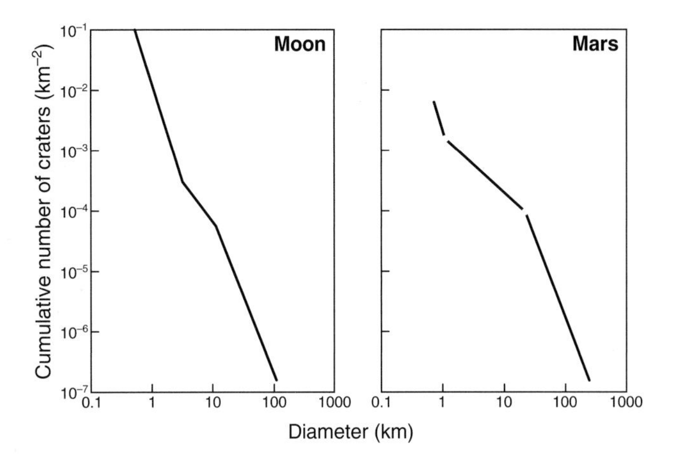

Figure 6.1 Size–frequency crater curves for Mars and the Moon compared.

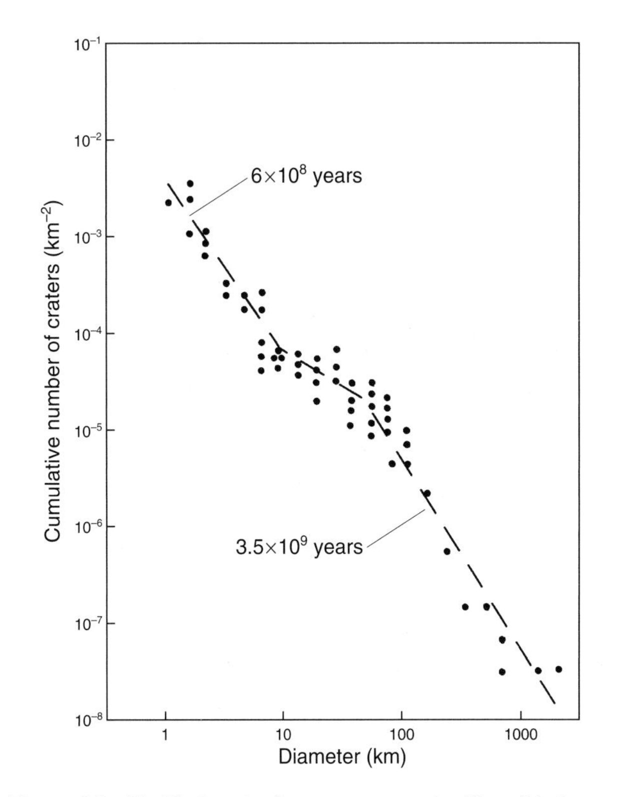

Figure 6.2 Modified crater frequency curve for Mars (Hartmann 1973).

had survived from an early period when the cratering and obliteration rates were most intense, producing a saturation curve. The middle segment on the size–frequency curve (that with the lower slope) he suggested was representative of an equilibrium curve produced during the same early era, during which time the formation and destruction of craters exactly matched, the smaller craters being more readily destroyed than the larger. To explain the third segment on the curve, Hartmann postulated a phase of impact by smaller bodies that post-dated the previous two, probably after a significant (unspecifiable) hiatus.

In the mid-1970s, an alternative explanation for the cratering statistics was provided by Jones (1974) and Chapman & Jones (1977). They suggested that, during the early period of intense bombardment, rates of crater formation were much the same throughout the complete size spectrum; subsequently they declined to leave a crater distribution close to saturation for all sizes (thus far, their model is similar to that of Hartmann). They then went on to hypothesize that there was a phase of intense obliteration that destroyed all craters less than about 5 km diameter and most in the 5–30 km class; it also greatly modified the larger ones. The new population of < 5 km craters was formed after this event.

Regardless of which of these and similar hypotheses comes closest to the truth, it does appear that two populations of impactors affected the inner regions of the Solar System. Thus, a recent statistical study by Barlow (1988) identifies an older population, which billions of years ago gave rise to the multi-slope distribution curve and which represents the phase of intense bombardment that affected Mercury, the Moon and Mars (producing the heavily cratered plains that cover about 60 per cent of the latter's surface). She also identifies a younger population that is primarily recorded in the more lightly cratered plains regions of these same planets.

Naturally, if cratering were the sole process of crater destruction, then the 5–30 km crater population ought to have approached the lunar slope value of −2. The fact that it does not strongly implies that other crater-destructive processes must have been at work; the more obvious of these would have been volcanism, fluvial and aeolian activity. There is ample evidence that intense volcanism was widespread during the early geological history of Mars, and the presence of fluvial channels on the upland plateau implies that the early Martian atmosphere may have been considerably denser. Furthermore, the varied state of preservation of the channel networks themselves is most readily interpreted to have been the result of obliteration by the same process as that which modified the cratering record. Most estimates point to a date of 3.8 billion years ago for the decline in impact flux (e.g. Carr et al. 1984).

Of the various attempts at providing an absolute Martian timescale, those of Neukum & Wise (1976), Soderblom et al. (1974), Hartmann et al. (1981) and Carr (1981) are most widely credited. Table 6.1 shows estimated absolute ages for selected regions of cratered plains using the scheme of Carr (1981), who takes 1 km-diameter crater statistics and calibrates them in billions of years against the ages from Hartmann et al. (1981). The large errors inherent in such methods are shown by the wide range of possible ages quoted.

Table 6.1 Ages of Martian plains as derived by Carr (1981).

Plains region	No. of craters Best estimate <1 km/10^6 km^2	Likely age Years (billion)	Range
Mare Acidalium	830	1.2	0.2–1.7
Sinai Planum	970	1.4	0.4–3.0
Utopia Planitia	1270	1.8	0.6–2.3
Noachis Planitia	1740	2.5	0.9–3.6
Amazonis Planitia	1940	2.8	1.0–3.7
Syrtis Major Planum	2053	2.9	1.2–3.7
Chryse Planitia	2100	3.0	1.2–3.8
Lunae Planum	2400	3.5	1.7–3.8
Hellas	2640	3.8	2.9–3.9
Hesperia Planum	2710	3.9	3.0–3.9

Martian impact basins

As with the Moon, the oldest recognizable geological features on Mars are the circular impact basins. Of these, Hellas – at 1600×2000 km arguably the largest impact basin so far discovered within the Solar System – Argyre and Isidis are the most obvious on spacecraft images, but many other ring structures with diameters greater than 250 km have been identified by Wood & Head (1976) and Schultz et al. (1982) (Table 6.2). Whereas lunar impact basins are characterized by several rings of scarp-like peaks, those of Mars have a rather different morphology. The rim morphology is best seen around the Argyre basin, where a belt of closely spaced blocky massifs extends from about 300 to 800 km from the basin's centre (Fig. 6.3). The incomplete ring of the Isidis basin is closely similar, and that surrounding parts of Hellas is broader and lower. Some Martian basins have several rings. Schultz et al. (1982) mapped six for the Chryse basin, four each for the Ladon and Aram basins, and two each for Argyre and Isidis. Hellas, Isidis and Argyre also have concentric fractures around their peripheries, in which respect they resemble the experimentally produced Snowball impact structure and some circular lunar

Table 6.2 Impact basins of Mars (diameter >250 km).

Name	Latitude	Longitude	Diameter (km)
Hellas	−43.0	291.0	2000
Isidis	16.0	272.0	1900
Argyre	−49.5	42.0	1200
South Polar	−82.5	267.0	850
Chryse	24.0	45.0	840
Renaudot, S of	38.0	297.0	600
Ladon	−18.0	29.0	?550
Sirenum	−43.5	166.5	500
Hephaestus Fossae	10.0	233.0	500
Schiaparelli	−3.2	343.5	470
Huygens	−14.0	304.2	495
Le Verrier, W of	−37.0	356.0	430
Antoniadi	21.7	299.1	400
Nilosyrtis Mensae	33.0	282.5	380
Nr Newcomb	−22.5	3.0	380
Schroeter	−02.0	304.0	310
Al Qahira	−20.0	190.0	300
Nr South	−73.0	344.0	300
Herschel	−14.6	230.2	320
Newton	−40.0	158.0	280
Holden	−25.0	32.0	260
Newcomb	−24.0	358.0	250

basins. The marked difference between lunar and Martian basins is undoubtedly explained by differences in the target materials, in terms of both their volatile content and subsurface layering.

Although rim ejecta facies are seldom well developed, peripheral areas of enhanced erosion may focus on original circum-basin ejecta that became saturated with water at some stage in the distant past, when the atmosphere of the planet was far denser than at present. The volatile-saturated deposits are presumed subsequently to have undergone sapping.

Martian multi-ring basins have all suffered a complex history of burial and exhumation, which has led to several of them going unrecognized until recently. However, their ultimate recognition has led not only to a fuller understanding of the early impact record but also to an appreciation of the way in which basin rim massifs have apparently controlled subsequent geological developments. Thus, the older ring structures commonly exhibit extensive systems of narrow valleys on their rim massifs, whereas large outflow channels may originate along one of their component rings. Such an association is unlikely to be fortuitous and it implies that, although their original form is now obscure, the deep-seated concentric fracture belts generated during basin excavation and subsequent tectonic readjustment have exerted an important control on geological features.

The degree to which such control has affected global-scale Martian geology has been discussed by Schultz et al. (1982), who suggested that the Chryse basin, although experiencing an early stage of lava infilling, was subsequently reactivated (perhaps by regional Tharsis-related volcanism) such that there was a complex interplay between it, the adjacent Aram and Ladon basins, major outflow channels and Valles Marineris. In particular, there is a curving belt of disruption that extends from the Aram basin, through the northern rampart of the Ladon basin and the offset canyons associated with the central regions of Valles Marineris, to Kasei Vallis. This appears to be a manifestation of an arcuate zone of instability along the outer rim of Chryse; if so, it could be taken to imply that basin structure, resurfacing, flood volcanism and subsequent basin reactivation may all be interrelated. However, recently acquired gravity data throw doubt on the impact origin of Chryse.

The Mars Global Surveyor MOLA experiment has shown that the Hellas basin is the deepest structure on Mars: it descends more than 9 km below datum. Furthermore, MOLA established that the mountain massifs formerly thought to mark the perimeter of the structure, and having a diameter of ~2300 km, actually lie on the inner slopes of the topographic basin. The new data reveal that there is a marked topographic peak that

Figure 6.3 Viking mosaic of the Argyre impact basin. Note the blocky rim massif. The large crater on the rim is Galle. JPL mosaic P-17022; centred at 14.79°W 43.43°N.

lies roughly 2 km higher than the main ring, with a diameter of ~4000 km. As has been observed by Smith et al. (1999a), Hellas shows striking similarities with the lunar South Pole–Aitken basin: they have similar volumes, diameters and distributions of ejectamenta.

The huge volume of material associated with this giant basin represents a major redistribution of ancient cratered terrain at the time of basin excavation and must account for much of the high-standing topography of the southern cratered hemisphere. Although it is not possible to ascertain the relative importance of impact melt, ejecta and structural uplift, calculations based on the new topographic model indicate that, if all of the peripheral rim materials were put back inside the basin ring, its surface would lie some 600 m above the datum level. In addition, the MOLA map shows quite clearly that the annulus of ejecta and massifs associated with basin formation expresses itself along the line of dichotomy. The somewhat asymmetric distribution of the rim massifs may indicate that the incoming bolide that generated this massive feature approached Mars at a relatively low angle, heading S60°E (Tanaka & Leonard 1995).

Morphology of impact craters

It was clear even from the early Mariner 6 and 7 images that Martian impact craters were shallower than those on the Moon and had a more Earth-like aspect than had been anticipated. This is a function of Mars' complex erosional and depositional history, with partial infilling of the interiors by windblown material and of rim destruction by subaerial degradation and sapping. Wall slumping, similar to that seen in larger lunar craters, leads to cavity enlargement, but the transition from simple to complex crater morphology occurs within the diameter range 3–8 km, as opposed to about 20 km on the Moon (Pike 1979, 1980a). Certainly, the larger Martian craters are shallower than their lunar counterparts but deeper than those on Earth. Since gravity is the predominant factor in controlling this parameter (Pike 1980a), this is hardly surprising, but one further factor that may have led to an enhancement of the enlargement process is the presence of volatiles within the Martian regolith (suggested by the fluidized ejecta patterns shortly to be described) and probably attributable to the presence of an atmosphere much denser than the present one. By analogy with the terrestrial Prairie Flat crater, Boyce & Roddy (1978) indicated that, if large volumes of volatiles were locked up in the regolith, Martian craters should be proportionately shallower than those on both Mercury and the Moon. Furthermore, many craters within the size range 30–45 km contain central pits (Wood et al. 1978) or peaks with summit pits on them. Wood et al. assert that the presence of ground ice, which would volatilize from the core of a crater's central uplift, could explain this particular characteristic.

Pike (1980a) notes that morphology changes with increasing crater size: for craters with 3–4 km diameters, flat floors are typical; between 4 and 5 km, cavities become shallower and central peaks appear; at around 6 km scalloped rims are typical, and around 8 km, wall terracing develops. He suggests that shallow depth of excavation and some unspecified rebound mechanism, rather than centripetal collapse and deep-seated slippage, billions of years ago gave rise to the central peaks and, in turn, engendered rim collapse. Pike (1980b) also notes that the transition from simple to complex craters is not the same on the ancient heavily cratered terrain as on the less heavily cratered plains, thus the onset diameter for complex craters is about 10 km on the latter but only 3 km on the former. This pattern must be terrain dependent, whereby stronger materials typical of the younger plains support larger bowl-shape cavities than does the weaker heavily impacted ancient crust.

Fresh craters are of bowl shape, have sharp rims and well defined ejecta deposits, and often contain central peaks, albeit small ones; furthermore, they are always superimposed on valley networks and the more degraded craters. Older craters show increasing degrees of degradation, beginning with the removal of the bowl shape and the development of a flat floor, and any central elevation is removed. Ejecta blankets are found incised by valley networks. At higher levels of erosion, the rim of the crater may be removed entirely, allowing material from outside to wash or blow in. In the most advanced stages of ageing, crater walls may become breached and the craters themselves deeply buried by sedimentary materials (this sequence is shown in Figs 6.4–6.7).

An analysis of 264 craters in the region of Sinus Sabaeus and Margaritifer Sinus by Craddock et al. (1997) indicated that, in general terms, craters become enlarged by 7–10 per cent initially. They also showed convincingly that larger craters become enlarged more than smaller ones – the reverse of what would be expected – and suggest that a ~10 m-thick aeolian deposit must have been laid down on crater interiors after the initial stage of degradation had ceased. This lends support to the notion that early Martian climatic conditions were very different from the present.

Because of extensive modifying processes on Mars, the original forms of most large impact craters are poorly preserved, some craters being almost palimpsestic (visible from above but without relief, rather

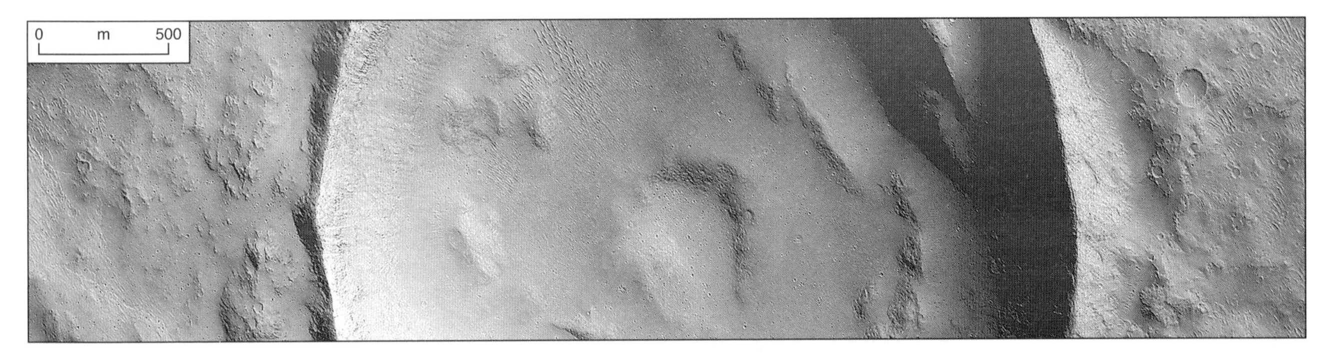

Figure 6.4 7.3 km-diameter impact crater in Hesperia Planum, showing deposits of sedimentary material in the interior, boulders on the rim slopes and stratification in the walls. MGS image PIA02019.

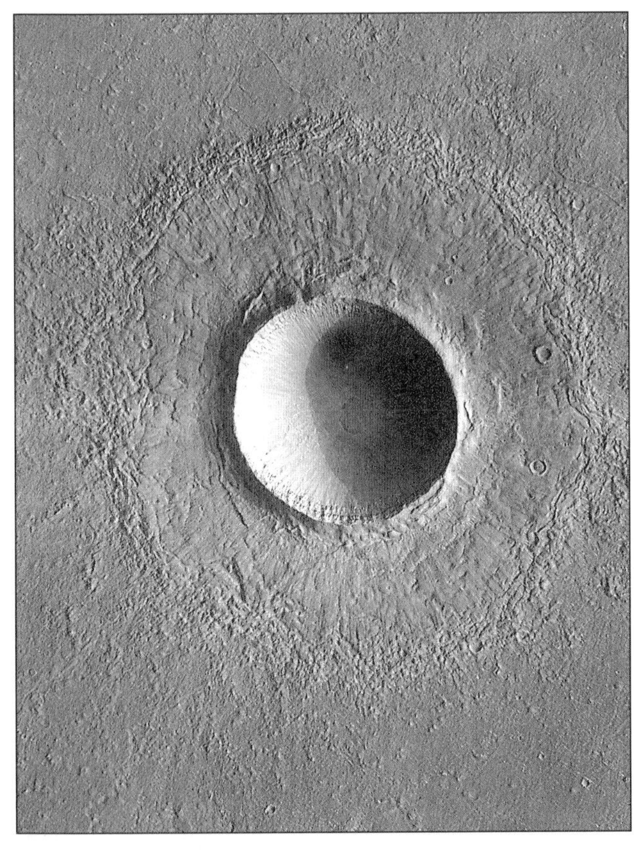

Figure 6.5 Small fresh impact crater in Elysium Planitia. Note the sharply defined rim, simple structure and radially grooved ejecta deposit surrounding it. MGS image MOC2-161 (PIA02084).

Figure 6.6 Degraded impact craters at latitude 35°N. The large crater, 65 km across, has strongly gullied walls, a floor with at least two levels partially filled with aeolian debris and/or volcanic flows and little or no evidence of a surrounding ejecta blanket. Viking orbiter images 192s19–192s21 (parts), centred at 24.5°N 301°W.

like a crop circle). However, their associated ejectamenta are often quite well preserved, their distribution and form having provided important clues to the constitution of the planet's regolith and to its variability from place to place.

Crater ejecta morphology

The ejecta patterns surrounding Martian impact craters are most distinctive and they contrast strongly with those typical of the Moon and Mercury. Lunar craters are typically surrounded by consecutive zones of continuous hummocky ejecta, often with coarse blocky debris along the rim, discontinuous dune-like ejecta, or radial ridges and grooves, and, beyond this, discontinuous ejecta that merge with a zone of secondary cratering and bright rays.

On Mars the situation is different. At less than a diameter of about 5 km, crater ejecta blankets appear to have been laid down with ballistic trajectories; however, above this size, most ejecta patterns show

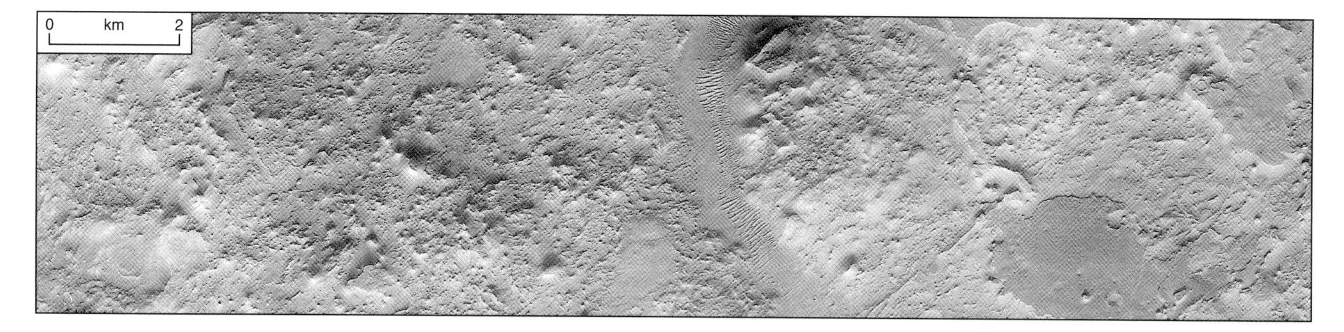

Figure 6.11 Narrow-angle image across Auqakuh Vallis and a highly degraded area of intercrater plains in eastern Arabia (PIA02043; centred at 296.5°W 27.5°N).

Over very large regions the cratered surface is incised by branching valley networks and may be very well defined or barely discernible. Branching valleys are typical of less heavily impacted areas between large craters, whereas stubby valleys frequently are seen cutting through the crater ramparts and extending part way across their floors. These ancient valleys are probably of fluvial origin and were formed at a time when the Martian atmosphere was denser and richer in volatiles than it is now. Their variable state of degradation strongly suggests that, like craters, they were subject to early attack by meteorites during the stage of heavy bombardment. On this basis, the better-preserved networks are probably the younger; they are fully described in Chapter 11.

Intercrater plains are defined as the less heavily cratered regions that occur between the large craters and on their floors. In places the level floors of these craters are clearly made from eroded layered deposits; elsewhere they are covered by plains units crossed by curvilinear ridges; then again, some floor-mantling units are marked by hundreds of very small pits. The latter are particularly difficult to accommodate in the light of how active the surface of the planet has been. The intercrater plains units partially submerge the plateau surface over large areas and are more fully described in Chapter 8. However, the eroded nature of this terrain, the apparent presence of what appear to be volcanic dykes, and fields of windswept dunes, are illustrated in the narrow-angle image (Fig. 6.11) that was taken across the channel Auqakuh Vallis.

Volcanoes of the cratered plateau

Although the most extensive central volcanism is associated with the northern hemisphere, there are several ancient volcanic structures in the south, particularly around the Hellas basin. Plescia & Saunders (1979) referred to these as highland paterae. The most ambiguous is Amphitrites Patera, which lies close to the southern border of Hellas. It comprises several 100 km-diameter rings, with little vertical relief, and associated radiating ridges, some of which extend onto the Hellas plains. Hadriaca Patera is very similar, although somewhat better defined, and has at its summit a caldera 60 km across. The flanks have a smoothed appearance and are incised by radiating channels. The low shield has a diameter of about 300 km. It is very reminiscent of the mantled parts of Alba Patera, and may have been built mainly from pyroclastic deposits.

Tyrrhena Patera has been studied more closely than either Hadriaca Patera or Alba Patera, particularly by Greeley & Crown (1990). Situated northeast of Hellas, there are two sets of ring fractures surrounding the summit, the inner one defining a 50 km-diameter region in which there is an off-centre depression; a broad channel leads off from this. Although there is a lava unit on the southwest flank, the lower parts of the shield are highly dissected and embayed by younger units; they have a higher albedo than these. This led Pike (1978) and Greeley & Spudis (1981) to surmise that it was composed of ash. Smooth units surrounding the shield were also interpreted by Greeley & Crown to be pyroclastics; they envisaged Tyrrhena Patera to have been built from hydromagmatic[1] eruptions, more or less contemporaneously eroded by water, wind and mass wasting. A similar origin can probably be supported for several other structures around Hellas; they are discussed further in Chapter 8.

Developmental history of the early crust

The oldest rocks that can be discerned in the upland hemisphere are the upstanding remnants of impact basin rims. These are assigned to the Lower Noachian epoch of Martian history. The most characteristic of

1. Eruption involving interaction between hot magma and groundwater. Formerly the term "phreatomagmatic" was used.

these are perhaps the massifs of Nereidum and Charitum Montes, which represent the elevated rim materials produced during the Argyre impact. These blocky mountains form an annular belt up to 700 km wide and in places 1–2 km above the interior of the basin. Similar massifs would have been produced for each of these early impact events and the associated and widely dispersed basin ejecta would have covered the surrounding cratered crust. The latter, now completely buried by younger deposits, doubtless represented the primitive material formed by solidification of molten materials from within Mars, which subsequently was bombarded by myriad meteorites to generate the ancient cratered terrain.

During the Middle Noachian, cratering was still intense and the Martian crust was battered by impacts that generated not only cavities but ejectamenta that constructed a complex interdigitating succession of rocks. Subsequently, mainly in Upper Noachian times, intercrater plains were emplaced widely; these now overlie or embay older cratered crust. The smooth surfaces of these plains, the presence of wrinkle-type ridges and the occurrence of volcanic flow lobes suggest that these were generated by early volcanism. The development of volcanism at this early stage is confirmed by the presence of several very large low-profile paterae, particularly surrounding the Hellas basin. These may have been of the explosive type, throwing out enormous quantities of ash, some as ash-flows, as well as extruding fluid lava. At this time the atmosphere of Mars was apparently denser than it is today and the climate may have been significantly milder, since many valley networks with well developed tributary systems were incised into the inter-crater areas and the rims of large impact craters. These are assumed to have been formed by fluvial processes. The recent description by Malin & Edgett (2000) of widespread stratified units – presumed to be sedimentary – within several equatorial impact craters (e.g. Gale, Henry, Holden, Becquerel and Trouvelot), also inside Candor Chasma, within regions of Margaritifer Chaos and at Hellas, appears to imply that both lacustrine and shallow marine sedimentation was widespread during Noachian times. The material involved may well prove to be a mix of volcanic, aeolian and lacustrine sediment. Early Mars is proving to have been very different from our early views of it.

Additional information about the early crustal development has been forthcoming from the MGS magnetometer experiment. Although the present field is known to be very weak and it is understood that the interior of Mars has cooled and solidified (therefore generating no magnetic field today), in mid-1997 Mars Global Surveyor's magnetometer began recording patches of rocks within the upland hemisphere that retained a magnetic imprint. These appeared likely to be bodies of igneous rocks that had risen to near the surface, cooled and taken on the magnetization of the time. However, they did not seem to be arranged into any kind of recognizable pattern. Then, purely by chance – the need to make 1000 rather than 100 aerobraking passes as the probe orbited Mars because the solar panel arm had become a problem – the magnetometer team had considerably more data than had been anticipated. During that period, it became clear that some of the magnetic patches began to coalesce into a pattern, so much so that, across a large region of the southern hemisphere, irregular magnetic stripes measuring about 100 km in width and up to 2000 km in length were recognized (Plate 5).

The most intense magnetic striping occurs in the region of Terra Sirenum, where field intensities of ~1500 nT were recorded (Acuña et al. 1999). This corresponds to an area with a very high density of impact craters larger than 1 km in diameter, implying that it represents very old crust. Interestingly, a magnetic field was not found to be associated with any of the major impact basins (Hellas, Argye, Isidis), nor with the plains north of the line of dichotomy, nor with any of the major volcanic constructs. This implies that the global field had declined to virtually zero prior to their formation, which was generally considered to be during early Noachian times. If this is so, then the dynamo that generated Mars' global magnetic field had ceased to operate by around 3.9 billion years ago. Furthermore, the presence of striping within the rocks of Terra Sirenum indicates that this is one of the oldest exposed regions of the planet's surface.

7 The plains of Mars

The greatest continuous expanse of plains on Mars occurs north of the line of dichotomy, where units of Late Hesperian to Amazonian age predominate. Similar plains are found within both Hellas and Argyre, and east of Tharsis volcanic plains outcrop in the Tempe volcanic province. Others surround both polar regions. The extensive outcrops of Late Noachian to Early Hesperian intercrater plains within the heavily cratered southern hemisphere are very different. There is much evidence to suggest that volcanism has played an important part in their history, but of equal importance in the geomorphological context are impact, aeolian, fluvial and alluvial processes.

In Valles Marineris, the outcrop of stratified rocks in the sidewalls of canyons attests to a long period of pre-canyon plains deposition in the equatorial regions. These deposits are a mixed sequence of volcanic and sedimentary rocks of considerable but unknown thickness. Elsewhere, the active degradation of upland plains has led to subsequent deposition of extensive plains deposits north of the line of dichotomy and, of course, to burial or partial mantling of cratered terrain south of this line. After the successful deployment of Mars Global Surveyor, high-resolution images of these plains areas yielded a totally new perspective on their detailed structure, epitomized by the strange ridged terrain shown in Figure 7.1. Never before has Mars been scrutinized in such detail, with images sampling narrow strips of terrain that are as varied as the surface of Earth. This picture provides an example of how strange Mars looks at this new resolution. The fill, whether it be sand or dust, is probably hardened to form a surface strong enough to have bright wind-blown ripples and small impact craters on it. It is also possible that some mid-latitude northern hemisphere plains may have been laid down in large lakes or even open seas, which existed on Mars in the distant past.

Plains polewards north and south of about latitude 30° exhibit dramatic albedo changes over quite small distances, as well as over tens of kilometres. This phenomenon appears to be a function of circumpolar wind activity, which sweeps clear some parts of the

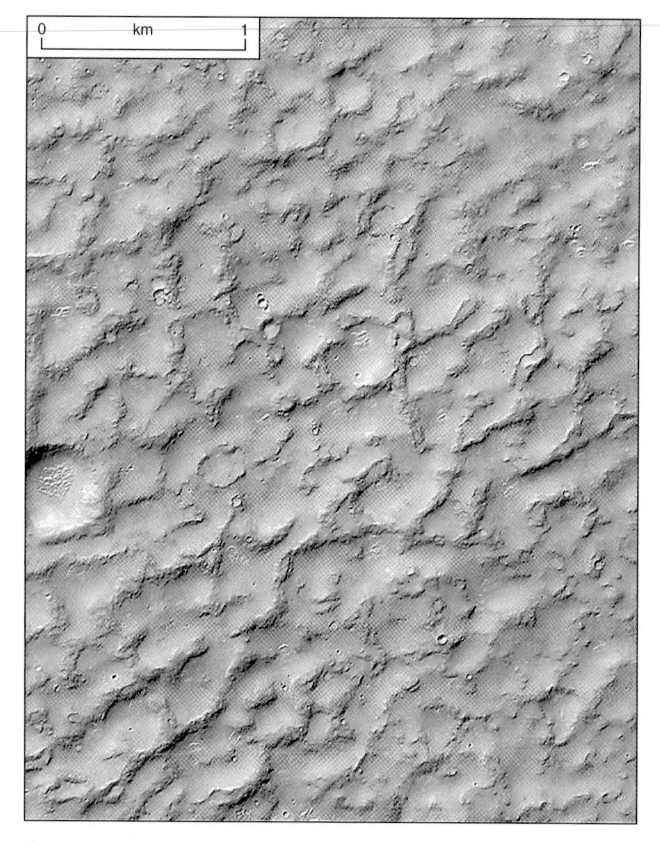

Figure 7.1 Martian plains 210km southwest of Gusev crater at the new resolution available to Mars Global Surveyor. MGS image PIA01699.

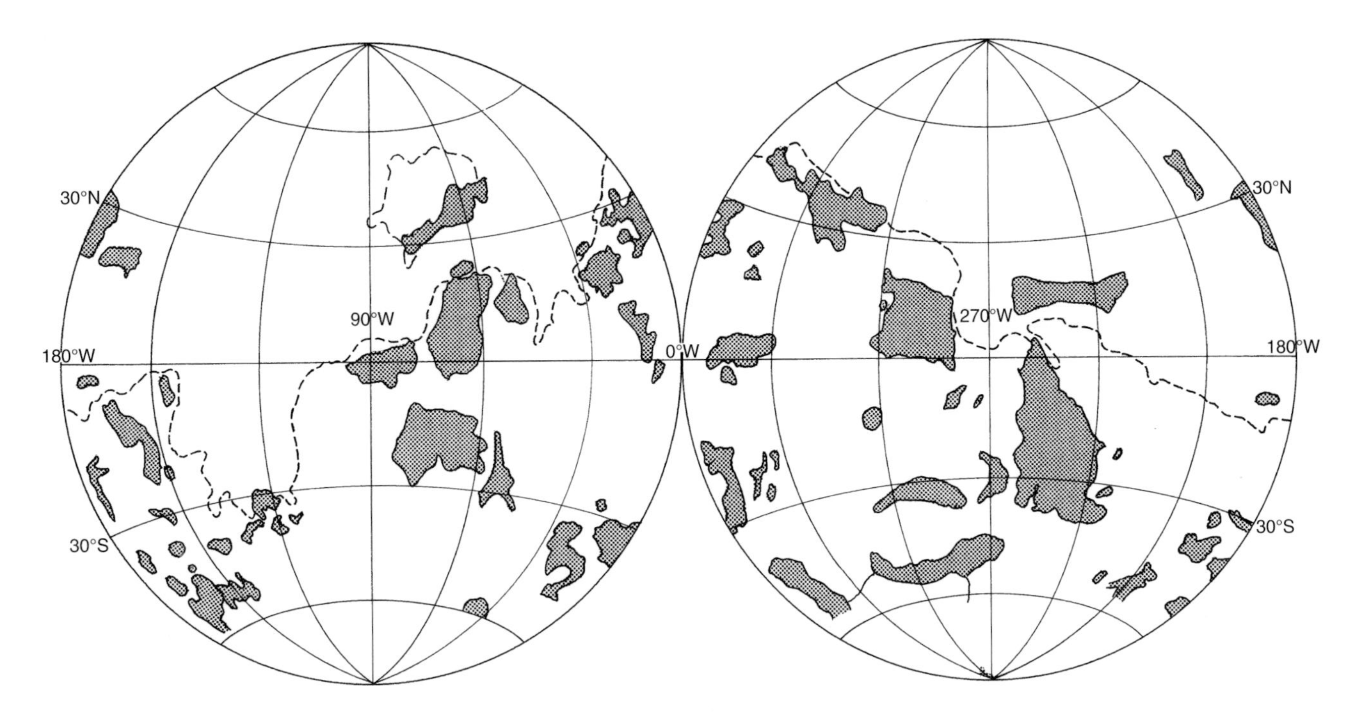

Figure 7.2 Distribution of ridged plains on Mars. The dichotomy boundary is shown by the dashed line.

plains surface but leaves others coated in aeolian debris. Climatic changes may also be responsible for the cyclic deposition seen in the young laminated deposits exposed near the poles, the regularity of alternating low and high albedo layers exposed in polar canyon walls signifying cyclical deposition. Precessional variations have been invoked to account for such cyclicity.

Noachian and early Hesperian plains

Resurfacing of the Martian southern hemisphere cratered surface followed the formation of the ancient cratered crust. In a long period of deposition on the plateau surface, deposits were emplaced that either subdue or bury the underlying terrain. These plateau plains have smoother surfaces than the heavily cratered areas and during the 1980s were classified on the basis of albedo patterns into smooth and mottled plains, the latter forming in response to the deposition of a patina of wind-sorted sediments (Greeley & Guest 1987). Both types probably consist of interbedded lava flows and sedimentary deposits.

Spudis & Greeley (1978) estimated that the aerial extent of these ancient plains is about 29 million km^2; of this, about 36 per cent is covered by intercrater plains with ridged surfaces (Greeley & Spudis 1981). Figure 7.2 shows the extent of such ridged plains units on Mars. Noachian plains are seen to extend widely in an easterly direction across Noachis Terra, to outcrop in Memnonia and across the southern part of Sirenum Terra. Cratering studies suggest that these Noachian units are among the oldest plains units (Scott & Tanaka 1986), which is supported by the recently recorded magnetic striping in the Sirenum Terra area. Younger ridged plains of Hesperian age occur in the western hemisphere in a broad outcrop about 1000 km wide on the eastern flank of the Tharsis Rise and covering an area of about 4 million km^2 in the eastern hemisphere; a similar expanse is to be found in Hesperia Planum.

Upland depressions, such as Syrtis Major Planum, the western part of Amazonis, and impact basins such as Hellas, Argyre and Isidis, also show development of plains units. Crater ages derived for those in Syrtis Major Planum yield 3.6 billion years (Hartmann et al. 1981); plains in Hesperia Planum and Lunae Planum are also of Lower Hesperian age. The ridges characteristic of many of these old plains are similar in many respects to lunar wrinkle ridges, with an asymmetric profile and anastomosing course. Although they are widespread on the intercrater plains units, they may also occur within large impact craters that appear to have been inundated by lavas and mantled by aeolian deposits (Fig. 7.3). The development of such ridges, which are invoked by many to be evidence for a volcanic origin, is not necessarily a volcanic process; however, as is known from lunar experience, such ridges usually form in resilient rocks such as basalts.

More direct evidence for Noachian volcanism may

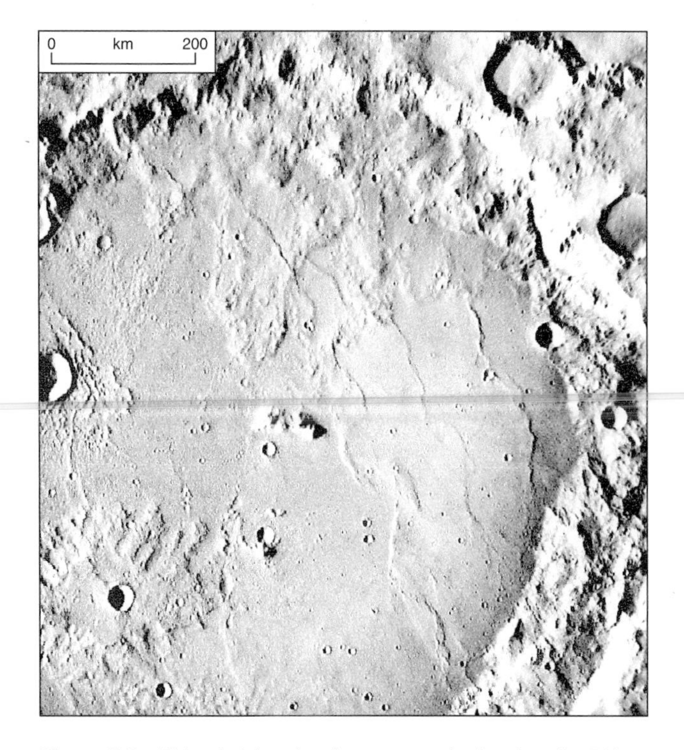

Figure 7.3 Ridged plains development on the interior of a 70 km-diameter crater. Viking orbiter image 432s25; centred at 20°S 180°W.

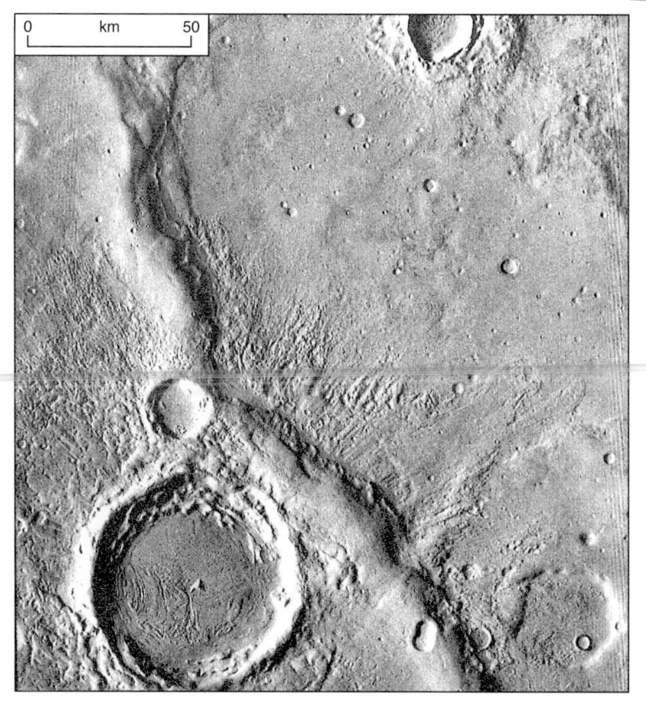

Figure 7.4 Wrinkle ridges and volcanic flows, crossing the plains near Hesperia Planum. Note the asymmetric profiles of the ridges and their sinuous courses. Viking orbiter image 417s05, centred at 35.62°N 246.84°W.

be found in such regions as Protonilus Mensae, where there are quite large areas without ridging and with a smoothed geomorphological signature (Fig. 7.4). Aeolian mantling may account for some of the smoothing, but there are many other landforms that are difficult to interpret as other than volcanic flows, flow lobes and either exhumed dykes or spatter ramparts. Such evidence can be augmented by noting (Schultz 1977), that there is a close correlation between the occurrence of ridged plains and floor-fractured impact craters, which are widely believed to have been modified by volcanic activity. An early episode of volcanism (prior to 3.9 billion years ago) would certainly alleviate the crater extinction problem discussed on pp. 49–50. Furthermore, as has been observed by Carr (1984), high effusion-rate volcanism might be expected during the period of intense bombardment and brecciation, and of elevated accretional energy dissipation, which must have been characteristic of the period around 4 billion years ago.

Ridges developed within the younger plains of Hesperian age show more definite evidence of a relationship between regional and local tectonic regimes: ridge segments typically are aligned over large areas. In both Solis Planum and Hesperia Planum, flow fronts and lobate flow terminations accompany the ridges (Fig. 7.4), whereas on the western side of the latter there are extensive smooth-facies ridged and lobed plains units associated with the highland volcanoes Hadriaca and Tyrrhena paterae. There is also a broad swath of such plains around Hellas. Narrow flows, small cratered domes and what appear to be either exhumed dykes or spatter ridges are discernible on the borders of Isidis, and there are two very prominent low volcanic shields with attendant caldera structures on the surface of Syrtis Major Planum. High-resolution images of the region adjacent to one of the calderas clearly reveal a development of tube-fed volcanic flows. Crater statistics suggest an age of 2.6 billion years for flows associated with this volcanic plains complex (Meyer & Grollier 1977).

Less heavily cratered Upper Hesperian ridged plains outcrop in Chryse Planitia and south of Elysium. The landing of Viking 1 in the former locality indicated the presence of what appear to be vesicular basaltic blocks on the surface of the plain and confirmed that there were basalt weathering products in the surface soils. The subsequent landing by Mars Pathfinder at the mouth of Ares Vallis confirmed the presence of volcanic blocks, including basaltic and andesitic types, fragmental rocks that may be conglomerates, and red drifts showing varying degrees of oxidization and weathering.

Studies of similar ridges on the east flank of the Tharsis Rise by Phillips & Ivins (1979) strongly suggest that they are tectonic in origin; however, the characteristic morphology of ridges and arches on plains units – a feature they share with their lunar counterparts – appears to support the notion that they were formed in rocks with considerable intrinsic strength. This appears to rule out ancient impact breccias, lavas of low viscosity being more likely. Their morphology, the extensive distribution of plains with ridged surfaces, and their association with various kinds of volcanic structures, support the idea of an early phase or phases of flood volcanism that modified the ancient cratered terrain widely, perhaps as early as 3.9 billion years ago.

Channels, plains and volcanism

South of Tharsis, in the regions of Thaumasia Fossae and Aeolis, the many valley networks that incise the intercrater plains appear to have formed after north–south Tharsis-related faults and northeast-striking graben (Brackenridge 1987). The plains are cut by most faults, but not all, and some of the graben do not displace the plains units, suggesting that graben faults may have been forming while plains were still being emplaced. Low-albedo volcanic flows were extruded after both types of faulting had taken place, many of the lavas partially covering tributary channels. Sections visible in some remnants of eroded plains exhibit a light/dark internal stratification, which suggests that dark volcanic flows are interbedded with lighter sediments. Furthermore, valley heads tend to occur at the base of a prominent low-albedo volcanic unit that is sandwiched between a lighter surface stratum and an older one beneath (which could be a sill). Brackenridge suggested that the volcanic horizon could have acted as an aquiclude for heated volatiles that escaped as springs along its base. If this was so, then valley development could be explained by hot-spring activity associated with the emplacement of a sill into ice-rich fragmental material.

Elsewhere in Aeolis, underlying structural control is shown by a strong preferred northwest trend in the many channel networks present. Moreover, channel-wall interbedding of light (sedimentary or volcaniclastic) and dark (volcanic) layers, implies a strong correlation between volcanism and channel development. The more degraded of these intercrater plains deposits are most readily explicable in terms of groundwater release from a frozen subcrust, the former being instigated by early mafic post-bombardment volcanism. One of the major problems here is deciding how such a reservoir of groundwater may have accumulated near the surface of Mars during its early history. Jakosky & Carr (1985) infer from calculations of the pre-Tharsis obliquity of Mars that enhanced obliquity would have instigated ice condensation at low latitudes (where it is currently unstable when in contact with the atmosphere). If this assertion is valid, then here is a mechanism for ice deposition in the cratered highlands at an early stage in Martian history.

Hesperian flow plains

The most extensive occurrence of Hesperian flow plains is peripheral to the major volcanic provinces of Tharsis and Elysium (see Figs 3.4, 3.7). There is also a major outcrop of somewhat ambiguous volcanic deposits in Malea Planum, which may be fluvially modified volcanic flows. The oldest lavas outcrop around Tempe Terra, Memnonia and Ceraunius Fossae, and their eruption marked the first of several major volcanic episodes that resurfaced huge areas of the northern lowlands. Extensive flow plains also were erupted from near the crest of the Syria Rise and now cover large areas of Syria and Solis Planae at the western end of Valles Marineris. Particularly extensive lobate flows of Mid- to Late Hesperian age are found east and northeast of the base of the Olympus Mons shield and on the western side of Tempe Terra.

A major sequence of flows emanates from beneath the low pile of Alba Patera, where broad sheetflows extend at least 1500 km from the summit. Similar flow plains originated from centres now situated beneath younger volcanoes such as Arsia Mons and Uranius Patera, where occasional rimless depressions and discontinuous spatter-type ridges are aligned along what are assumed to be linear source vents or fissures, most of which have been buried by their own flows, or younger ones. Flood-type lavas of Late Hesperian to Early Amazonian age also occur around Ceraunius Fossae, where they flood fractures incised into older highland terrain.

Flows are typically 60–120 m thick, often composite, and have relatively featureless upper surfaces. Many can be traced for hundreds of kilometres and several may coalesce to form broad overlapping sheets. Individual flows are immense and have volumes in excess of 400 km^3. Flow channels can often be discerned and there are many low domes and small depressions, which are presumably the sources of some flows. However, the sources of such lavas are generally obscured by younger flows and it can only be assumed that they issued from fractures or linear vents that were buried by their own products.

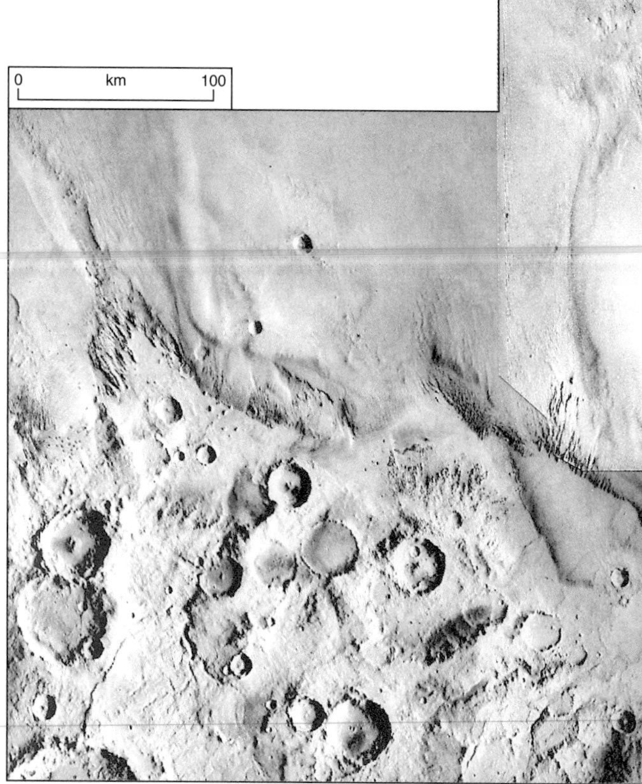

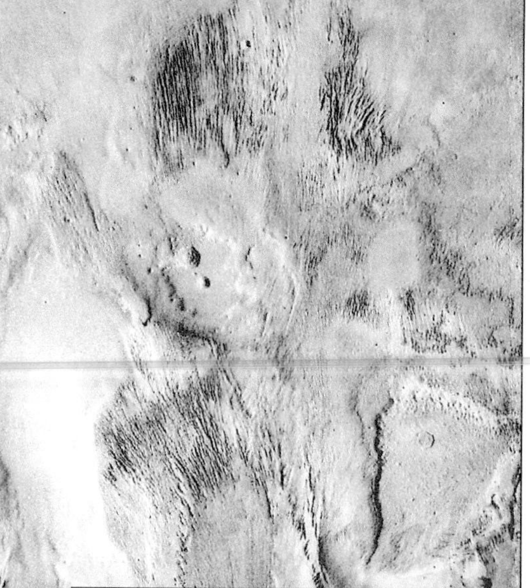

Figure 7.5 Diverse morphology of volcanic units in the Amazonis region. The yardang development shows a different orientation from unit to unit. Note the wind-etched rocks seen on the interiors of the large impact craters towards the foot of the mosaic. Viking orbiter images 635a83 and 635a84; left-hand image centred at 0.33°S 177.73°W.

The Medusae Fossae plains

Unusual plains units are located in a region that has the volcanoes Apollinaris Patera, Biblis Patera and Olympus Mons at its apices. In general these deposits have higher relief and albedo than most plains materials, but are less cratered and hilly than the highlands. They outcrop in a broad but discontinuous zone, which runs east–west along the lowlands/highlands boundary. Here a series of discontinuous flat-lying sheets, each about 100 m thick, has smooth and gently undulating or etched surfaces and a total thickness of at least 3 km. Some of the higher sheets show a development of yardangs along plateau edges, which suggests they are not competent (of significant intrinsic strength) lavas but relatively friable rocks (Fig. 7.5). In morphology they bear a striking resemblance to some terrestrial ignimbrites, as was first noted by Malin (1977).

Scott & Tanaka (1982) indicated that the deposits cover an area of the about 2.2 million km^2 and have an estimated minimum volume of 3.85 million km^3. Seven units are distinguishable, and at four locations there is a marked thickening of the sheets, suggesting possible eruptive foci. Early work by Scott (1969) among terrestrial silicic volcanics, allowed him to highlight features common to both:

- rounded patches of smooth, high-albedo (non-welded) materials that overlie low-albedo jointed (welded) flows
- local complementary joint sets in some (welded) materials
- thick flow sheets of great lateral extent that subdue the underlying topography.

One other interesting feature of this region is the general absence of volcanic domes, which are common in terrestrial silicic igneous provinces. Their absence does not rule out a pyroclastic origin, since theoretical considerations indicate that there should be wider dispersion of pyroclasts on Mars than on Earth for the same mass eruption rate (Head & Wilson 1981). Thus, the absence of domes may be a function of the lower expected relief and an inability of Viking images to resolve low relief features.

In the Basin and Range region of the western USA, ignimbrites are closely associated with block faulting where highly volatile silicic magmas were available for energetic eruption. The high concentration of NW–SE faults and graben within the Amazonis–Aeolis region (e.g. Medusae Fossae) shows that there was indeed extension in the Martian crust at this time and it is possible that faulting resulted in roof failure in a large magma chamber located beneath the deposits, producing extensive ashflows. Calculations by King & Riehle (1974) suggested that, if such ashflows were generated, they would remain in a fluidized condition for three to nine times the period typical on Earth, making it possible for flows to travel great distances from their source.

In my opinion, the circumstantial evidence for a pyroclastic origin is considerable; however, other workers have suggested that the ashflows could equally well be a thick sequence of aeolian deposits transported and trapped along the highland/lowland boundary (Lee et al. 1982, Thomas 1982). Higher-resolution imagery of the Medusae Fossae region obtained by MGS confirms that aeolian erosion has been very active in the region, but does little to distinguish the true nature of the deposits. Thermal emission mapping by Ivanov et al. (1998) revealed unusually high emissivity in this region, implying that the surface deposits have low density; they could not distinguish between an aeolian or ashflow origin.

As can be seen in Figure 7.6a, which covers an area measuring only 3.0×4.7 km, much of the region is mantled in sediment of smooth texture, but there is ample evidence that more resilient layers have been (and possibly still are being) stripped and wind-etched into yardangs. It is difficult to decide whether or not the prominent dome-like landforms seen on the image are small volcanic blisters or sediment accumulations around butte-like remnants of the stripped unit. The very high-resolution image (Fig. 7.6b) shows beautifully the stripping and etching of the friable plains deposits. Several areas show families of dunes that are currently undergoing erosion; many have somewhat ragged edges and rather rounded crests, and others evidently are now active, having good crescentic morphology and strongly demarked slopes (Malin et al. 1998).

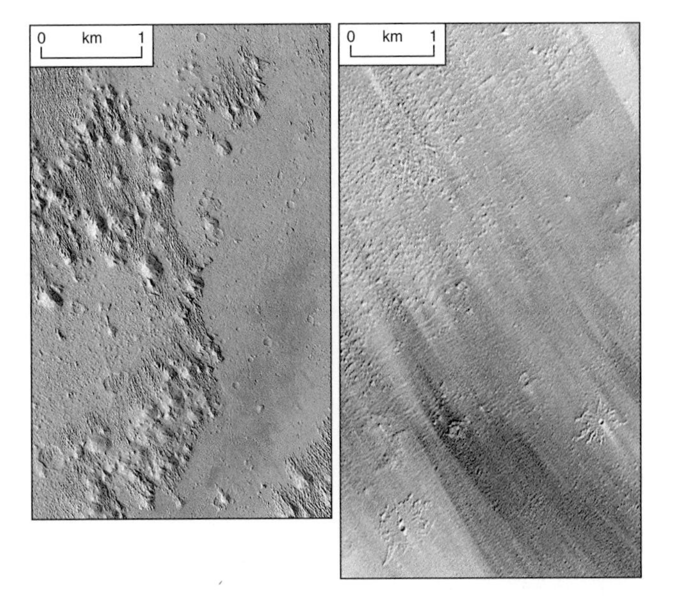

Figure 7.6 MGS high-resolution images of the Medusae Fossae region. **(a)** Wind-etched surface and smooth region of sediment deposition. Yardangs well developed and many rounded landforms that could be small volcanic domes. (PIA00801) **(b)** High-resolution image showing fine details of yardang development and morphology of edge of stripped Medusae Fossae units (PIA01698).

Tempe Terra plains province

Although there is a general lack of specific evidence regarding the source vents of the very extensive Hesperian flowplains lavas, this is not the case with those in the Tempe volcanic province. This interesting complex developed within the uplands that form a northwestward continuation of Lunae Planum, on the opposite side of Kasei Vallis. The Hesperian volcanic sequence occupies an area of 3.4 million km^2 and comprises three distinct kinds of terrain: rugged hilly, faulted, and smoother uplands, which are an extension of the Lunae Planum plateau (Scott 1982).

The geological relationships in this region are as follows. Blocks of the Noachian cratered plateau are mantled in younger lavas and what may be pyroclastic rocks, and are embayed by younger Hesperian volcanic plains, sometimes heavily mantled with aeolian debris. Distinct narrow lava flows can be seen on some of these plains units. Debris aprons also have accumulated, often hugging the boundary between the fractured plateau and the adjacent plains. On the surface of the plateau is a surprising variety of smaller volcanic landforms, several major volcanotectonic structures with little or no relief, and steeper-sided volcanic mountains, which may be dissected shield volcanoes. Details of several of these structures have been given by Scott & Carr (1978), Underwood & Trask (1978), Wise (1979), Plescia (1980) and Scott (1982). The major ring structures are somewhat similar to Alba Patera, although with even less relief; the largest is about 250 km across.

The many smaller-scale volcanic features, and in particular the substantial number of low shields on the resurfaced parts of the fractured plateau, bear a striking resemblance to those that developed during the plains-type volcanism of the Snake River Plains (Idaho), which produced widespread volcanic plains from source vents arranged along rifts. Thus, small low shields have well defined summit pits that often are aligned in the direction of the SW–NE fracturing.

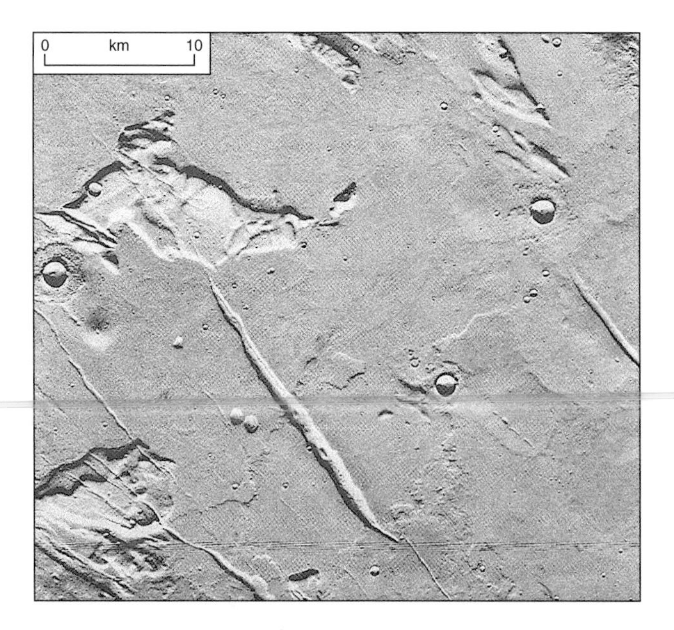

Figure 7.7 Plains-style volcanic province of Tempe Terra. Note the low shields and row of aligned axial depressions, and a single circular vent. A sheetflow is seen to pre-date the SW–NE faults. Also, a volcanotectonic feature traverses the region in a SW–NE direction. Viking orbiter image 627a26; centred at 36.22°N 86.71°W.

There are also many elongate vents and thin sheet flows, some of which appear to have originated in fissures (Fig. 7.7). Plescia (1981) estimated that low shields account for about 75 per cent of the constructional landforms present. Their obvious development along rift faults and the presence of fissure-fed sheet flows and elongate depressions reveal a close genetic relationship between volcanism and crustal extension, as they do in the Snake River Plains. What appear to be absent from the Tempe plains are tube-fed flows, so common a feature in the Snake River province; however, it is possible that such flows were emplaced but that their roofs have not suffered collapse.

In addition to the low shields, there are several steeper-sided landforms with diameters in the range 5–10 km. These are probably composite cones and they have no discernible associated lava flows. Such features are not found in the Snake River Plains and their occurrence in Tempe may imply a more extended range of volcanic style, involving more viscous silicic magmas than are common in the Snake River province. Supporting evidence for such a hypothesis is provided by a large dissected patera structure located in the northeast of the plateau, which, by analogy with highland paterae elsewhere on Mars, may have developed during a phase of explosive volcanism generating ashflow and airfall pyroclastic deposits.

Plains deposits of Hellas

The Hellas impact basin, which had a major impact on the ancient cratered crust, shows a widespread development of plains units, some of which extend well beyond the rim (Fig. 7.8). The basin itself was produced in Noachian times, when the rim and adjacent cratered plateau materials were uplifted, possibly because of an oblique impact. Subsequently, there was sustained modification, during which, volcanism, channel formation, erosion and deposition occurred. The basin was first flooded by fluid basalt-like lavas in Late Noachian times, which were deformed with the production of plains ridges. The fact that wrinkle ridges did form suggests a thick lava sequence, possibly >1 km thick (Tanaka & Leonard 1995). Contemporaneously with this activity, it seems that explosive volcanism from Amphitrites and Peneus paterae spread out extensive sheets of pyroclastic material that has subsequently become quite deeply eroded. Other volcanic plains units were generated from volcanoes such as Hadriaca and Tyrrhena paterae, their deposits generally external to the basin's inner ring.

Subsequently, the interior of Hellas was mantled by a variety of materials that were later modified to give what Greeley & Guest (1987) term "a dissected floor unit"; this has been subdivided by Tanaka & Leonard (1995) into units including degraded deposits, smooth material, an "interior" deposit, and dunes. These Early Hesperian units represent a sequence of predominantly fine-grain aeolian materials, 1–2 km thick, within which large numbers of pedestal craters and knobbly outliers can be identified, attesting to considerable erosion, except where protected by ejecta. The relatively subdued ridging on the eastern side of the basin's interior appears to be a function of flooding associated with Harmakhis, Reuil and Dao Valles. During Late Hesperian and Amazonian times, production of sediments from rim materials was accomplished by a mixture of fluvial activity and groundwater sapping (Fig. 7.9). At the same time, aeolian activity emplaced sedimentary materials on each of the floor, rim and surrounding rim massifs. Subsequently (and currently), these have been etched, pitted and otherwise modified by wind activity, which is currently extremely strong in this region of the southern hemisphere. The orientation of yardangs and dunefields allows current wind vectors to be established, there being a general N–S or NNE–SSW trend on the western rim and floor, and a NE–SW windflow on the eastern rim and beyond.

During Amazonian times, there was significant fluvial activity on the basin's eastern rim, with the activity along major channels such as Dao and Harmakhis

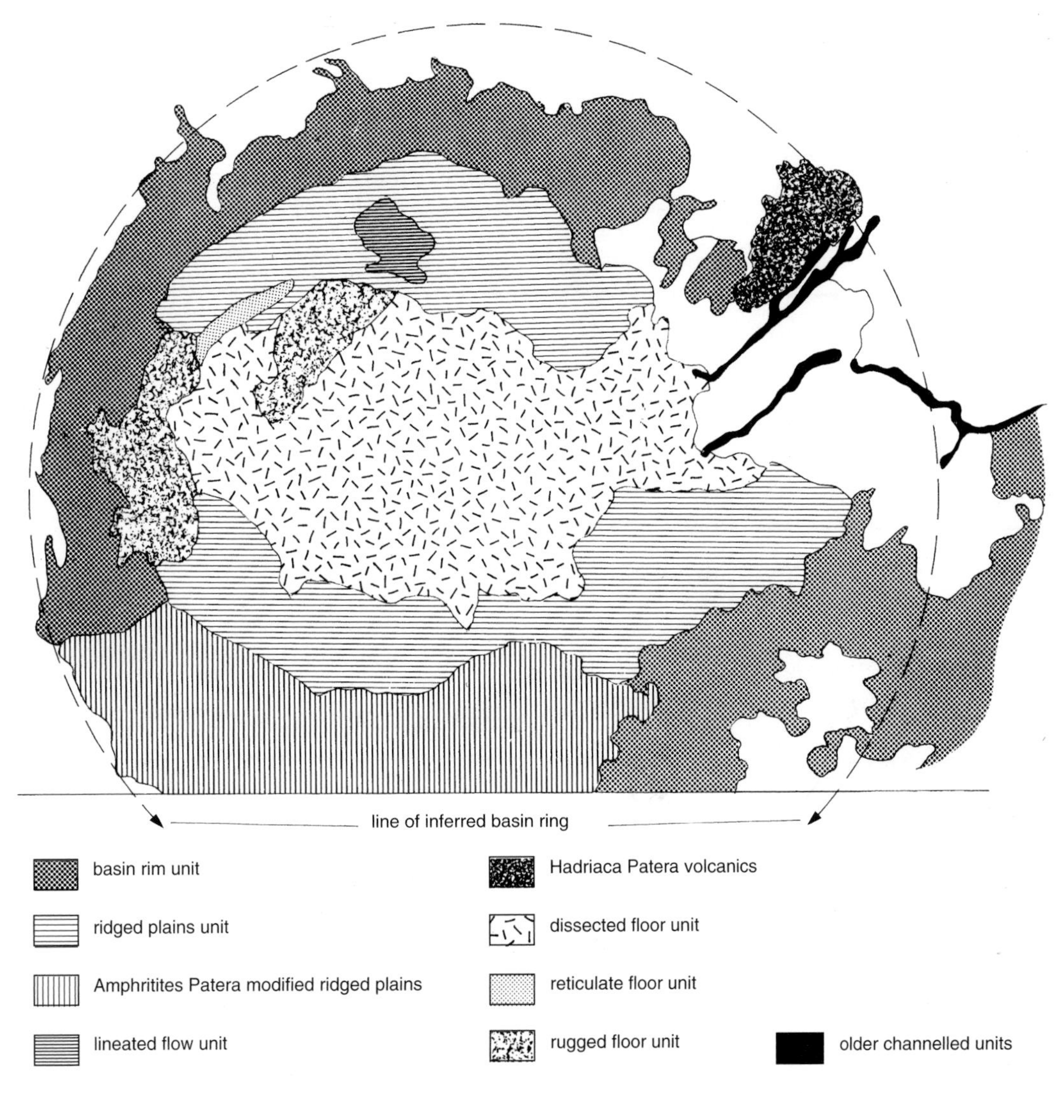

Figure 7.8 Geological sketch map of units in the Hellas region.

Valles. These spread out sediments widely on the eastern floor, which were later modified by deflation and collapse. The geomorphological signature of these younger materials suggests that both fluvial and periglacial activity may have played a part in their production (Crown et al. 1990). Within this plains mantle there is a lineated floor unit and a channelled plains rim unit, which may be only slightly modified areas of the original sedimentary mantle. The latter has subdued topography and low remnant mesas and narrow channels, whereas the former shows a development of both straight and curved lineations within smooth plains material and may have formed by tectonic modification of mantling material.

Other Hellas plains have a reticulate pattern of ridges, rugged on the kilometre scale, and others are punctuated by knobs a few kilometres across (Fig. 7.10). Extensive ridged plains extend beyond the southern rim and include low-relief volcanic structures, including Amphitrites and Peneus paterae and several unnamed ring structures with radiating ridge patterns.

The northern plains

Introduction

The very extensive northern plains were emplaced during the Late Hesperian and Amazonian epochs. Despite having the largest contiguous area of any of the plains developments on the planet, in some ways they show the fewest easily identifiable landforms, in general having a muted and softened character. This notwithstanding, there are enough features to justify hypothesizing about their formation. Plains development apparently entailed deposition of a complex

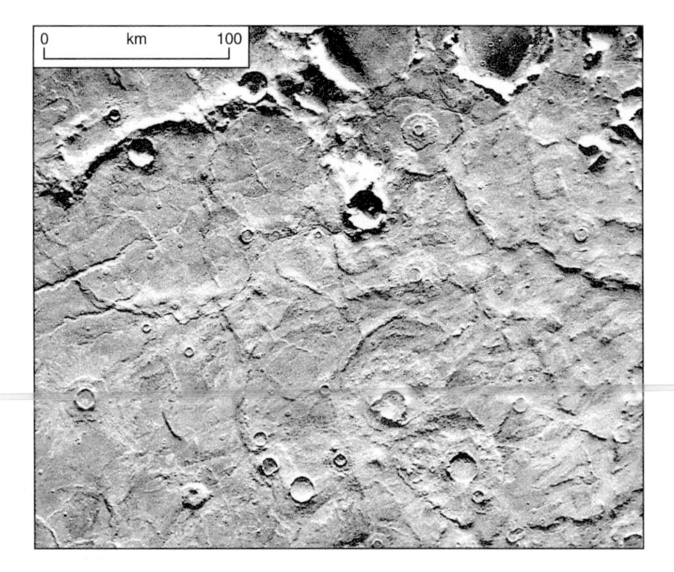

Figure 7.9 Plains units emplaced along the southern rim of Hellas. Viking orbiter image 361s17, centred at 54°S 317°W.

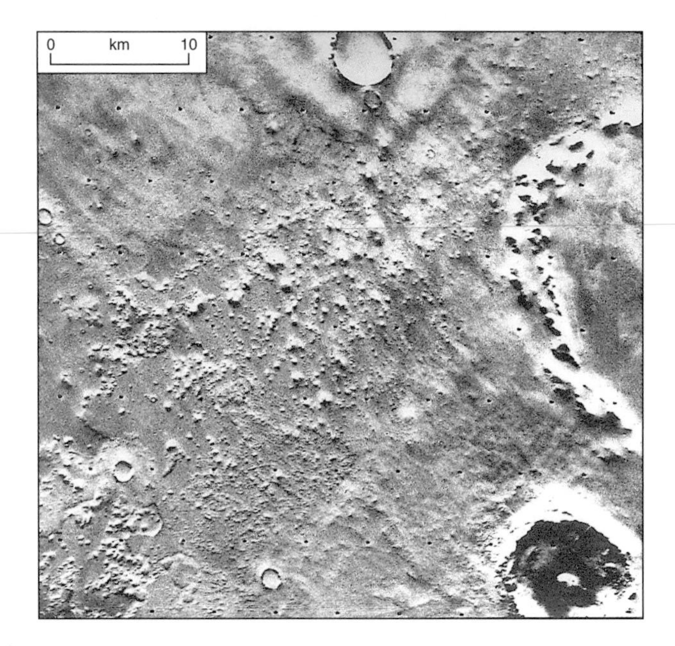

Figure 7.10 Knobby plains development in the northern plains, north of the volcanic shield of Hecates Tholus. Viking orbiter images 086a36–47, centred at 33.35°N 212.44°W.

sequence of lava flows, aeolian and alluvial sediments, and this may have been aided and abetted by widespread glaciation. Resurfacing and modification of the northern lowlands continued through the Amazonian epoch, a strong argument having been put forward for extensive continental glaciation (Kargel et al. 1995). Sedimentation and volcanism produced diverse plains deposits that are divisible broadly into smooth plains and knobby plains. In places, the former are composed of aeolian deposits, whereas some knobby plains appear to have been produced by stripping of older units. Recently obtained MGS images have thrown new light on the detailed structure of some small areas.

As we have seen, the dominant characteristic of Mars' topography is the striking difference in elevation between the northern and southern hemispheres. As the recent MOLA results have shown, this difference is also manifested in the surface roughness: a large percentage of the northern hemisphere is given over to the Vastitas Borealis Formation (of Amazonian age), which is flat and smooth, even at a scale as small as 300 m (Smith et al. 1998). This supports the hypothesis that this region is composed of either sedimentary or volcanic rocks, or both.

Northern lowland plains in mid- to high latitudes

One of the more distinctive geomorphological imprints seen on the lowland plains is a polygonal fracture pattern that divides the plains into a series of giant blocks, anything from 2 km to 10 km across (Fig. 7.11). In two short papers, McGill (1985a,b) proposed that such patterns had developed in relatively thin layers of sedimentary rocks, where contraction and compaction had occurred over the underlying topography. In this way some curvilinear fracture patterns would represent outlines of impact craters buried beneath the sediment cover. A study by Lucchitta et al. (1986) confirmed this suggestion. There are several reasons why

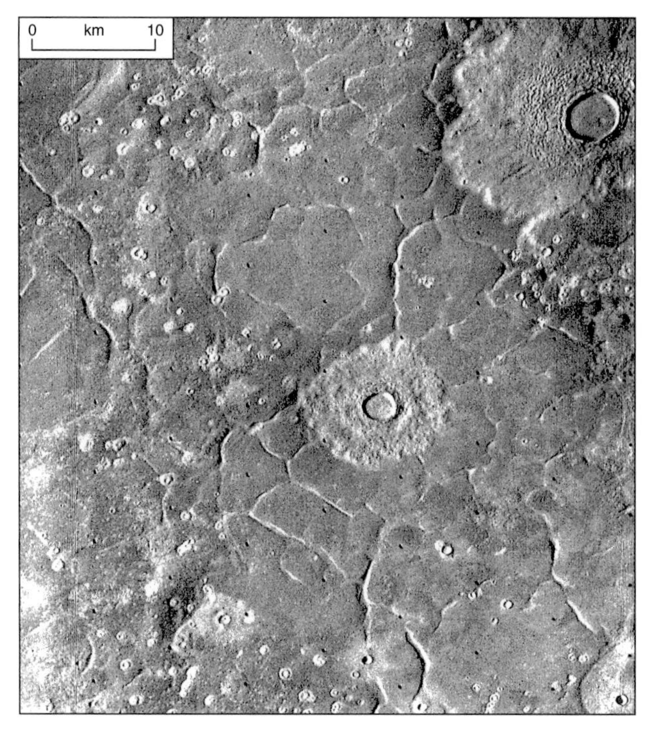

Figure 7.11 Polygonally fractured mottled terrain in Acidalia. The polygons evidently pre-date many impact craters, whose ejecta stand proud as pedestals that overlie the plains surface. Viking orbiter image 026a32; centred at 44.96°N 19.19°W.

the idea seems plausible: first, polygonally fractured deposits tend to be located in low-lying areas that apparently received an influx of sediments; secondly, the fracture patterns are concentrated near embayments in the upland hemisphere where major outflow channels debouch; and thirdly, crater ages indicate the channels and fractured plains to have similar ages.

McGill & Scott-Hills (1992) suggest that the sedimentary material in which the features have formed was transported by outflow channels from the southern highlands and rests on a surface of old crater rims, mesas and knobs. They estimate that the Utopia Planitia polygonally fractured terrain covers an area of 3.4 million km^2, is 600 m thick and has a volume of 1.9 million km^3. There is a similar expanse in Acidalia Planitia. However, Pechmann (1980) prefers to consider them as tensional fracture patterns developed in the plains by regional doming. What is clear is that the pattern of cracking implies uniform tension in all directions within the plane of the surface. Closely similar patterns (in form, not scale) occur on Earth where horizontal extensional stresses develop because of cooling or desiccation.

Scott & Tanaka (1986) assign a genetic relationship and an Upper Hesperian age to both the Chryse outflow channels (which debouch into Acidalia Planitia) and the polygonally fractured plains in Acidalia. Greeley & Guest (1987) assign a similar age to the fractured plains of Elysium and Utopia Planitia. However, the prominent Elysium channels appear to be slightly younger (Early Amazonian) and connected with the Elysium shield volcanism. This anomaly may be more apparent than real, since the Elysium channel deposits may be simply obscuring an older channel system that was directly responsible for the Hesperian plains.

Other distinctive geomorphological features occur on the plains within a few tens of kilometres of the lowland/upland boundary. Like much of the polygonally fractured ground, these are located near the mouths of the Chryse channels that debouch into Acidalia Planitia and below the highland scarp in western Utopia. One particularly interesting set of features comprises sinuous ridges 0.5–1 km wide, set in elongate depressions (Fig. 7.12). Lucchitta et al. (1986) observe that they bear a striking resemblance to the ridges that are generated near the mouths of Antarctic ice streams and on ice shelves, particularly where shoals are developed. For this reason they suggest that the Martian uplands were bordered by frozen materials that had mechanical properties similar to those of modern Antarctic ice shelves. If their suggestion is accepted, then the northern lowland materials represent outflow channel sediments that were laid down in a partially frozen ocean. Further support for a glacial

Figure 7.12 Sinuous ridges and rounded hills in Utopia Planitia. The cratered uplands are located to the left of the image. Viking orbiter image 608a06, centred at 45°N 278°W.

origin comes from Kargel et al. (1995), who view the sinuous ridges as possible eskers. The fact that some ridges are located within troughs may indicate that these are tunnel channels. However, Parker et al. (1986) compared the ridges to terrestrial lacustrine or shallow marine coastal spits and barriers, which imply a predominantly liquid environment in which waves are responsible for the features seen.

The knobby plains comprise extensive units with a softened aspect, often showing impact craters that are partially infilled with debris that has been modified or stripped, and surfaces crossed either by curvilinear ridges that give what is termed thumbprint terrain or by lines of small knobs (Fig. 7.13).

De Hon (1988) showed that in many locations close to the mouths of major outflow channels (e.g. Maja Vallis, Ladon Vallis) there are relatively featureless plains units that appear to represent lacustrine deposits. Temporary ponding in channels would have been expected under Martian conditions, with the result that local sedimentation would have taken place both within channels and near their mouths. Flood-plain sediments are widespread down stream of channel mouths, and lacustrine deposits have been identified in such areas as Chryse Planitia, Terra Sirenum and Lunae Planum. Possible deltaic sediments are revealed on Viking images in Amazonis and Elysium

Planitia on Viking imagery, and it is to be hoped that higher-resolution images from MGS may help establish the true nature of these units.

High-latitude plains

The character of plains units in higher latitudes (above 35–40°N) is very different from those elsewhere. These plains are rather poorly understood and appear to be blanketed in aeolian debris that softens the topography over vast areas. Little or no direct evidence is found for primary volcanic deposits, although such may well be present beneath the sedimentary cover. Landforms developed within these high-latitude plains include surface mottling, ridges, grooves and knobs. Impact craters at these latitudes tend to have encircling pedestals of raised ejecta and, where aeolian burial is deeper, are entirely blanketed in sediment, giving them the appearance of volcanic domes. Mottled plains are particularly enigmatic and they almost encircle the planet between latitudes 50°N and 70°N. As we have seen, mottling is largely attributable to the relatively high albedo of impact crater ejecta compared with intercrater areas.

Fortunately, Mars Global Surveyor has been able to obtain several high-resolution images of these enigmatic regions. Figure 7.14 shows a typically softened terrain, among which impact craters in a variety of stages of both burial and stripping can be seen. Some craters and their attendant ejectamenta are almost completely mantled in sediment (possibly aeolian but could be glacial), looking almost like small domes, whereas others appear to have been almost stripped of the mantling material, which has a pitted appearance. Higher-resolution images appear to show boulders along the rims of stripped impact craters (Fig. 7.15).

Figure 7.13 Thumbprint and knobby plains development in Arcadia. Viking orbiter image 319s49; centred at 46°N 158°W.

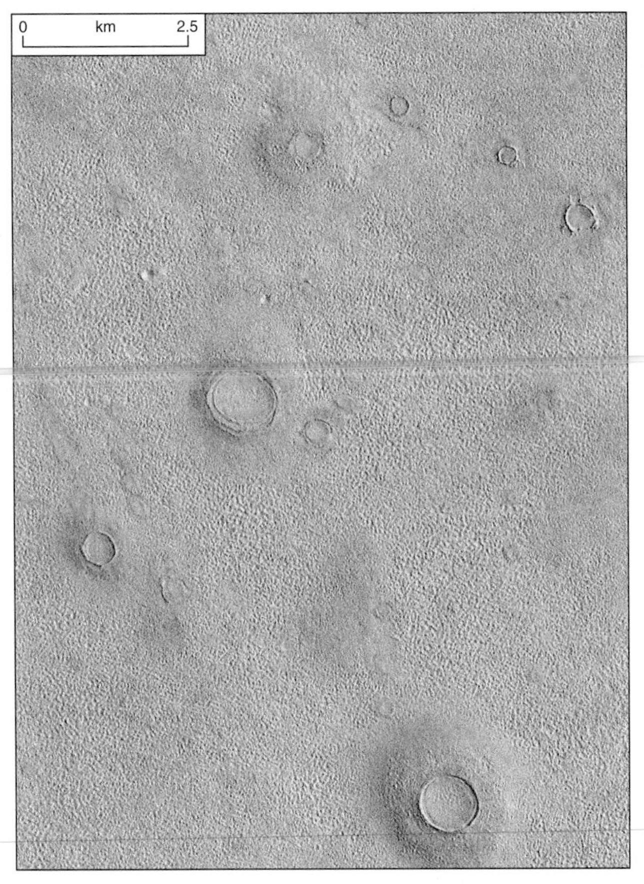

Figure 7.14 Softened topography of the northern plains. Impact craters can be seen in various stages of having their sedimentary mantle stripped off. MGS image PIA02073.

Although there has been considerable volcanic input into these Hesperian plains – there was major volcanism in the Tharsis and Elysium regions at this time – there has also been much aeolian modification. Periglacial processes, tectonism and sediment compaction may all have been involved and certainly there has been extensive stripping of mantling deposits. Because densities of impact craters greater than 5 km diameter are surprisingly uniform for these extensive plains units, it would appear that, however they were formed, they were produced relatively quickly.

Other plains deposits

Flat-lying surficial deposits resulting from mass wasting are located both within and beyond the mouths of

Figure 7.15 MGS image taken close to the Martian terminator. The low lighting conditions enhance the fine detail of the rough and bumpy northern plains. MGS image PIA01697.

fretted channels in the Deuteronilus Mensae region and east from here. These overlie older channelled terrain and have a very low incidence of impact craters, indicating a relatively young age. Plains formed from this material may be up to 600 km wide and are marked by low-albedo sinuous intertwining patterns that become increasingly mottled to the west. They appear to be fluvial deposits with channels marked by sandbars and relict islands; mottled regions may represent deposition from the ponded terminus of an ancient fluvial system. The fact that such channels are restricted to the upland/lowland interface suggests that only here was there a downchannel slope large enough to allow for transportation of the sedimentary debris down stream. Other, recently identified, units can conveniently be considered here. Malin & Edgett (2000), using high-resolution MGS imagery, identified widespread well stratified deposits within the cratered southern hemisphere. These flat-lying plains deposits vary in type, but where all variants occur together the sequence is (in descending order): massive units, dark-toned layered units, light-toned layered units. This succession, or parts of it, occurs as remnants within several large impact craters, among the intercrater plains, in the equatorial canyon-chaotic terrain system, and around Hellas, and must have been formed at much the same time as Noachian-age cratering was in process. Thus, the upper ancient crust of Mars, rather than being simply impact brecciated like that of the Moon, is well stratified over large regions, probably to a depth of at least 10 km. This points to the likely existence of lakes and shallow oceans, and of subaqueous activity during early plains formation.

8 Volcanism on Mars

The global distribution of Martian central volcanoes (see Fig. 3.6, p. 21) is not linear like that on Earth, because of the lack of a segmented lithosphere; thus, Mars' large shield structures are presumed to have formed above long-lived mantle hotspots. Plains volcanism, characterized by fissure eruptions, growth of low shields and hydromagmatic activity affected the southern and northern hemispheres, but the most recent Amazonian phase apparently did not affect the southern hemisphere.

The earliest recognizable major outbreak of volcanism produced the Upper Noachian plateau plains (Chs 6, 7); these were succeeded by the even more extensive Lower Hesperian ridged plains. Both were emplaced largely by flood lava eruptions and outcrop on the upland southern hemisphere of Mars. Centralized volcanic activity appears to have begun with the formation of the enigmatic structure, Amphitrites Patera, during the Lower Hesperian epoch, and this was followed in Upper Hesperian times by the generation of a few large, low-profile ash or mixed lava and ash volcanoes close to the Hellas basin, also at Hecates Tholus in the region of Elysium, and in Syrtis Major Planum. The characteristic hydromagmatic activity of this period attests to the rather different climatic conditions affecting Mars at that time, as a result of which significant volumes of volatiles became entrained in magmas rising through the subcrust.

Towards the close of Hesperian times, another episode of flood volcanism began constructing the volcanic plains of the northern hemisphere. This was accompanied by the growth of major volcanoes in the region of Elysium at Albor Tholus, and in northern Tharsis, where activity became focused on Alba Patera, the largest central volcano on the planet. While some explosive activity continued, there was a gradual change towards effusive volcanism, with the eruption of huge volumes of low-viscosity mafic magmas. Finally, there was the growth of vast shield volcanoes along the crest and margins of the Tharsis Rise, the growth of Elysium Mons, and finally the construction of Olympus Mons, west of the crest of Tharsis.

The largest number of volcanoes and the youngest individual structure are located in the Tharsis province, the region of a major tumescence in the planet's crust. Roughly the size of Africa south of the River Congo, it extends 4000 km from north to south and 3000 km from west to east, and, on average, at least 10 km above datum. The northern slopes range from 0.2° to 0.4°, but in the south are only half this. This asymmetry is largely because it straddles the line of dichotomy: the gradients steepen towards the lower hemisphere. Its extent is also greater towards the north, probably since volcanic flows extend further in that direction of steeper slopes. Strangely, most of the large shield volcanoes are sited either near the crest of the rise or on its northwestern flank.

The most prominent shield group, the Tharsis Montes, comprises Arsia, Pavonis and Ascraeus Montes. These are approximately 700 km apart, aligned SW–NE along the crest of the rise. Since even beyond the Tharsis volcanic province major fractures pursue the same trend northeastwards and southwestwards, it can be assumed that they have developed along a major fracture zone, now buried by the products of volcanism. Several smaller shields and steeper-sided volcanoes (tholi) lie close to the continuation of this line northeast of Ascraeus Mons; others lie to the east and west (Fig. 8.1). Olympus Mons lies 1200 km northwest of Tharsis Montes. It is the youngest of the Martian shields, whereas slightly farther north of Ascraeus Mons, on the extreme edge of Tharsis, lies

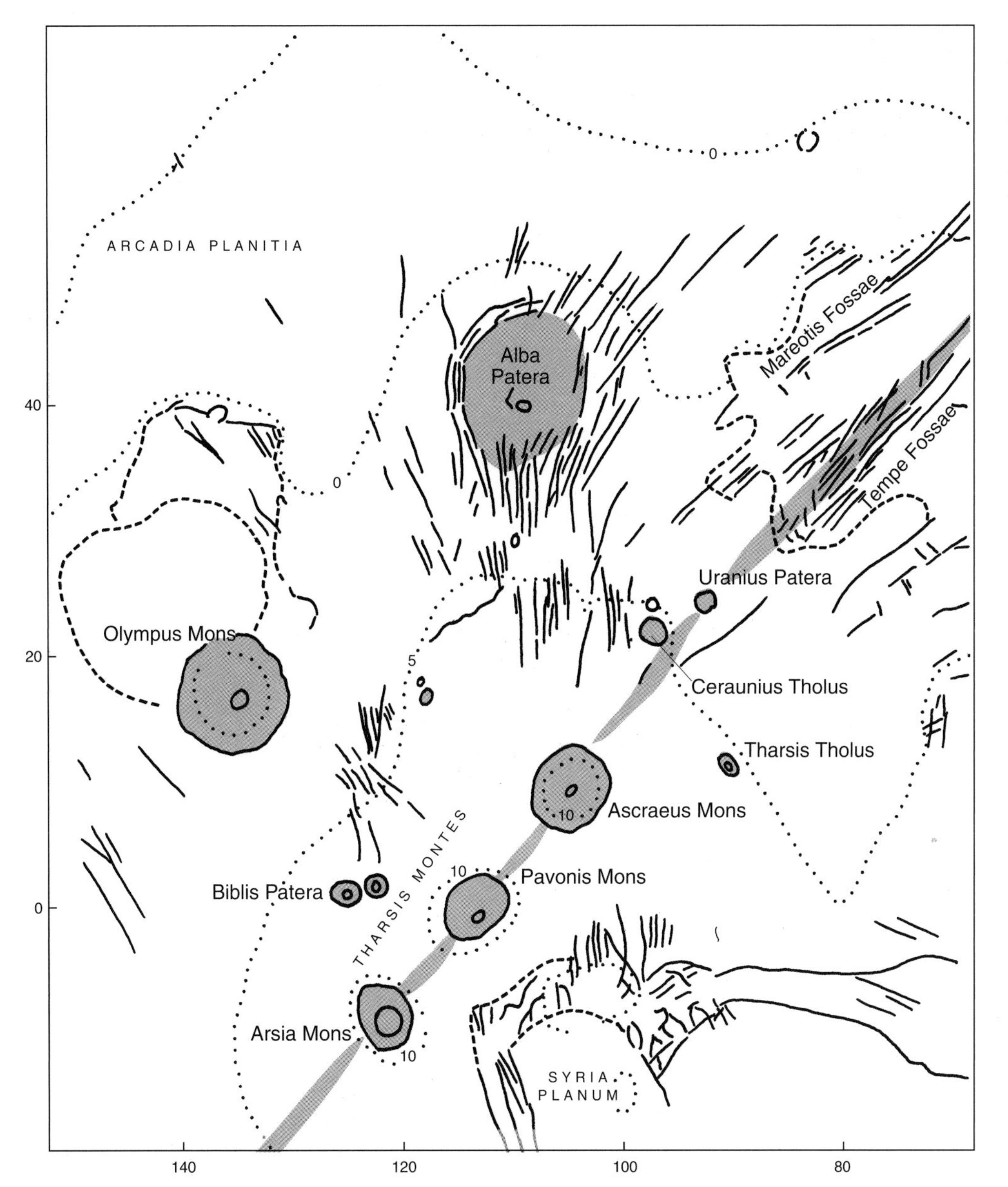

Figure 8.1 Map of the western hemisphere of Mars between 60°N and 60°S, showing the position of the main Tharsis volcanoes (grey areas) and principal fractures and ridges; contour values in km (modified from Plescia 1979).

the vast Alba Patera, an older low-profile volcano surrounded by prominent circumferential fractures and a massive mantle of pyroclastic rocks and lava flows.

Elysium Mons, Hecates Tholus and Albor Tholus are located in Elysium, another rise with a diameter of approximately 2000 km and a mean height of 5 km. The Elysium group is somewhat different in that its development involved both lavas and pyroclastic rocks.

Several major volcanoes are located around the Hellas impact basin; of these, Peneus, Amphitrites, Hadriaca and Tyrrhena paterae are almost certainly mixed lava and ash volcanoes. The only other major volcano is Apollinaris Patera, located southeast of Elysium on the lowland hemisphere at 10°S 185°W, that is, just north of the line of dichotomy. Its morphology suggests an origin in both effusive and explosive activity. Other structures are low shields in Syrtis Major and volcanotectonic features in Tempe.

Without returned samples there can be no radiometric dates as there are for the Moon; therefore, only crater counting can establish a relative timescale. Plescia & Saunders (1979) provided crater counts for each of the major volcanoes, divided into their types, with Lunae Planum as a datum (Table 8.1). Absolute ages, derived by putting these data into the chronologies of both Neukum & Wise (1976) and Soderblom (1977), give an idea of the sequence of events in Martian geological evolution. However, individual structures did not form at any one instant but evolved often over very lengthy periods, episodes of activity being interspersed with periods of inactivity measurable in tens or hundreds of millions of years.

Table 8.1 Number of craters greater than or equal to 1 km in diameter per million km^2 for individual Martian volcanoes, compared with absolute ages derived from the chronologies of Neukum & Wise (1976) and Soderblom (1977) (after Plescia & Saunders 1979).

Volcanic centre	Number of >1 km craters per 10^6 km	Implied absolute age in 10^9 yr (Soderblom 1977)
Olympus Mons	127	0.03
Ascraeus Mons	110	0.1
Pavonis Mons	350	0.3
Arsia Mons	780	0.7
Apollinaris Patera	990	0.9
Biblis Patera	1400	1.3
Tharsis Tholus	1480	1.38
Albor Tholus	1500	1.4
Hecates Tholus	1800	1.7
Alba Patera	1850	1.7
Jovis Tholus	2100	1.95
Hadriaca Patera	2100	1.95
Elysium Mons	2350	2.2
Tyrrhena Patera	2400	2.25
Uranius Patera	2480	2.3
Uranius Tholus	2480	2.3
Ceraunius Tholus	2600	2.4
Ulysses Patera	3200	3.0
Tempe Patera	4300	3.4
Lunae Planum	2500	2.3

Types of volcanic structure

The large volcanoes have been classified into three main types, depicted in Figure 8.2:

- *Shields* Built from thousands of individual flows, have summit calderas and overall low profiles characterized by steeper upper regions and gentler lower flanks.
- *Tholi or domes* Similar to the former but have somewhat steeper slopes that may be a function of more viscous lava, lower eruption rates or increased pyroclast content.
- *Paterae* Of two kinds: lowland paterae (e.g. Alba Patera, Uranius Patera), which are lava shields in the northern hemisphere characterized by extremely low profiles and complex summit calderas; and highland paterae (e.g. Tyrrhena Patera, Amphitrites Patera), which are located mainly in the southern hemisphere, have very low profiles and summit caldera complexes and also may be incised by channels (Plescia & Saunders 1979).

Most of the highland paterae are believed to be mixed lava and pyroclast edifices. A fourth group, volcanotectonic depressions, may also be recognized, including certain suspected volcanic structures in Syrtis Major, in the region south of Hellas and among the volcanic plains around Tharsis.

In addition to the major landforms, other volcanic structures include dykes, spatter ridges, craters, chains of craters, small domes and, of course, a variety of lava and ashflow types. Details of some of these landforms are gradually being revealed by the high-resolution imagery being sent back by Mars Global Surveyor.

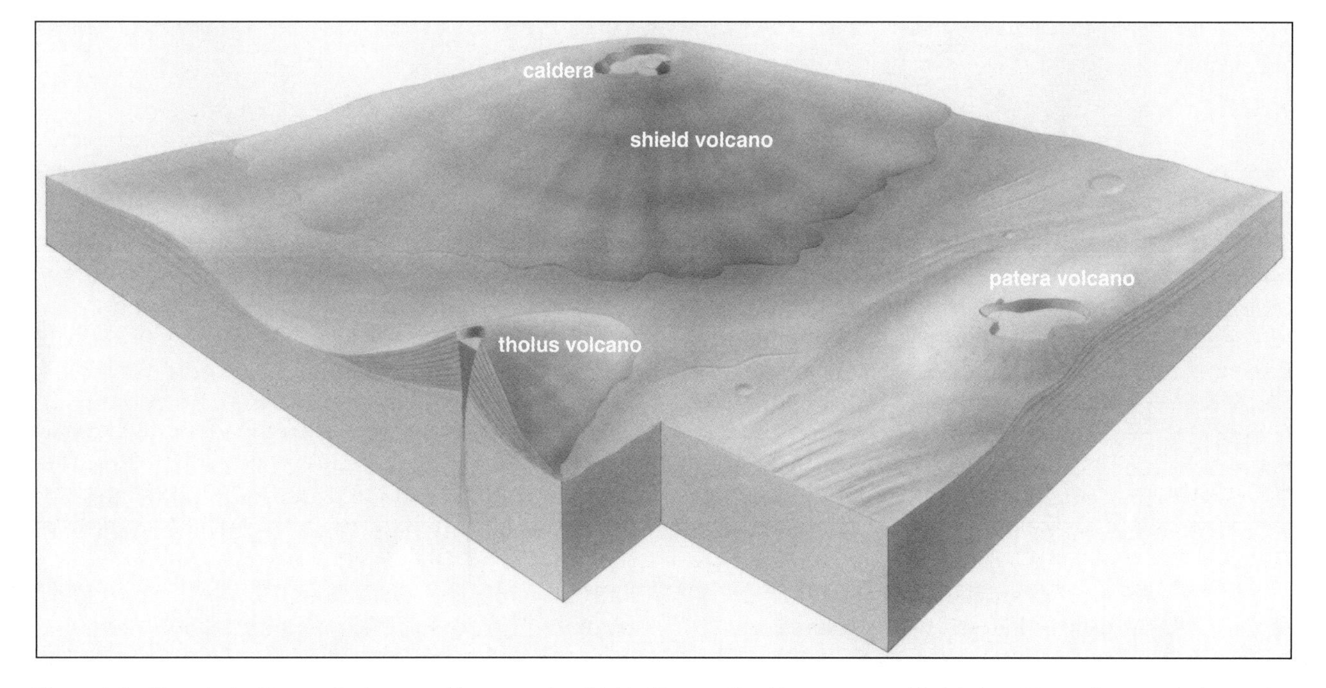

Figure 8.2 The principal types of volcano on Mars: massive shield with summit caldera, steeper-sided tholus with smaller summit depression and low-profile patera with summit caldera.

Controls on volcanic processes on Mars

Whereas the earliest central volcanism may have been controlled by deep-seated fracturing produced during the excavation of Hellas, this is less likely to have been the case for the activity in Tharsis and Elysium. More likely, volcanism here was in some way related to the growth of the major crustal upwarps that exist there or to the internal processes that generated them. In an attempt to explain the gravity data for the Tharsis region, Sleep & Phillips (1979) proposed a model that assumed the Martian crust to be thinner beneath Tharsis than elsewhere and the mantle less dense. On the basis of such a model, they were able to postulate isostatic compensation under Tharsis at depths of only 300 km. The notion of a lesser density would be expected in a region of active volcanism. However, the Sleep & Phillips model is not universally accepted, as will be seen from the fuller discussion of Tharsis and its formation on pp. 102–104.

The ascent of magma on Mars is likely to follow a course similar to that on Earth; however, there is evidence to suggest that Martian partial melts may be iron-enriched or even ultrabasic. Also, because of the lower surface gravity, diapiric ascent rates are anticipated to be lower than on Earth, meaning that diapiric bodies would rise to shallower depths on Mars. This would encourage the intrusion of volcanic dykes twice as wide as terrestrial ones, enabling much higher mass effusion rates for Martian magmas, perhaps up to five times those typical of Earth. Given the same style of cooling-controlled effusive activity, lava flows on Mars could be up to six times longer than terrestrial ones. A full treatment of this aspect of volcanism and those described below can be found in Cattermole (1996).

Various lines of evidence support the notion that Mars is a volatile-rich planet, and water has been found in all of those Martian lavas so far analyzed. Because eruption-cloud instability occurs at lower mass-eruption rates than on Earth (for the same volatile content), pyroclastic flow formation is more likely for basic magmas than it is on Earth. For the same magma chemistry and volatile content, eruption velocities on Mars would be at least 1.5 times those characteristic of Earth, and lava fountains feeding such flows would rise to at least twice the height of terrestrial analogues. Furthermore, because of the suspected extensive subsurface volatile reservoirs on Mars, the likelihood of hydromagmatic eruptions is high.

On Earth, for disruption of magma to occur, thereby breaking it up into pyroclasts, the volatile content (assuming H_2O to be the dominant gas) must exceed 0.07 wt per cent; on Mars, the critical value is less, at 0.01 wt per cent. This fact, together with the low atmospheric pressure, facilitates pyroclast formation. Unless Martian magmas are characterized by being unexpectedly volatile poor (and there is no evidence for this), low discharge-rate effusions should be characterized by Strombolian type of activity. The abundant presence of both CO_2 and H_2O on Mars when volcanism was at its peak strongly favours most eruptions being explosive, regardless of their chemistry; even basaltic eruptions would therefore be subject to vigorous fire fountaining.

Because of the low atmospheric pressure, eruption clouds on Mars would be expected to rise much higher than on Earth, largely because of less atmospheric drag. Add to this the lower gravity of Mars and it is not surprising that eruption plumes would typically be predicted to rise three times higher and spread out over significantly wider areas than on Earth. In consequence, airfall tephra layers would be thinner than those of Earth but more widespread.

Highland paterae of the Hellas region

Several of the most ancient volcanic structures on Mars are located near to the borders of the Hellas basin. Plescia & Saunders (1979) termed these highland paterae and, on the basis of crater counts, all appear to have formed 3.7–3.1 billion years ago. Potter (1976), Peterson (1977) and King (1978) made the first studies of these volcanoes, which appear to be a manifestation of the earliest explosive volcanism on the planet. Pike (1978) suggested that this group of highland paterae were similar to terrestrial ash shields. Subsequently, several investigators have applied models of pyroclastic flows to deposits found in eastern Hellas, and it has been established that long-runout pyroclastic flows are a better explanation of the observed morphologies than ash fall deposits are, and are quite capable of producing the very extensive channelled deposits seen. The inferred transition from explosive to effusive activity witnessed at both Tyrrhena and Hadriaca paterae has important implications for Martian volcanic and climatic history.

The stratigraphy of the Hellas basin has been redefined by Tanaka & Leonard (1995), who interpreted the asymmetric distribution of basin rings to imply an oblique impact event during Middle Noachian times. In the Late Noachian the basin was infilled by a variety of deposits, including lavas, and probably both volcaniclastic and aeolian sediments, whereas burial of the region north of Malea Planum is seen as a product of the massive production of pyroclastic material from two ancient volcanoes, Amphitrites and Peneus paterae, that sit on the southern rim of the basin (Fig. 8.3).

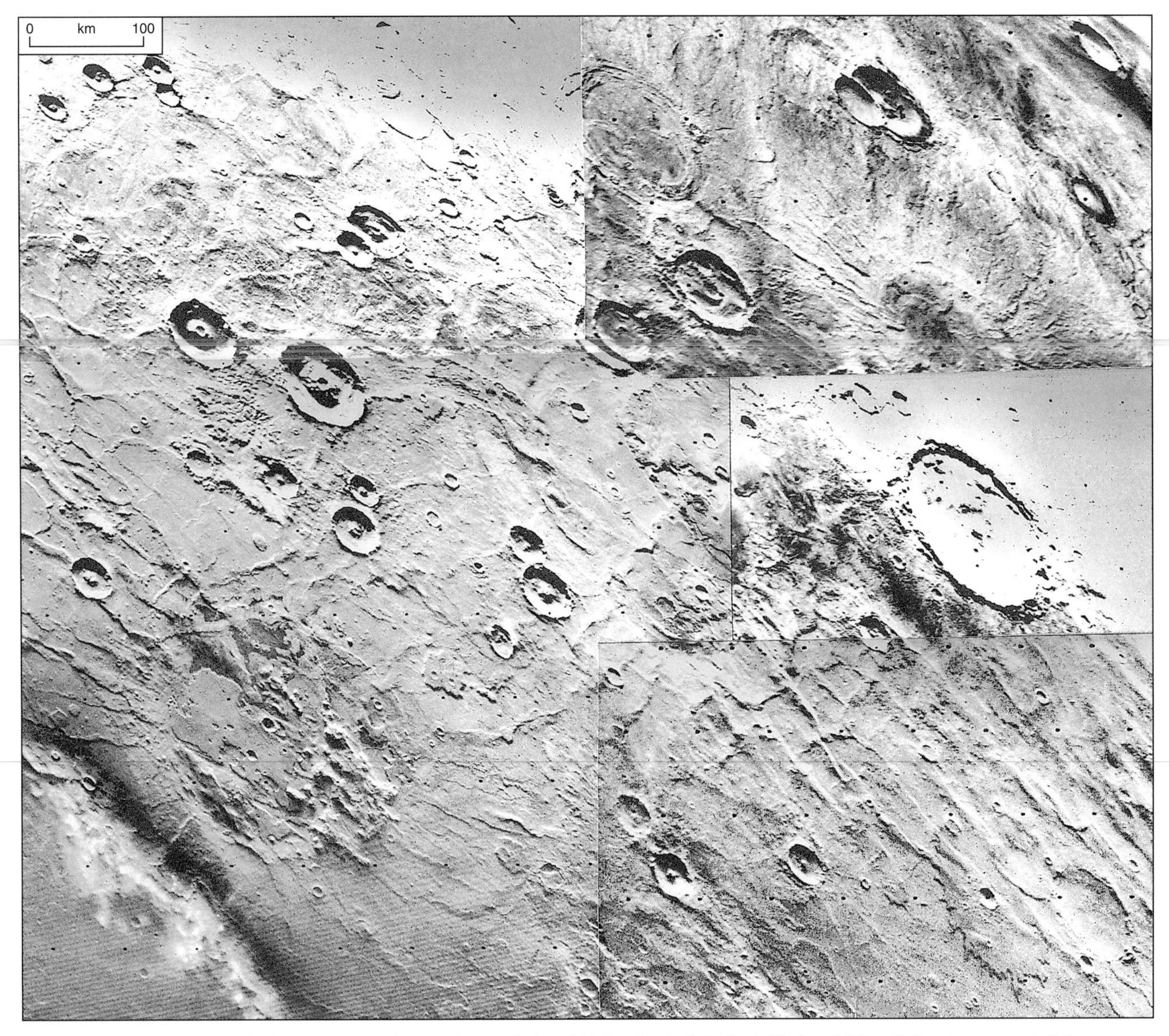

Figure 8.3 Ancient volcanic ring structures close to the south rim of Hellas (north is to the left). Amphitrites Patera has a central depression with prominent radiating ridges and valleys, and to its west is another large ring feature with concentric graben, Peneus Patera. The gap on the right-hand side of the mosaic is explained by very poor lighting at these latitudes during the pass in which the region was imaged. (Viking orbiter image 056b35; centred at 303°W 58°S.)

Peneus and Amphitrites paterae

Amphitrites and Peneus paterae, located as they are on the southern basin rim, are surrounded by the ridged plains of Malea Planum which, north of the volcanoes, are incised by narrow channels. The paterae attain an elevation of 4 km, the adjacent plateau sloping down to around 1 km southeast of Amphitrites Patera and to 2 km below datum at the perimeter of Hellas itself. The area of volcanic rocks in this region is at least a million km^2.

Peneus Patera is over 100 km across, having an elliptical shape, and is composed of a 10–30 km-wide zone of annular faulting that surrounds a central depressed floor. This is deformed by wrinkle ridges in places. Its overall form implies that it is a major volcanic structure whose development entailed voluminous outpourings of magma, as either lava or pyroclasts. Amphitrites Patera is somewhat more ambiguous, comprising several 100 km-diameter ring structures with associated radiating ridges, some of which extend into the basin itself. The rings appear to have negligible vertical relief. However, the overall structure does suggest some kind of volcanic origin, and recent mapping by Tanaka & Leonard (1995) has distinguished a 300 km-diameter low-relief shield surrounding the ring faults. The channels that radiate out from the centre of the volcano appear to terminate at the foot of this shield. The rather degraded form of the volcano flanks suggests that they may be made up principally of relatively friable pyroclastic rocks, the lesser depth of the central depression perhaps implying a lower volume of magma than that associated with the deeper Peneus volcano.

The slopes to the north of both volcanoes are highly dissected and show lobes and tongues suggestive of either lava or pyroclastic flows, or even lahars. Average slopes lie within the range 0.2–0.9° over distances of 400–500 km. Tanaka & Scott (1987) interpreted these deposits as pyroclastic rocks; Tanaka & Leonard (1995) found nothing to change that view. Furthermore, the lengths of the flows, if such they are, and the slopes are in accordance with the constraints for long runout materials discussed by Greeley & Crown (1990) and Crown & Greeley (1993); however the distances are too great for those suggested for airfall deposits. Certainly, there is a strong likelihood that such eruptions would have been triggered by interaction between magma and groundwater.

Hadriaca Patera

Hadriaca Patera, located on the northeast rim of Hellas, has a flat-floored summit caldera 77 km in diameter (Fig. 8.4). The northern and eastern parts of the caldera backwall are rather indistinct and may have been covered by pyroclastic materials or by overflow of volcanic flows. On the eastern floor are several small volcanic constructs that could be domes or cinder cones, or could represent remnants of some kind of volcanic flowsheet. Hadriaca appears to have little vertical relief (although more than Amphitrites), but there is possibly as much as a 2 km height difference between the caldera and the foot of the channelized shield that extends 300 km from the caldera rim before merging with the surrounding plains. To the southeast it has been eroded by the Dao Vallis channels and is embayed by the low-lying plains that were laid down during the Hesperian and Amazonian periods.

The flank slopes range from 0.01° in the northeast to 0.47° in the south and west (Crown & Greeley 1993). The smooth ridged and channelled morphology of the flanks, which appear to be devoid of lava flow lobes, is rather reminiscent of the mantled parts of Tyrrhena Patera, located 870 km to the northeast, and, to some extent, of Alba Patera, located in northern Tharsis. Such morphology indicates friable materials.

A large channel, which begins its southwestward course in a large depression on the lower southeast flank, eventually merges with the floor of Hellas, some 800 km distant. The stratified and dissected deposits of the flanks have been incised by many outflow channels in whose walls layering is sometimes revealed. The channels are 3–4 km wide, typically, and over 100 km long. Zisk et al. (1992) indicated their depth to be 200–300 m. Typically they have flat floors and in this respect are dissimilar to lunar sinuous rilles, which are U-shape and are known to have been formed by evacuation of volcanic-flow tubes. The channels on the flanks of Alba Patera generally occur on ridge crests, not on trough floors. Thus, channel morphology on Hadiaca Patera is more similar to that of channels on Tyrrhena Patera than to either those of lunar rilles or those on Alba Patera.

Tyrrhena Patera

Tyrrhena Patera has been studied in more detail, having benefited from somewhat better image coverage than other structures in this region. Located northeast of Hellas, it is markedly eroded (Fig. 8.5). At the summit are two sets of ring fractures, the innermost being 50 km across. Within it is an off-centre caldera depression. A prominent southwest-striking channel leads away from the caldera and appears to be volcanotectonic in origin; two others originate lower down the flanks. According to altimetric data (Zisk et al. 1992) the summit rises ~1.5 km above the adjacent plains and has a maximum slope of ~3°.

Greeley & Crown (1990) identified five geological units: two older ones that form a basal shield unit extending 340 km to the south and over 300 km to the north and west of the summit caldera, and a slightly younger summit shield unit about 200 km across. The margins of both are embayed and dissected by younger units, and it was these eroded deposits with a relatively high albedo that were attributed by Pike (1978) and Greeley & Spudis (1981) to volcanic ash. On the southwest side of the volcano is a younger fan-shape flow unit, the apex of which focuses on the caldera and which is composed of narrow lava flows, often with levees and flow channels. At the summit there is a caldera-filling unit, believed to be composed of ponded lavas produced at a late stage in the volcano's cycle.

The radial texture and etched appearance of much of the basal and summit shield deposits (Fig. 8.5) was interpreted by Greeley & Crown to be the result of erosion by water, wind and mass wasting of early ash deposits that were generated contemporaneously by hydromagmatic eruptions here and at other centres surrounding Hellas. Such activity is seen as the result of eruptions through water-charged regolith that is believed to have been widespread in the southern

Figure 8.4 The highland patera, Hadriaca Patera, showing the 77 km caldera, gullied flanks and adjacent channels. Viking orbiter image 625a18; centred at 30°S 266°W.

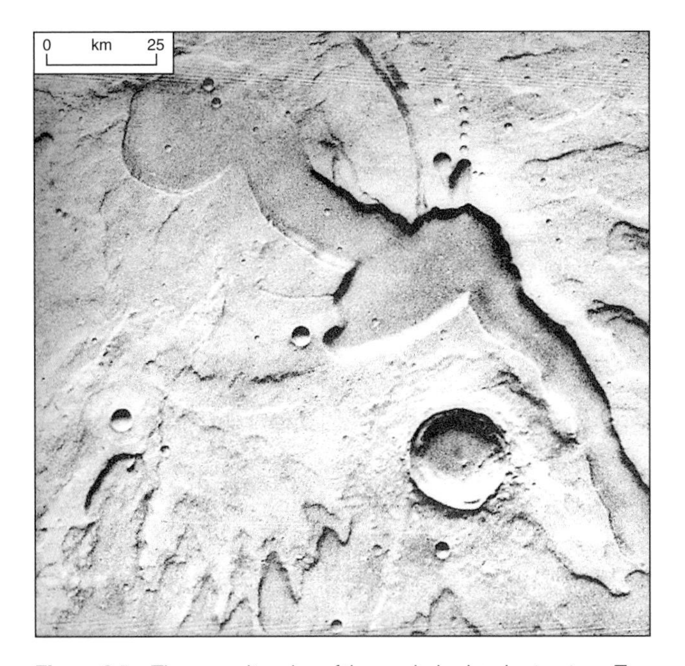

Figure 8.5 The summit region of the eroded volcanic structure, Tyrrhena Patera (north is to the left). The central ring-faulted depression has been filled by later deposits and subsequently incised by volcano-tectonic depressions or channels. Viking orbiter image 445a53–445a54; centred at 21.8°S 253°W.

hemisphere and whose former presence is indicated by the extensive fluvial channels incised into Noachian plateau plains (Greeley & Guest 1987). Subsequent to ash deposition, internal heating appears to have increased, whereupon flood lava eruptions emplaced the surrounding plains, which partially buried the flanks of the volcano. The sinuous channels that originate close to the summit may have continued to supply lavas to the lower flanks, after infilling most of the summit caldera.

Zisk et al. (1992) asserted that the absence of correlation between channel width and flank slopes on the volcano implies that erosion by gravity-driven flows was not responsible for cutting the channels. They suggested that other factors, such as spatial or temporal variability of groundwater release by sapping, or physical differences in the volcanic units forming the volcano's flanks, probably controlled their geometry and location.

Volcano morphology and early volcanism

Gulick & Baker (1990) assessed the relative importance of contributions from lava, volcanic density flows and fluvial erosion in the genesis and modification of channels on Martian volcanoes. In most cases a combination of these phenomena appears to have been involved, with erosion by groundwater sapping and surface runoff apparently dominating the latter stages of channel evolution. In the cases of both Tyrrhena and Hadriaca paterae, the channels are erosional and have been cut into the rocks making up the volcano flanks. These rocks are unlikely to be lava flows (their morphology, texture and susceptibility to erosion are inappropriate) and more likely to be either volcanic ash or ashflow deposits.

Applying the same criteria to Tyrrhena Patera that Mouginis-Mark et al. (1988) used on Alba Patera, Greeley & Crown conclude that the ash deposits cannot be airfall in origin because of their dispersion 300–600 km from the caldera region. Ashflow eruptions could, however, account for the observed distribution. Theoretical analysis of terrestrial ashflow eruptions has indicated that large flows may have been emplaced at velocities of up to 300 $m\,s^{-1}$ (Sparks et al. 1978), and initial velocities of 400–600 $m\,s^{-1}$ may be possible (Sparks & Wilson 1976). Initial flow velocities of 325–450 $m\,s^{-1}$ would be required to emplace the basal shield unit, and an initial velocity of around 250 $m\,s^{-1}$ could have emplaced the summit shield unit of Tyrrhena Patera. If the smooth plains units are also accepted as being ashflow deposits, initial velocities of around 650 $m\,s^{-1}$ would be necessary. An ashflow origin therefore seems highly plausible.

Crown & Greeley (1993) applied gravity-driven flow models to Hadriaca Patera and found that the flank materials also can be most convincingly attributed to the emplacement of pyroclastic flows. That such materials were preferred to ballistically deposited ash deposits is a reflection of the inability of non-welded ashes to support the steep-sided scarps typically found on Hadriaca. Both magmatic and hydromagmatic[1] eruption models were tested and found applicable. Assuming the former, the requisite eruption rates (10–100 million $kg\,s^{-1}$), ejection velocities (400 $m\,s^{-1}$) and volatile contents (1.5–3.0% H_2O) are parameters derived for terrestrial Plinian activity. In the Martian case, the conversion efficiencies are comparable with experimental results, and the permeability of the Martian regolith allows for the storage of large volumes of water and for its rapid transportation (flow rates 10^3–$10^4\,m\,s^{-1}$). Thus, either style of activity could transport the observed flank deposits to the requisite distance from the vents. The change from explosive to effusive style (from flank deposits to caldera-filling flows) could be a manifestation of depletion of volatiles in the proximity of the centres with time, rather than to planet-wide changes.

The question of whether or not such flows could become welded under Martian conditions was also

1. Involvement of volcanic activity with ice.

addressed by Crown & Greeley (1993). The well developed layering within the pyroclastic sequence at both Hadriaca and Tyrrhena paterae certainly suggests that either compaction or welding of the deposits must have occurred. King & Riehle (1974) evaluated the potential for welding of shards on Mars by modelling emplacement temperatures of pyroclastic flows as a function of runout distance from source vents. Taking those results, Crown & Greeley argued that Martian pyroclastic flows would remain close to their original temperatures for considerable distances after eruption; for instance, the particles in a 10 m-thick flow travelling at $10\,m\,s^{-1}$ would not be hot enough to weld together at a distance of ~125 km from the source vent. The low heat-loss characteristics of such flows appear to indicate that welding may well be common and that differences in erosion style and channelling may be in part a function of the degree to which parts of a single unit are welded, or one flow is and another is not.

King & Riehle (1974) showed that compaction of terrestrial silicic flows occurs for emplacement temperatures of greater than 600°C for 10 m-thick flows and 575°C for 40 m-thick flows, whereas Riehle (1973) showed that minimum temperatures for welding of terrestrial silicic flows are 625°C for 10 m-thick flows and 575°C for 40 m-thick flows. The distance for which welding could continue would be dependent upon the loss of heat in the eruption column (if the flow was produced by column collapse) and the loss of gases prior to emplacement.

On the basis that eruption volumes were comparable with large terrestrial eruptions (100–1000 km^3) – probably an underestimate from what we know about Mars – Greeley & Crown estimated that between 100 and 1100 basaltic eruptions could account for the entire edifice of Tyrrhena Patera. On the additional basis of calculations relating the rate at which subsurface water could accumulate to magma volume per eruption, they conclude that the volume of water required to drive the proposed explosive volcanicity could, in fact, accumulate rapidly (in tens of years) and that hydromagmatic activity is quite consistent with the climatic changes suggested to have affected Mars during this early epoch (Clifford & Zimbelman 1988). The fact that climatic conditions changed as time progressed may explain why ash eruptions do not appear to have characterized the younger shields of Tharsis. In other words, by the time these major shield provinces were active, there had been significant depletion of the volatiles that played such an important role in the Hellas province during earlier times.

Shield volcanoes of Tharsis and Elysium

Shield volcanoes are broad, gently sloping cones that usually they have a shallow caldera at their summit. Smaller pits and spatter cones are often concentrated along lateral rift zones. Terrestrial shields show a range of sizes from small low-profile shields such as those of Iceland to the very large volcanoes of the Galápagos Islands and Hawaiian–Emperor chain. Those of Mars are of even greater size, some being more than an order of magnitude larger than any terrestrial counterpart. They appear to have been constructed largely from fluid basalt-like lava flows. However, the discovery of rocks with andesite chemical composition at the Pathfinder landing site implies that somewhat more viscous flows and pyroclastic rocks may have a played a more important part in the volcanic activity of both Tharsis and Elysium than hitherto has been supposed. Terrestrial lava shields are sited within some of the most volcanically active regions on Earth and are a manifestation of volcanicity associated with long-lived mantle hotspots. The same appears to be true of their Martian counterparts.

Although Martian shield volcanoes share many similarities with those of Hawaii, they are very much larger, were active over extremely long periods and show quite a different global distribution pattern. The first and second of these differences imply a structural and thermal stability of the Martian lithosphere that has probably never been matched on Earth, and the third is evidently a reflection of the absence of plate tectonics on Mars. Substantial gravity anomalies are associated with the Elysium rise (450 mgal) and with Olympus Mons (2750 mgal); the Tharsis Montes also show strong central positive anomalies and surrounding gravity lows, but their position near the centre of the Tharsis Rise makes it difficult to separate their specific contribution from that of Tharsis itself (Smith et al. 1999b). Alba Patera also shows a positive anomaly of 430 mgal. The fact that all five of the Tharsis centres deviate significantly from isostatic equilibrium implies a young age and mechanical support by a strong crustal layer.

The central volcanoes of Tharsis

The Tharsis Montes epitomize the large volcanoes of Mars and bear a striking morphological resemblance to terrestrial large shields. By implication, therefore, they are likely to be the products of long-term eruption of large volumes of fluid basalt-like lavas. The most northerly shield, Ascraeus Mons, rises to the greatest height (26 km) and has the largest relative height difference with respect to the surrounding plains (17 km); both Pavonis and Arsia Montes reach about 20 km above Mars datum. Each is 350–400 km in diameter, with a summit caldera complex significantly larger than any known terrestrial analogue. However, flank slopes are relatively gentle, averaging less than 5°, and the summit region and lower flanks tend to be less steep than the mid-flank zone. The prominent radial texture seen on Viking imagery is attributable to hundreds of narrow (<3 km) lava flows, many of which have apical channels; a large proportion of these can be traced up slope to the rims of the summit calderae, and a further significant proportion appear to have emanated from prominent embayments made up of many coalescing pits and situated adjacent to the calderae in the southwest and northeast sectors of each shield. Those near the summit of Arsia Mons are very large and striking, the embayment on the southwest side having erupted a vast shoulder of fan-like flows. Successively older flows are exposed as the distance from the main shields increases, a characteristic common to many Martian shields.

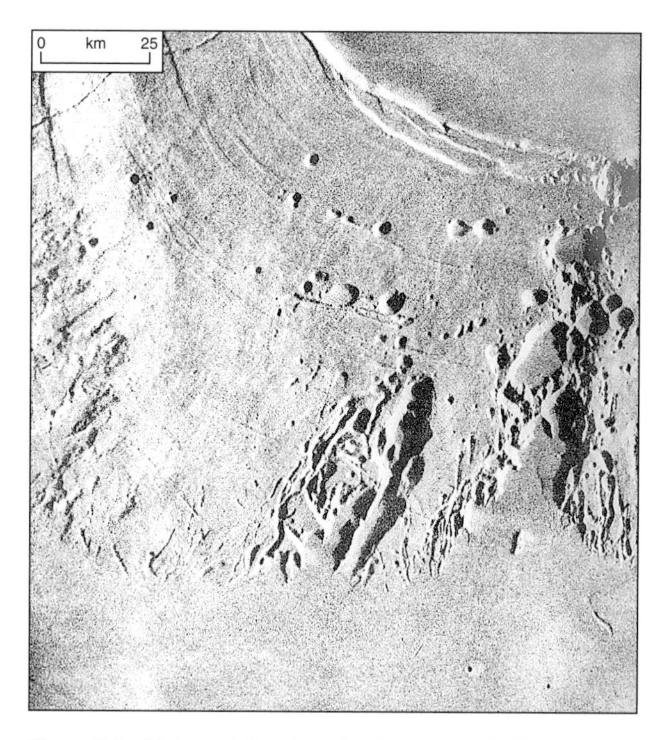

Figure 8.6 High-resolution view of embayment on SW flank of Arsia Mons. Note the coalescent pits and annular graben. Viking orbiter image 204a08; centred at 10.4°S 120.9°W.

Arsia Mons has a simple 120 km-diameter caldera bounded by arcuate faults. A row of low domes connects the embayments in each wall, suggesting eruptive activity associated with the major SW–NE fracture line continued after the latest caldera subsidence had taken place. However, as is the case with all three Tharsis Montes, little or no evidence of intra-caldera constructional volcanism is to be found. On the caldera rim are many graben, which extend outwards from the rim for about 60 km and whose spacing ranges between 1 and 12 km; the narrow (0.5–1.3 km) lava flows on the northwest slopes pre-date the faults, having been emplaced prior to fracturing of the caldera rim. The lavas themselves can be traced to a source in a major graben depression situated on the rim, approximately 12.5 km from the edge of the modern floor (Fig. 8.6).

Figure 8.7 Summit region of Ascraeus Mons, showing the nested summit caldera complex, radial flows and flank embayments. Viking orbiter images 224a88–224a91; centred at 11.1°N 104.2°W.

Arsia Mons's caldera is relatively shallow in compared with that of Ascraeus Mons, whose deepest pit lies 3.15 km below the rim (Mouginis-Mark 1981b). It is but one of eight major depressions that form its nested caldera complex (Fig. 8.7). Study of the summit region by Mouginis-Mark (1981b) shows that some of the collapse events were preceded by major slumping of the caldera backwalls, these being approximately contemporaneous with the formation of circumferential graben. Originally the summit must have boasted several smaller pits, but the latest collapse saw the production of the large 40 km-diameter depression that now unifies the caldera.

Along the entire caldera rim it is possible to discern many narrow lava flows and some sinuous channels, which radiate down the volcano's upper flanks. The flows are generally less than 1 km wide and 10–20 km in length, whereas channels are 100–200 m wide and up to 18 km long. Schaber et al. estimate that most flows are no more than 10 m thick (Schaber et al. 1978). Because these flows and channels can be traced right up to the caldera backwall, it is clear that effusive activity continued at the summit until the final collapse occurred and it proves that late-stage explosive activity was not important in the shield-building process. However, source vents for flows are not visible, even on high-resolution Viking images, neither can traces of the flows be discerned on slump terraces within the caldera. This implies, first, that each collapse depression experienced resurfacing after its formation, and, secondly, that the source vents for the lavas were originally located farther up the shield than they can now be traced (Fig. 8.8).

MOLA topographic data reveal that the summit of Arsia Mons is regionally flat, indicating that the main volcanic structure postdates the regional Tharsis topography (Smith et al. 1998).

Pavonis Mons has a single 45 km-diameter caldera, about 4.5 km deep, surrounded by arcuate fault terraces that define a shallow summit depression approximately 100 km in diameter. Annular grabens also occur lower down the northeast, east and southeast slopes, beginning about 120 km from the summit. These are similar to those developed on both Ascraeus and Arsia Mons. Several sinuous rilles can be seen to cross these faults and are considered to be major lava channels incised by turbulent flow. Some of these apparently emerged from circumferential fractures, implying that the latter were the sources for the flows that subsequently poured down the volcano's flanks.

Rheological studies of flows associated with the main shield of Ascraeus Mons indicate that they were of low yield strength and low viscosity, consistent in the main with basalt-like composition (Moore et al. 1978, Zimbelman 1985). The average effusion rates calculated for these flows range between 18 and 60 $m^3 s^{-1}$, towards the lower end of Hawaiian and Icelandic rates. Although a basalt-like composition is implied by the morphology of the Tharsis Montes flows, they are not all exactly the same. On the upper parts of shield surfaces, out to radial distances of about 400 km, flows tend to be relatively narrow (less than 3 km), less than 150 km long (many being no more than 15 km in length) and often have a central channel, sometimes bounded by levees. Beyond this point they tend to widen (4–7 km) and become longer, some exceeding 400 km; central channels are common. At distances greater than 800 km, where regional gradients become lower, the flows broaden substantially

Figure 8.8 High-resolution mosaic of the south rim of Ascraeus Mons's summit caldera. The narrow summit lava flows are truncated by the caldera backwall, a characteristic they share with terrestrial lava shields such as Mauna Loa. Several flows are seen to be leveed; others show apical channels. Viking orbiter images 401b16–401b18; centred at 8.5°N 104.2°W.

and may be up to 50 km across at their distal ends. Such flows tend not to be channel fed; some are at least 650 km long. Interestingly, the greater the source height of Ascraeus flows, the shorter they are; this appears to be the case for all three shields, and has been noted at Alba Patera (Cattermole 1989a). This cannot be mere coincidence and almost certainly is a direct reflection of the greatest altitude to which lava can rise in response to the density contrast between the magma and rocks in its source region. Confirmation for this explanation comes from the observed concentration of shorter flows near shield summits, where eruption rates must have been less than on the lower flanks – a response to the much greater distance through which they had to be lifted before eruption.

It is hardly surprising, in view of the huge dimensions of the shield volcanoes, that instabilities were set up from time to time. The most dramatic evidence for this comes from the west-northwest flank of Arsia Mons, where a lobate slump measuring 400×350 km extends outwards from the base of the shield. Within the slide unit are hundreds of small hills; at its outer edge are closely spaced ridges that produce a strongly striated effect. Many striations crosscut impact craters and other small landscape features without in any way modifying their outline, as though having been superimposed upon them. This is a curious phenomenon and the suggestion has been made that the slide formed while the landscape was mantled by ice (Williams 1978). Not only would this have facilitated the flow of the debris, but its subsequent melting and removal would have superimposed the flow pattern onto the underlying topography. The strongly dissected terrain of the shield flanks above the slide's upper boundary suggests there was a zone of detachment along which the shield partly collapsed, slipping laterally as a gravity-assisted slide or series of slides that spread out over the surrounding plains.

It appears that the three Tharsis Montes followed a similar pattern of development. An initial stage of shield building was achieved by the gradual accumulation of fluid lavas, both from the summit area and from peripheral vents; then, after each shield had grown to its maximum height, effusive activity became concentrated along a major rift zone aligned in a SW–NE direction. Considerable lateral transport and supply of magma to vents and fissures along this zone over long periods of time resulted in repeated collapse of the shield summit region, forming major embayments and small satellite calderae in which some lavas became ponded. Eruptions from these rift-aligned sources built out substantial shoulders on the southwest and northeast flanks, from which many major eruptions disgorged high-volume flows. The great lateral extent of the latter implies that rates of eruption were high. Crater counts suggest that the construction of the main Arsia Mons shield terminated earlier than its two neighbours, but eruption from both SW–NE embayments and within the caldera continued into the more recent past (Crumpler & Aubele 1978).

Rising to a height of 27 km above Mars datum and at least 23 km above the surrounding plains, Olympus Mons is unquestionably the most spectacular of all Martian volcanoes. It has a diameter of at least 520 km and a volume over 50 times that of any terrestrial shield. Furthermore, it is surrounded by a huge scarp that in places is 6 km high, and an aureole of peculiar blocky terrain extending for up to 700 km beyond the scarp base (Fig. 8.9). It has a nested caldera complex 80 km across and has suffered multiple collapse, the most recent events having formed a 3 km-deep pit. Even on very high-resolution images of the summit region, evidence for caldera-floor constructive activity is not forthcoming. In contrast to Ascraeus Mons, where the largest pit was formed last, here the reverse is true.

Mouginis-Mark & Robinson (1992) estimate that at least 2.5 km of collapse was involved in the caldera-forming events, and that eight distinct events contributed to the final morphology of the structure. Tectonic features on the caldera floor indicate that an extensional stress field characterized the perimeter, but, at a radial distance of ~17 km from its centre, this changes to a compressional one. This fact is used to infer a depth of less than ~16 km for the underlying magma chamber.

The flank slopes are generally rather gentle (mean slope is 4°), but the shield has a somewhat sinusoidal profile since the middle part of the structure is steeper than either the summit region or the lower flanks. Also, the surface of the volcano is built from a series of 15–50 km-wide flow terraces separated by distinct breaks in slope and crossed by innumerable thin flows, which have a roughly radial disposition. The flows themselves are difficult to see close to the caldera region, but farther away long narrow flows (1–3 km wide), often with central channels and levees, are widely distributed and, except in the western sector, may be seen draping the face of the scarp before extending over the lower flatter ground (Fig. 8.10). The fact that these radial flows extend outwards for distances of about 350 km from the summit may mean that it is more realistic to consider 700 km as the real diameter of the volcano.

The scarp transects a broad pedestal of pre-shield material, about which little is known. This is cut by several faults, seen in section in the scarp backwall. These may be related to the generation of the scarp. Where the scarp has not been inundated by lava flows, there is evidence for several major landslides, which have a hummocky texture except near their margins where a ridge-and-trough fabric prevails.

Olympus Mons aureole

In some respects the remarkable Olympus Mons aureole, whose formation was either contemporaneous with or just subsequent to scarp development, looks

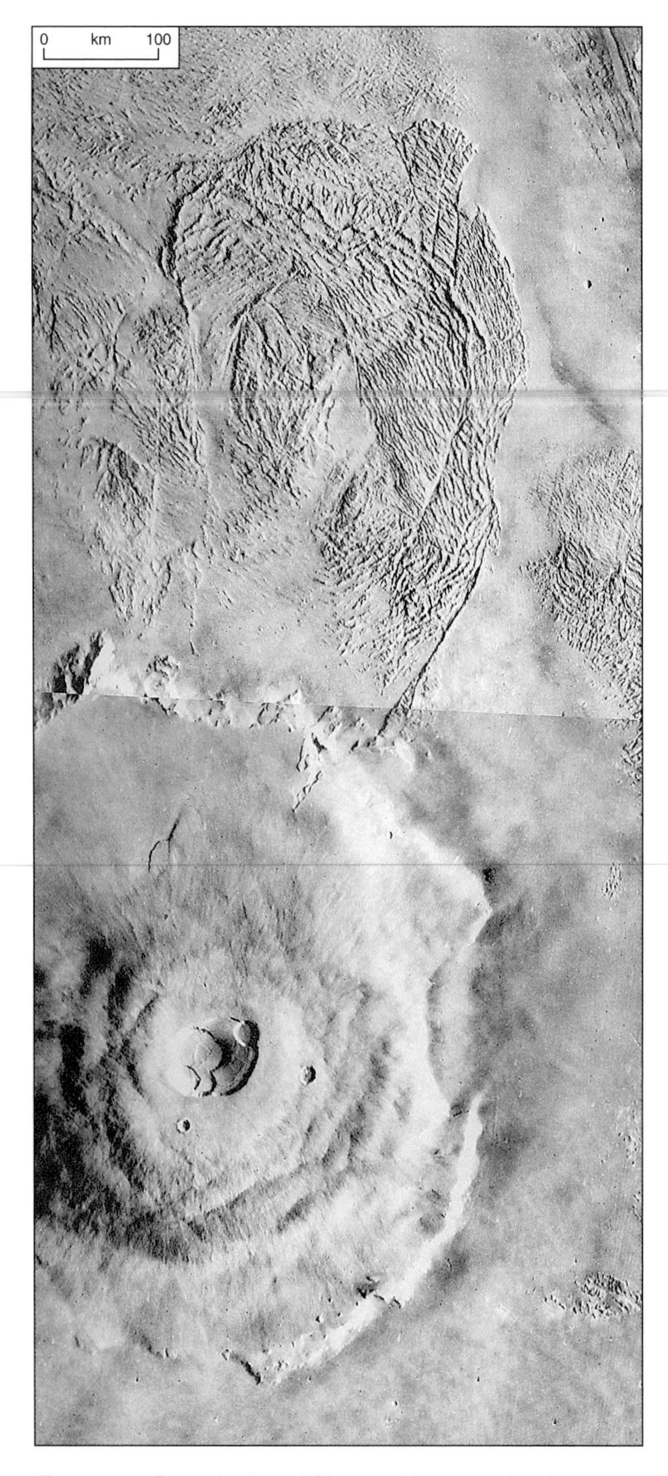

Figure 8.9 Synoptic view of Olympus Mons, showing the summit caldera, terraced flanks, basal scarp and aureole. Viking orbiter images 741a05 (upper) and 741a07 (lower); centred at 241°N 132.5°W; .

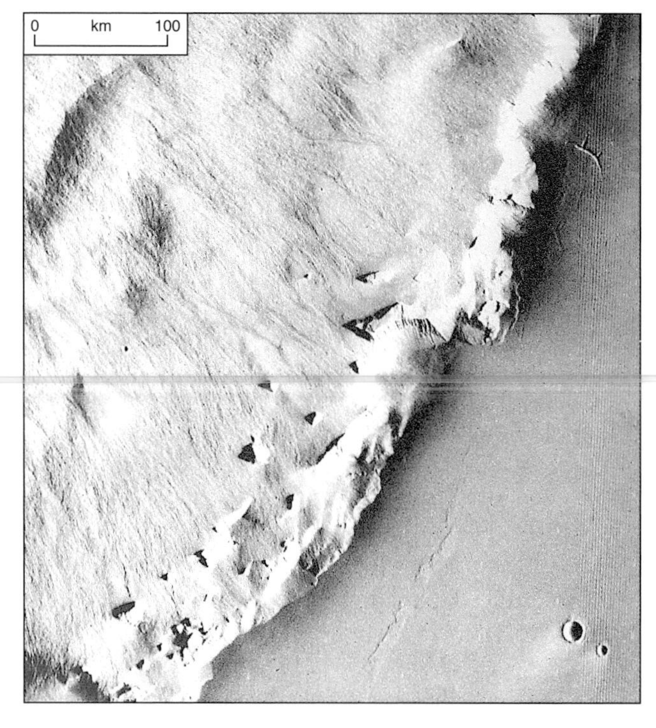

Figure 8.10 Narrow flows draping the flanks and scarp of Olympus Mons. Viking orbiter image 222a64; centred at 16.2°N 130.3°W.

like a massive landslide. It extends in places for at least 1000 km from the shield summit and comprises a series of terrain blocks made up from distinctively textured and closely spaced ridges (Fig. 8.11). The inner margin of each block is embayed by younger flows or aeolian deposits, whereas the outer boundary is more scarplike; the blocks form a complex of inwardly tilted prisms of intensely fractured terrain. The northwestern sector of the aureole is marked by a positive free-air gravity anomaly[1] of several tens of milligals (Sjogren 1979). Significantly, there are few superposed impact craters and, although smaller ones (<1 km) might be removed by surface creep or landslipping, the relative sparsity of 5 km craters (which are less likely to be removed in this way) implies the feature must be young (Schaber et al. 1978).

Explanations for this unique but enigmatic landform are as numerous as they are diverse. Carr (1973) suggested that the aureole might be the remains of an older shield volcano; Blasius (1976) suggested it was an unroofed intrusion. However, neither of these early suggestions appears valid, because both would necessitate much higher rates of erosion than are seen on

1. A discrepancy recorded between the hypothetical measurement of an ideal ellipsoid (Mars) and what is recorded. Those that are attributable to short-wavelength near-surface anomalies are termed free-air anomalies, those to longer wavelength deeper-seated anomalies are known as Bouguer anomalies.

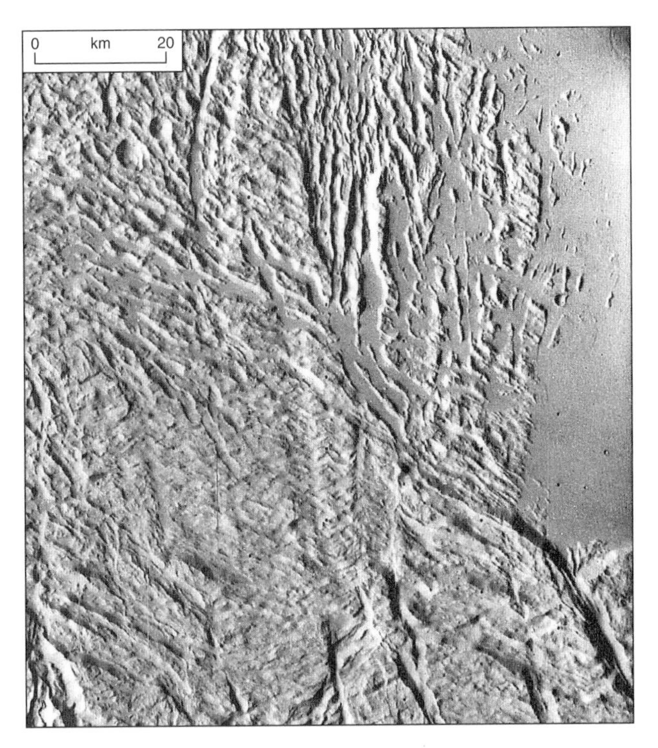

Figure 8.11 Olympus Mons aureole texture: a series of blocks, each composed of ridges interspersed with smoother plains units. Viking orbiter image 043b20; centred at 28.2°N 132.9°W.

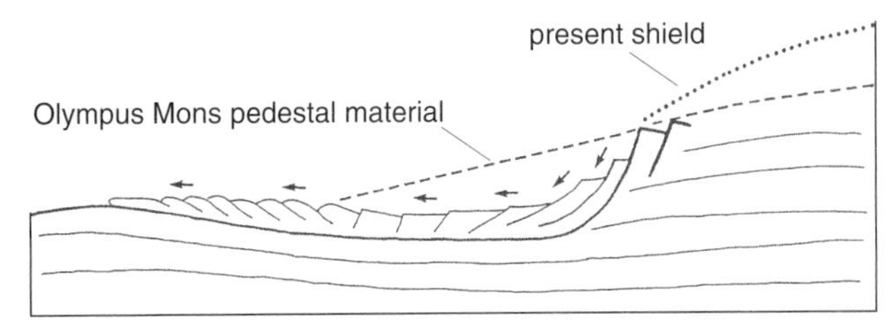

Figure 8.12 Gravity-sliding mechanism for generation of Olympus Mons aureole blocks (after Lopes et al. 1982).

Mars. King & Riehle (1974), on the other hand, proposed Olympus Mons to have been a composite volcano, a significant amount of construction having been accomplished by ash generated during extensive explosive activity. The aureole blocks are considered by them to represent the eroded remnants of lava and tuff sheets, the latter being deposits of nuées ardentes. Morris (1979) also suggested construction from tuff sheets, but in his hypothesis these were erupted from several vents distributed around the main shield. However, neither of these ideas seems entirely appropriate as there is a total absence of evidence for any pyroclastic deposits anywhere on the Olympus Mons shield or, indeed, on the Tharsis Montes.

In contrast, Hodges & Moore (1979) formulated the ingenious proposal that Olympus Mons bore similarities to certain Icelandic volcanoes, which erupted under a cover of ice. According to this theory, the height of the basal scarp represents the thickness of an ice sheet that once covered the region and beneath which eruptions took place. However, no evidence exists for subaerial activity within the confines of the aureole and, furthermore, neither is there any evidence for the ice sheet.

Various workers have considered the possibility of major thrusting or sliding. For instance, Harris (1977) suggested that the aureole lobes represent huge gravity-assisted thrust sheets, an argument pursued also by Morris (1981). Carr et al. (1977) and Lopes et al. (1980) have concluded that the aureole is the product of large-scale mass movement. The Lopes group invoked gravity-assisted rockslides on an immense scale, an interpretation supported by a comparison between the volume of material available before scarp formation and the present volume of the aureole materials (Fig. 8.12). Furthermore, the greater extent of the aureole on the northwest side, where the shield slopes gently down to the surrounding plains (compared with the southeast, where it rises towards Tharsis Montes), clearly implies gravity control. Mass sliding of this order could have been aided had there been a permafrost reservoir in the subshield pedestal. For this to have occurred, the pedestal would, of necessity, have to have included a significant proportion of either brecciated or relatively porous material, such as volcanic ash or tuff. Although there is no evidence for this, it is not impossible that explosive activity preceded the building of the main Olympus Mons shield. If this had been so, the increased heatflow associated with the rise of magma to form the main shield might have set off melting of subshield permafrost. This, in turn, might provide a catalyst for the sliding out of the aureole blocks. Although I favour some kind of large-scale sliding mechanism, neither this nor any other suggestion explains the gravity data. At the present time, therefore, it has to be admitted that no single hypothesis is completely satisfactory and an explanation of this enigma remains to be found.

MOLA topographic data indicate there to be several topographic depressions on the western flank of the volcano, immediately outside the edifice. These may reflect subsidence arising from the loading and flexing of the lithosphere by volcanic material. North of the volcano are two topographic highs: the nearest (centred about 600 km from the centre) comprises the disrupted aureole deposits; the farthest (centred at about twice the distance) is older terrain, including the Acheron Fossae region. If the more distal high has an origin in flexure, then a thickness of 75–100 km is implied for the underlying lithosphere. The nearest topographic high to the south of the volcano is found at a distance of 800 km, in the Medusae Fossae plains.

If this high too is flexural, then a thickness of 40 km is implied for the elastic lithosphere.

Elsewhere in Tharsis are smaller central volcanoes, all of which are older than the Tharsis Montes; some may be as ancient as paterae in the heavily cratered highland hemisphere (Plescia & Saunders 1979). Uranius, Ceraunius Tholi and Uranius Patera lie northeast of Ascraeus Mons (Fig. 8.13). The two tholi have steeper slopes (5° and 7°, respectively) than the patera (0.5°). Uranius Tholus has a diameter of 83 km and rises 3500 m above the surrounding lava plains. Its flat summit has a 14 km-diameter pit set towards the eastern edge of a larger but very shallow 32 km caldera, largely infilled with ponded lavas. It is a relatively old structure and, like the other two volcanoes, is older than the Hesperian flow plains that surround them. Ceraunius Tholus is somewhat steeper and larger (130×92 km) and has a 23 km-diameter summit caldera with vestiges of an older shallower pit on its north side. The flanks of both volcanoes are finely striated, the most obvious radial striations being narrow channels. Their general appearance is more reminiscent of the Elysian volcano Hecates Tholus (see p. 89) than of other Tharsis shields. It could be that Plinian-type eruptivity played a role in their development.

A prominent 2 km-wide channel runs from near the summit of Ceraunius Tholus down the north flank into an impact crater at the base of the shield. Carr (1974) suggested that this larger channel and several others like it may have been created by lava erosion, but Reimers & Komar (1979) argued that the general appearance of both coarse and fine channelling on both tholi is much more reminiscent of the terrestrial stratovolcano Barceno (Mexico), which, during 1952–3, experienced large-scale explosive activity that generated fast-moving density currents (base surges and nuées ardentes) that eroded arrays of channels in the volcano's flanks.

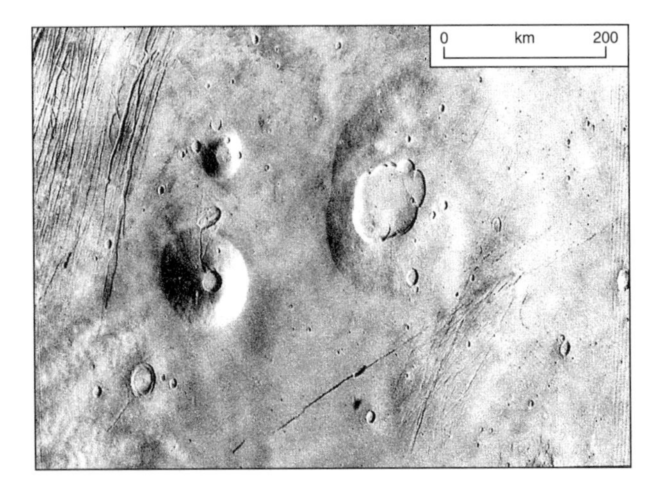

Figure 8.13 Group of three volcanic shields situated east of Ceraunius Fossae (left). Uranius Tholus, the smallest, has a flattened summit region, and to its south is the larger steep-sided Ceraunius Tholus. East of both is the volcano Uranius Patera, with a large summit caldera and much shallower flank slopes. Viking orbiter image 759a73; centred at 26.2° 94.1°W.

Biblis Patera and Ulysses Tholus, situated west of Pavonis Mons, both have simple calderas slightly over 50 km in diameter and many circumferential graben. The former measures 175×105 km and rises 4 km above the plains; the latter rises 3 km above the plains and has a diameter of 91 km. Both have flank slopes approaching 4°. Their general appearance is very reminiscent of the summit region of Arsia Mons. Jovis Tholus is a 55 km shield with a 27 km-diameter summit caldera and is set somewhat apart from the rest, midway between Olympus and Ascraeus Montes. Tharsis Tholus, east of Ascraeus Mons, rises 6 km above the plains and measures 155×120 km; it too has a caldera 62×46 km across. In view of their resemblance to the upper parts of the main Tharsis shields, all are presumed to be partially buried centres that grew prior to the Upper Hesperian to Lower Amazonian effusive activity, which saw the gradual building of major centres along the crest of the Tharsis Rise. Whether they ever grew to the vast dimensions of their successors is not known.

A recent study of southwest Tharsis by Edgett et al. (1997) has identified large dunefields located about 70 km northwest of Biblis Patera and about 500 km northwest of Arsia Mons. The materials from which these are made have low thermal inertias and high albedos, and are mantled by fine-grain materials (particle size <60 μm), Edgett et al. argued that to form aeolian dunes a supply of fine-grain material is required and that the only plausible source at the time they were laid down (the Late Amazonian) is volcanic ash or other explosively generated material. If this point is valid, it implies that explosive volcanism took place in the Tharsis region during Late Amazonian times.

The upper exposed part of Uranius Patera is traversed by well defined, narrow and often leveed flows and it measures 202×184 km across; it has flank slopes of just 0.5°. The floor of the nested caldera complex is crossed by several prominent wrinkle ridges, and the northwest part of the floor surface is inclined towards the centre of the depression, suggesting it subsided prior to the emplacement of the more or less horizontal ponded flows that occupy the centre. The very large size of the caldera implies that a large part of the shield has been buried by younger deposits.

The volcano Alba Patera has associated with it some of the most extensive volcanic flow fields found anywhere in the Solar System. Situated on the northern edge of the Tharsis Rise, its summit is located at 110°W,

40°N and it lies at the centre of an oval ring-fracture zone measuring 550×400 km. Crater ages suggest that its oldest flows pre-date those of Tharsis Montes, that its peak of activity may have occurred around 1725 billion years ago (it has an Hesperian to Amazonian age range), and its activity extended over a period of at least 1.5 billion years and perhaps as much as 2.8 billion years (Cattermole 1989a).

The summit of Alba lies 7 km above Mars datum and comprises a broad SW–NE trending ridge with the summit calderas situated towards the northeast end. The mean slope value for the patera is only about 0.5°, yet the volcanic flows extend at least 1350 km from the summit, giving it a diameter of 2700 km, an area approaching 2 million km^2 and a volume of at least 1.4 billion km^3. The extreme length of these flows implies both low viscosity and high volume for the individual flows. Quantitative analysis of these flows by Cattermole (1987) indicated that the large sheet and tube-fed lavas had low yield strength and viscosity and were erupted at very high rates. However, more recent work by Sakimoto et al. (1997), using a rather different fluid-dynamics model, suggests effusion rates may not be as high as originally estimated, more probably within the range 2–105 $m^3 s^{-1}$, and with viscosities between 100 and 1 million $Pa\,s^{-1}$. This suggests that the Alba lavas are basaltic rather than komatiitic[1] (one suggestion previously made to accommodate the supposed low viscosities involved). The chemical make up of the earlier pyroclastic flows remains a mystery.

MOLA profiles across the western flank of Alba reveal its shape to be somewhat asymmetric in a north–south direction; this asymmetry is in part related to the shape of the Tharsis Rise on which it is built. The southern flank has a slope of 0.77°, rolling over to 0.21° near the summit region. The northern flank is steeper (1.88°) and the lower flanks extend some 800 km farther to the north, where slopes typically are of the order of 0.15°. The northward-sloping summit of Alba runs roughly parallel to the slope of the surrounding terrain that forms part of the Tharsis Rise. This coincidence suggests that Alba volcano was built before the long-wavelength regional topography, being tilted northwards in association with the formation of the rise.

The summit region is the site of a double caldera complex, which became the focus of extended effusive activity after an early period of flood lava activity. Volcanic flows that originated from the younger caldera partially bury the older incomplete depression (Fig. 8.14). The former has at least five components, which together cover an area of 2065 km^2. The sizes of the individual collapse depressions decreased with time. Each is very shallow, the deepest being only 150 m, which contrasts markedly with the deeper pits on Ascraeus and Olympus Montes, and indicates a significant difference in the dimensions of their subjacent magma reservoirs. As is the case with the Tharsis Montes, narrow and often leveed flows can be traced right up to the backwall of the caldera, indicating that Plinian-style activity was not important during the later stages of patera growth.

Raitala & Kauhanen (1989) concluded that the

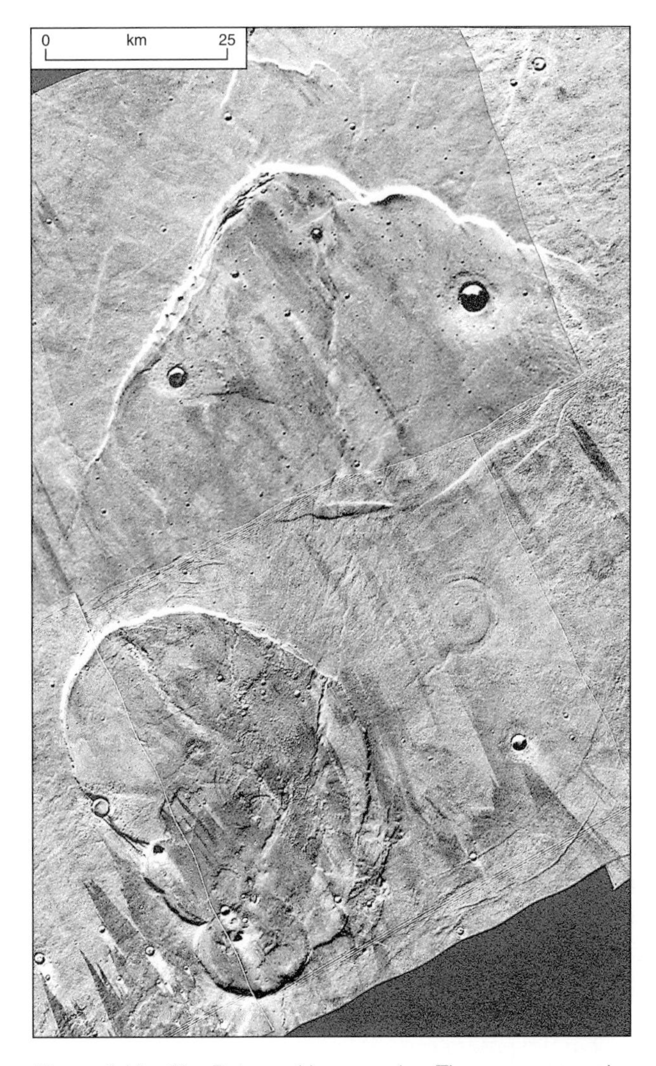

Figure 8.14 Alba Patera caldera complex. The younger complex measures 65 km×48 km and is composed of at least five coalescent depressions. Narrow summit flows can be traced to the caldera backwall and many are seen to be leveed. These late flows partly bury the floor of the older caldera to the northwest. Viking orbiter images 255s13–17; centred at 40.7°N 109.6°W.

1. A primitive ultramafic lava formed at very high temperature. Terrestrial komatiites are usually of Archaean age and many show a development of rapidly quenched olivine crystals known as spinifex, the crystal growths resembling the form of spinifex grass.

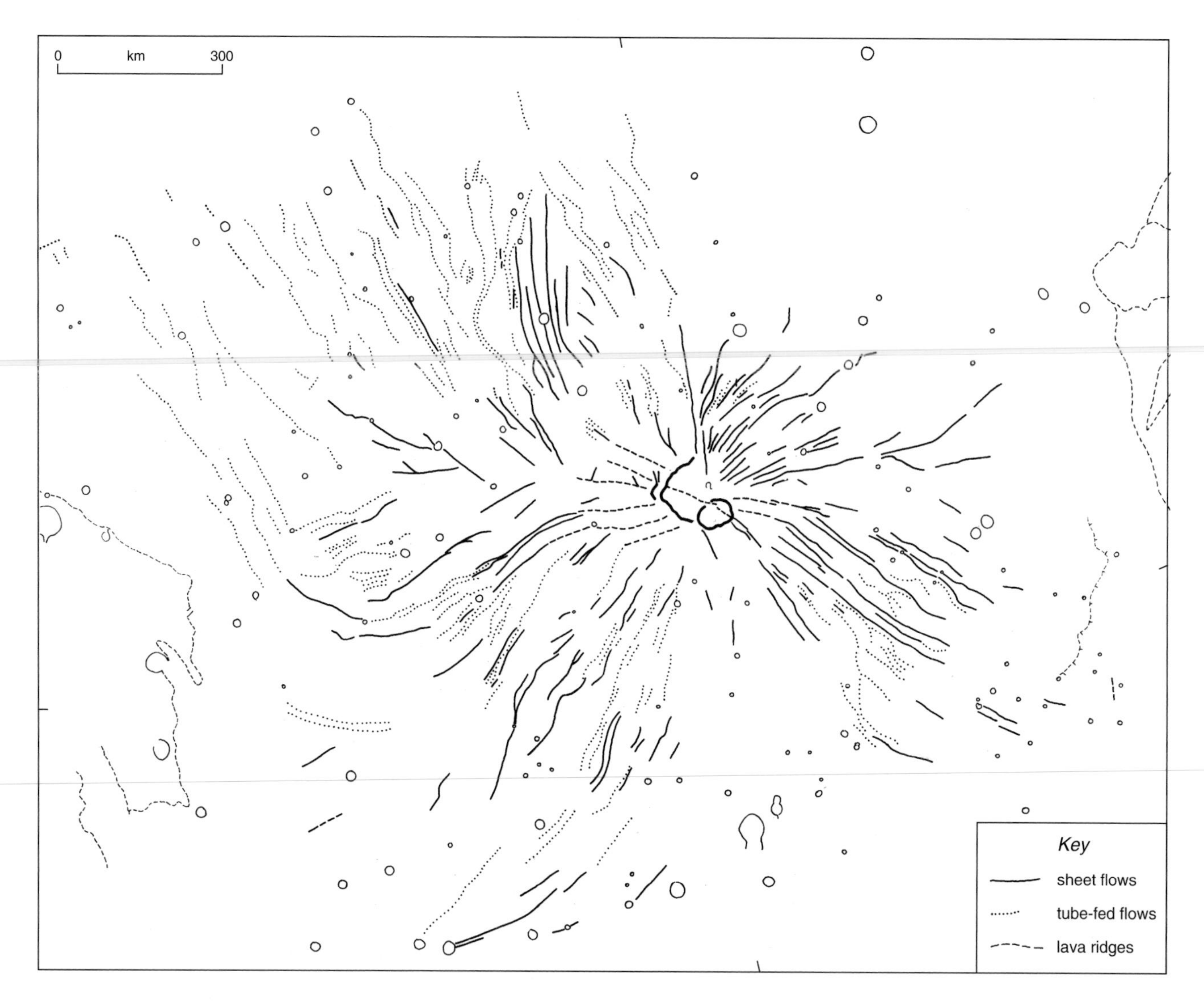

Figure 8.15 Distribution of different flow types at Alba Patera volcano. Although there is broadly radial pattern, massive tube- and channel-fed lavas to the west and northwest of the structure do not appear to have a summit-related origin.

formation of Alba's caldera was a result of collapse into a wide and shallow magma chamber, and that an assumed ratio of 5.0 between the diameter and the roof depth of the magma chamber implies an ascending mantle depth of around 100 km. The circumferential graben that characterize the volcanic edifice are thought to have formed in response to volcanic loading, and the centrally depressed region by the cooling and drying up of the upper asthenosphere towards the close of the volcanic cycle.

The general distribution of lava flows on the main part of the patera is radial about the summit, but flows are scarce over the uplifted block of Ceraunius Fossae, and in the northwest sector they are either absent or mantled by younger deposits (Fig. 8.15). On the lower ground to the northwest, large fields of tube- and channel-fed lavas have a generally NNW–SSE strike and appear not to have been related to summit activity. My own detailed geological mapping shows there to have been three principal episodes of patera growth (confirming earlier mapping by Scott & Tanaka 1980, 1986), together with the generation of several large discrete flow fields, the development of which may have been related to major fractures on the volcano's flanks. The earliest phase in the development of Alba involved the widespread emplacement of fissure-fed flood lavas of Lower Hesperian age, which now occupy distal locations. Subsequently, volcanism became more centralized, sheet flows and tube-fed lavas of large volumes being extruded, mainly from linear vents situated at or near the present summit, or from lower down the volcano's flanks (Fig. 8.16). These flows were responsible for the majority of patera construction. Quantitative measurements of these flows indicate volumes at least an order of magnitude greater than most Hawaiian flows, sheet flows having volumes in the range 1–110 km^3 and tube-fed lavas achieving volumes as great as 3500 km^3 (Baloga & Pieri 1985, Cattermole 1987).

A further characteristic of Alba Patera is the existence of anastomosing channel networks, mainly on the northern flank where lava-flow lobes are absent. Channel incision appears to have separated the two main stages of lava shield growth. Analysis of these by Mouginis-Mark et al. (1988) suggested, first, that the channels were of fluvial origin and, secondly, that they had been incised by a process of sapping induced by the release of non-juvenile water within relatively unconsolidated deposits on the volcano's flanks. They concluded that the fine-grain deposits themselves were of pyroclastic origin, but discounted an airfall origin on the basis that the channelled deposits extended too far (500–600 km) from the potential caldera source region and could have been dispersed that far only by extremely high eruption clouds. Since these are unlikely to have formed on Mars, they suggest that long runout pyroclastic flows dispersed the fine-grain material.

If this conclusion is accepted – and there is no *a priori* reason why this should not be so – then eruption of the volatile-rich ashflow materials can be shown to have postdated the emplacement of high-volume sheet and tube-fed lavas on the lower flanks, but to have preceded the final effusive phase of the volcano's evolution. As Mouginis-Mark and his co-workers noted, the existence of pyroclastics within the lava pile certainly could have played a part in explaining its very low relief and also the form and distribution of the circumferential fractures. The development of the volcano might well have followed the evolutionary sequence depicted in Figure 8.16.

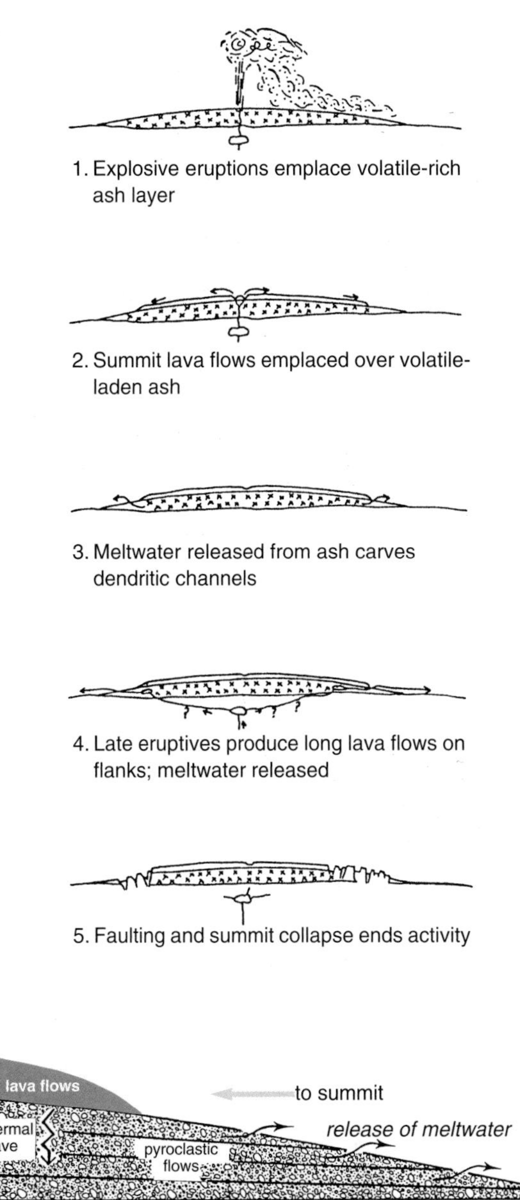

Figure 8.16 **(a)** Five-stage evolutionary model for evolution of Alba Patera. **(b)** Possible method by which channel networks were produced on smooth-texture flank deposits of Alba Patera. It is assumed that the non-welded ashflows would have become charged with volatiles, following which summit activity generated lava flows that partly buried the earlier units. A thermal wave passing through the lava pile into the underlying volatile-rich material would then generate meltwater that eventually emerged at the surface along bedding plains between successive ashflow layers. (After Mouginis-Mark et al. 1988.)

Because the Alba flows are so well preserved, it is possible to consider certain volcanological implications. First, it is clear from the flow morphology that Hawaiian-style effusive activity characterized the growth of the main shield and, secondly, that the volumes of individual flow units decreased with time. Now, since caldera collapse can take place only if a sufficiently large void is created beneath the summit, and because subvolcanic reservoirs must be full prior to the onset of eruption, only during the extrusion of lava or the injection of dykes into the volcano's substructure can such a space be created. Thus, the greater volumes of the earlier sheet and tube-fed flows imply that progressively lesser volumes of magma were required to trigger eruption as time proceeded. The implication here is that the magma flux rate beneath the volcano must have been greater during the earlier stages of its construction than during the later stages; that is, when the main part of the structure was emplaced during the early Amazonian epoch.

The very large dimensions of Alba flows have no real parallel on Earth and it is informative to consider the long-term magma supply rate to this huge volcano. On the assumption that all of the present topography of Alba is attributable to constructional volcanism, the lava pile has a volume of 4.16 billion km^3. Now, if volcanism occurred over a period of at least 2.3 billion years, this gives an overall magma production rate of $2000\,km^3\,yr^{-1}$. Such a figure is significantly larger than present day rates of magma supply to the volcanic pile of Mauna Loa and Kilauea. Calculations also show that several of the larger sheet and tube-fed lavas on the lower flanks must have released two orders of magnitude more thermal energy than Earth's annual release attributable to volcanism, and one order of magnitude more thermal energy than the annual conducted heatflow of Earth (Cattermole 1989a). This surprising conclusion leads inevitably to the implication that eruption of these very large volumes of magma must have constituted an extremely important, if not *the* most important, source of energy loss from the interior over geologically short periods during the Hesperian and early Amazonian epochs.

Volcanic shields and lava plains of Elysium

There are three shield-like volcanoes in Elysium, their morphology being significantly different from those of Tharsis. A fracture belt almost completely encircles the summit of Elysium Mons and outside this ring are many troughs trending WNW–ENE, most with flat floors and the general appearance of graben. For reasons not entirely understood, these pass northwestwards into a series of branching channels that extend for several hundreds of kilometres. The lower Amazonian lavas associated with Elysium central volcanoes cover an area of approximately 3 million km^2.

The largest of the Elysian volcano group is Elysium Mons, which has an overall diameter of 170 km and a single caldera 12 km in diameter. Its summit lies at about 12.5 km above the surrounding plains and it has slopes comparable with those of the large Tharsis volcanoes. Its overall geometry is asymmetric, with the summit caldera straddling a prominent ridge that trends NW–SE across the crest of a roughly circular shield structure. Although there are recognizable narrow flows on the main cone, the general appearance is rather smooth. There are many hummocks up to 5 km across and linear channels on the flanks (Fig. 8.17). In July 1998, Mars Global Surveyor obtained high-resolution pictures of the summit region; even at the higher resolution, there are no obvious lava flows visible on the volcano's flanks, but many craters certainly are visible, often without obvious ejecta. Many of these probably represent collapse pits formed

Figure 8.17 Synoptic view of Elysium Mons, showing 14 km summit caldera, textured shield and peripheral radial fractures. Viking orbiter images 544a46/58; centred at 25.6°N 213.7°W.

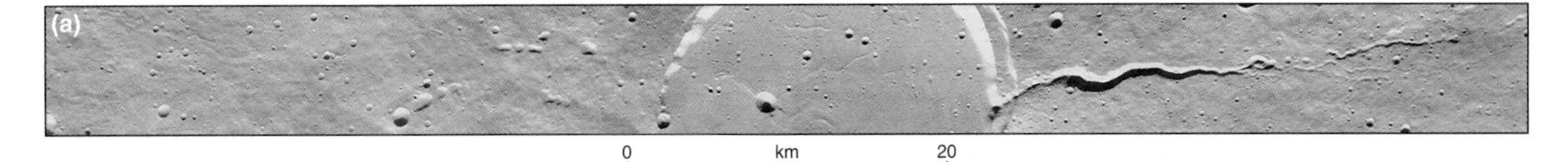

Figure 8.18 **(a)** A long crater chain traversing the shield (PIA01455; centred at 24.9°N 213.3°W); **(b)** A high-resolution MGS image of a part of the caldera wall (PIA01456; centred at 24.8°N 213.4°W).

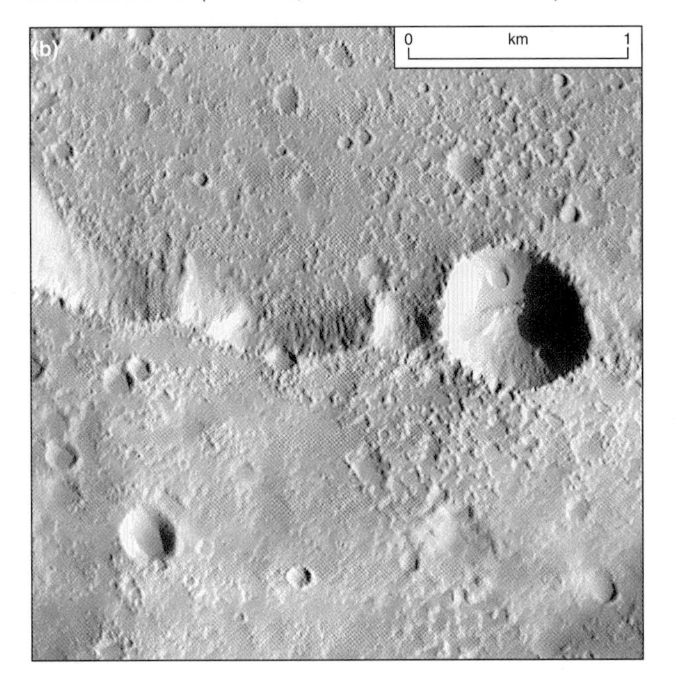

by withdrawal of magma during eruptions. A prominent channel runs due north–south and intersects the summit caldera (Fig. 8.18).

The other large Elysian structure, Hecates Tholus, is more interesting and has been studied closely by Mouginis-Mark et al. (1982). Situated north of Elysium Mons, it rises about 12.5 km above the adjacent plains and takes the form of a low shield 160×175 km across. At the summit is a nested caldera complex measuring 11.3×9.1 km. Unlike Tharsis shields, which are largely constructed from lava flows, none is exposed on Hecates Tholus. All that is visible is a complex of radial channels not unlike those seen on Ceraunius Tholus. However, Mouginis-Mark and his colleagues (1982) showed that the anastomosing courses and dendritic tributary networks were unlike channels associated with volcanic density currents. Furthermore, the absence of channels from the summit region makes it difficult to believe that their formation was associated with explosive volcanic activity. As a result, they suggest that the channels are fluvial in origin, having been incised into materials less coherent than the lava shield that they mantle.

Impact crater counts indicate that the channelled surface is actually younger than the youngest region of the Olympus Mons caldera (perhaps as young as 3 million years). The same group therefore proposed that this is a relatively young airfall deposit they estimate to be about 100 m thick, produced by an eruption cloud with a height approaching 70 km. Because a stable eruption column was apparently able to be sustained, the magma volatile content must have been about 1 wt per cent if the volatile was H_2O and more than 2 wt per cent if CO_2. Such a volatile component requires that the source magma be originated at depths of greater than 50–100 km if the volatile was CO_2 and 40–150 km if it was H_2O. Although this imposes mantle depths for CO_2-rich magmas, no such restrictions pertain if water was the volatile, and thus the possibility exists of absorption of permafrost or trapped groundwater as rising magma neared the Martian surface. In order to reach such an altitude, model calculations suggest that a volatile-charged Martian magma would need a mass eruption rate of around 107 kg s^{-1}. Therefore, we have yet another volcano in whose growth hydromagmatic activity has been important.

Albor Tholus lies 300 km SSE of Elysium Mons and, on the basis of crater counts, lies between Elysium Mons and Tharsis Tholus in terms of its age. A further 600 km to its south lies the relatively flat region known as the Elysium basin, which consists of a large depression about 3000 km across. Prior to the arrival of Mars Global Surveyor, there were two schools of thought regarding the origin of this region:

- The depression was once a massive lake some 1500 m deep and is now filled with sediments.
- The region is composed of fluid-lava flows that may have infilled an older lake and in places overflowed its margins to enter Marte Vallis.

High-resolution images confirmed that the basin surface consists of lava flows, the Viking appearance (dark plates with intervening light surfaces) now interpreted as huge rafts of solidified lava that floated above a molten lava lake (Fig. 8.19). The early centralized volcanism of the Elysium rise was followed by extrusion of extensive fluid lavas that ponded in the massive depression on the rise's southern side.

Apollinaris Patera

Apollinaris Patera is rather isolated, lying southeast of the Elysium group, at 96°S, 186°W. Crater counts indicate an age roughly half that of Alba Patera (Plescia &

Figure 8.19 MGS high-resolution image of huge lava slabs in the Elysium basin region. MGS image PIA01494; centred at 5.8°N 208.9°W.

Saunders 1979). It is a broad 400 km-diameter lava shield crowned by a 70 km summit caldera that has two different floor levels, a fact established quite clearly by photoclinometric work undertaken by Thornhill et al. (1993). The flanks are strongly striated and transected on the west, north and east sides by a prominent scarp. However, to the south a large fan with its apex at the caldera rim extends for about

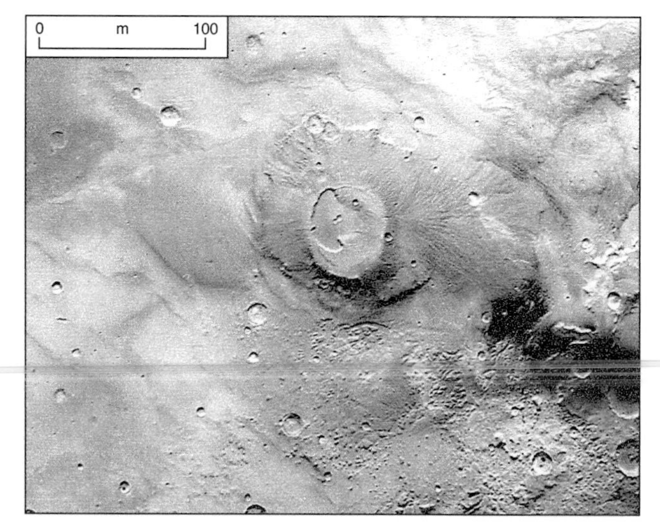

Figure 8.20 The isolated central volcano Apollinaris Patera, with its 100 km-diameter caldera (north is to the left). The shield surface is truncated by a prominent scarp, which is buried on the south side by a massive fan-shape deposit that spills onto the adjacent lava plains. This could be a lava fan, but it might also represent channelled pyroclastics. Viking orbiter image 639a92; centred at 9.1°S 186.2°W.

350 km and buries the scarp in that sector. On the surface of the fan is an array of what appear to be broad channels, some of which incise the caldera backwall (Fig. 8.20). The rather poor resolution of Viking imagery prevents detailed analysis and it is not possible to decide whether the fan-like deposit is built from lavas or from channelled pyroclastic deposits.

Central volcanism on Mars

Volcanic activity has been widespread in terms of both space and time. The most extensive form of early activity involved fissure-erupted flood lavas that built volcanic plains on a global scale. This activity emplaced the Upper Noachian plateau plains and was succeeded by the plains of Lower Hesperian age. This may have been related to formation of an early magma ocean, possibly driven by a multi-plume mantle convection pattern. The earliest centralized volcanism generated low-profile paterae, mainly in the southern hemisphere, a particularly prominent group of structures surrounding Hellas. These were active during Late Noachian (Apollinaris Patera) and Early Hesperian times (Hellas group of paterae) and, from morphological characteristics, their growth appears to have involved hydromagmatic eruptions and lava flows. Volcanoes also grew at Hecates Tholus and in the region of Syrtis Major Planum. The characteristic hydromagmatic activity of this period was a manifestation of the different climatic conditions that are

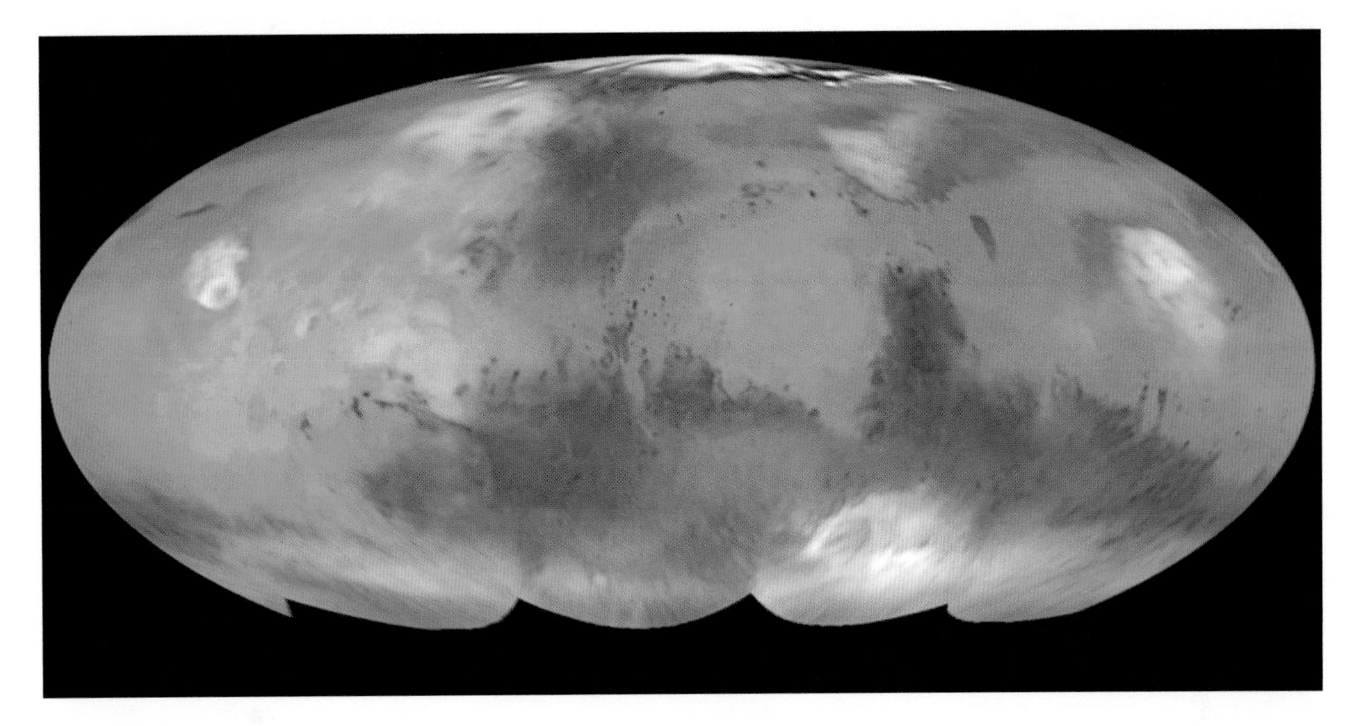

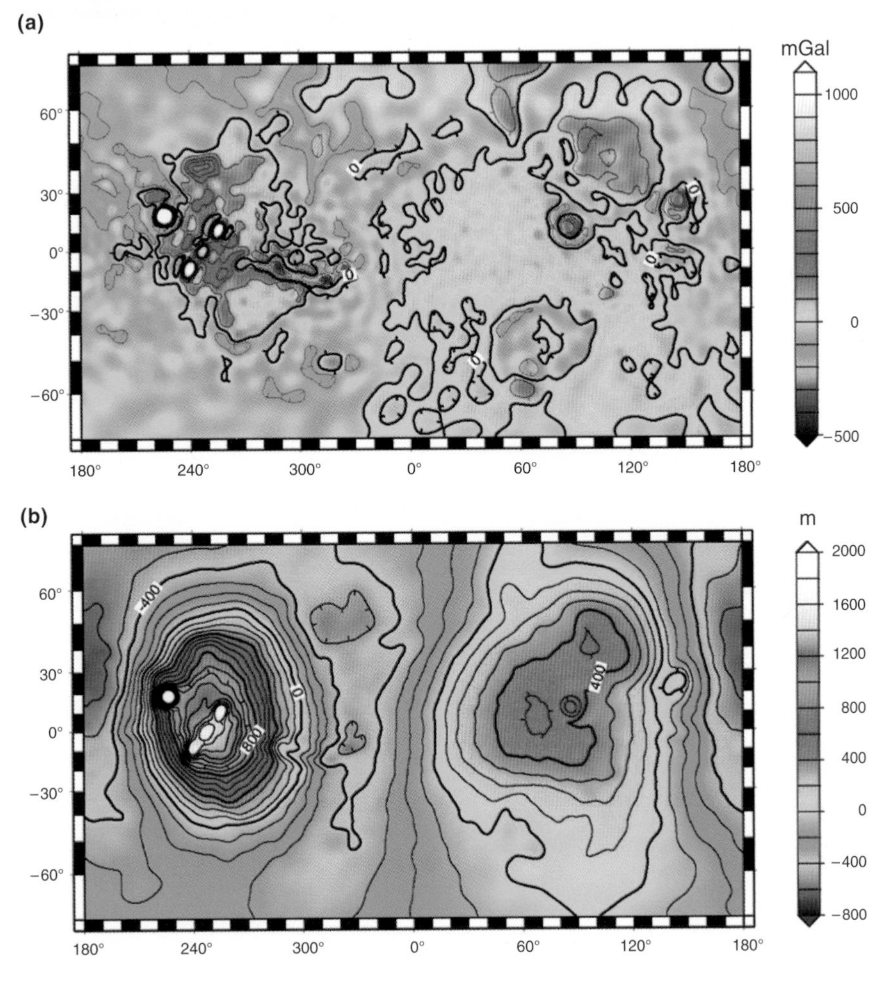

Plate 1 Hubble space telescope global view of Mars showing seasonal cloud distribution (PIA Image 01588).

Plate 2 (a) Free-air gravity for the MGS 75b gravity model. (b) Areoid anomalies; the estimated hydrostatic contribution (95%) of the planetary flattening has been removed. (After Smith et al. 1999b.)

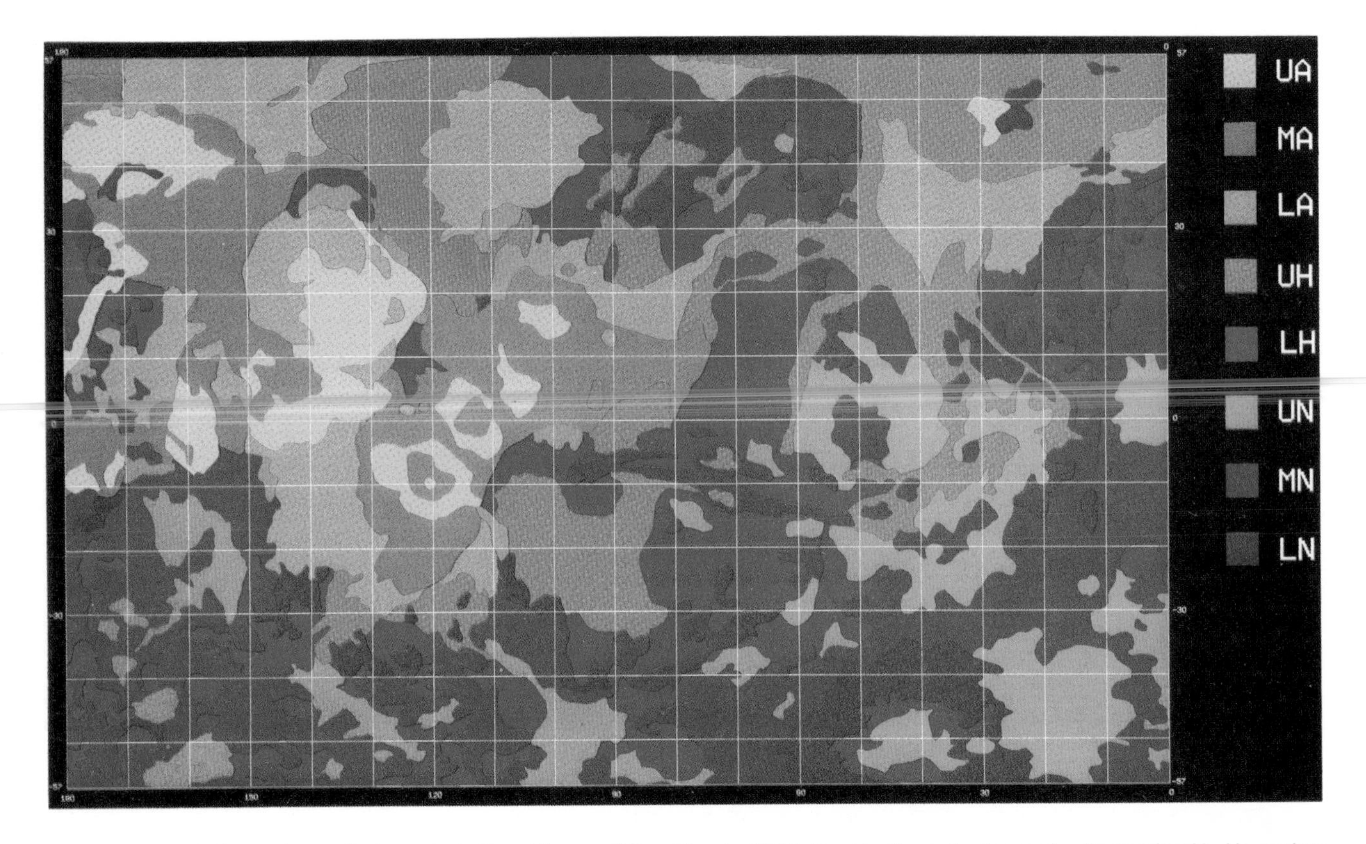

Plate 3 Above, below and opposite (upper): stratigraphic map of Mars (Tanaka 1986). Key: U = Upper; L = Lower; A = Amazonian; H = Hesperian; N = Noachian.

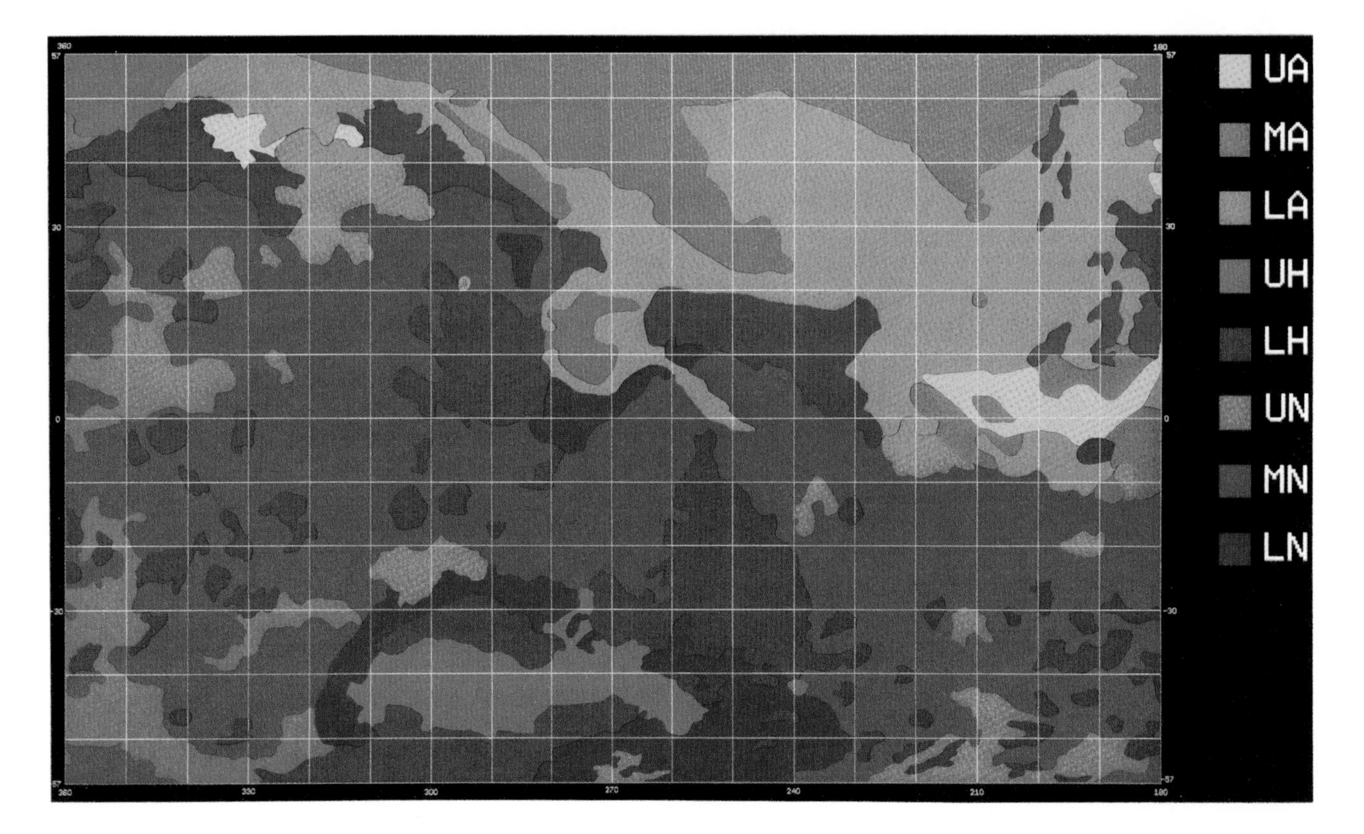

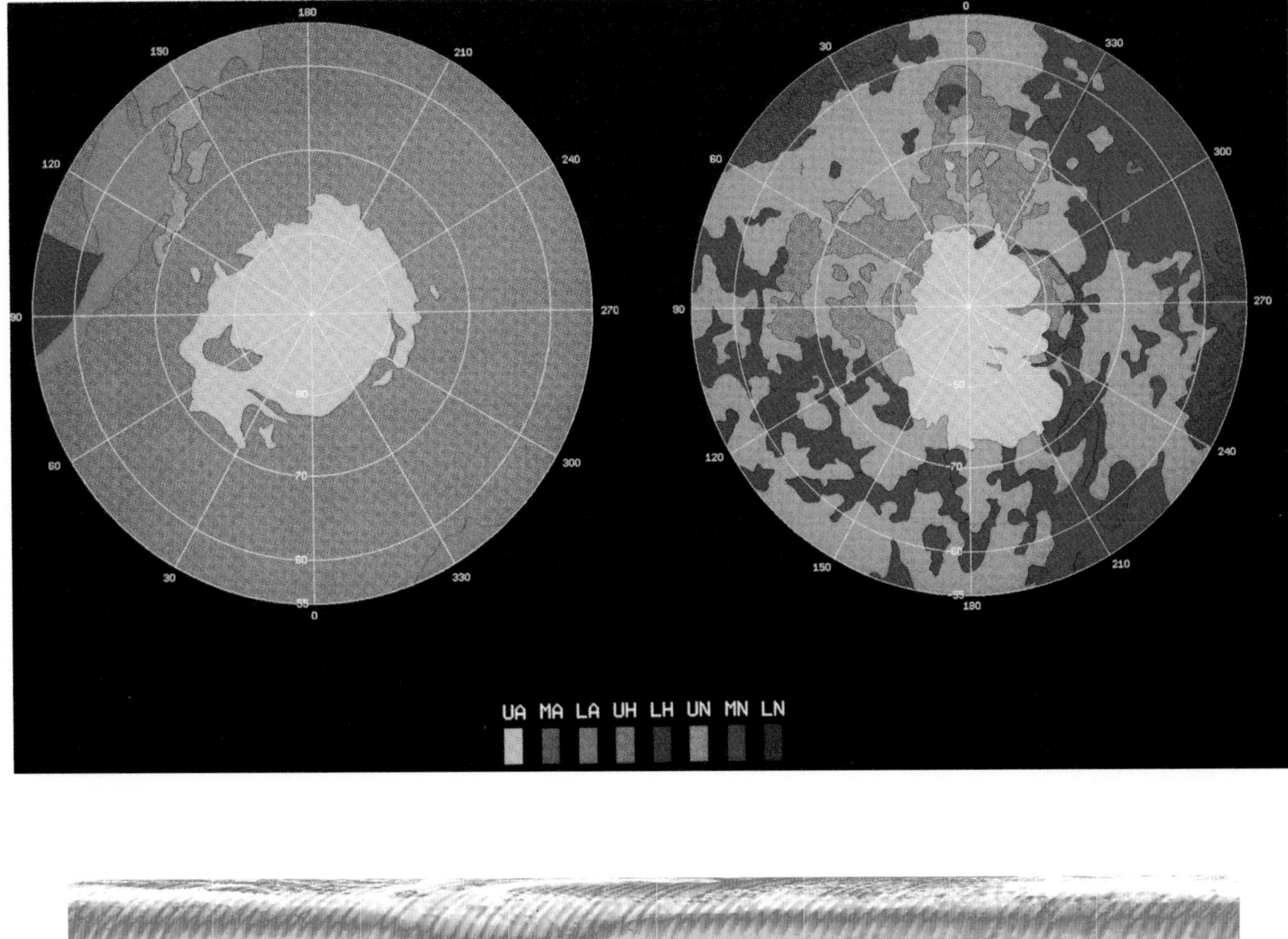

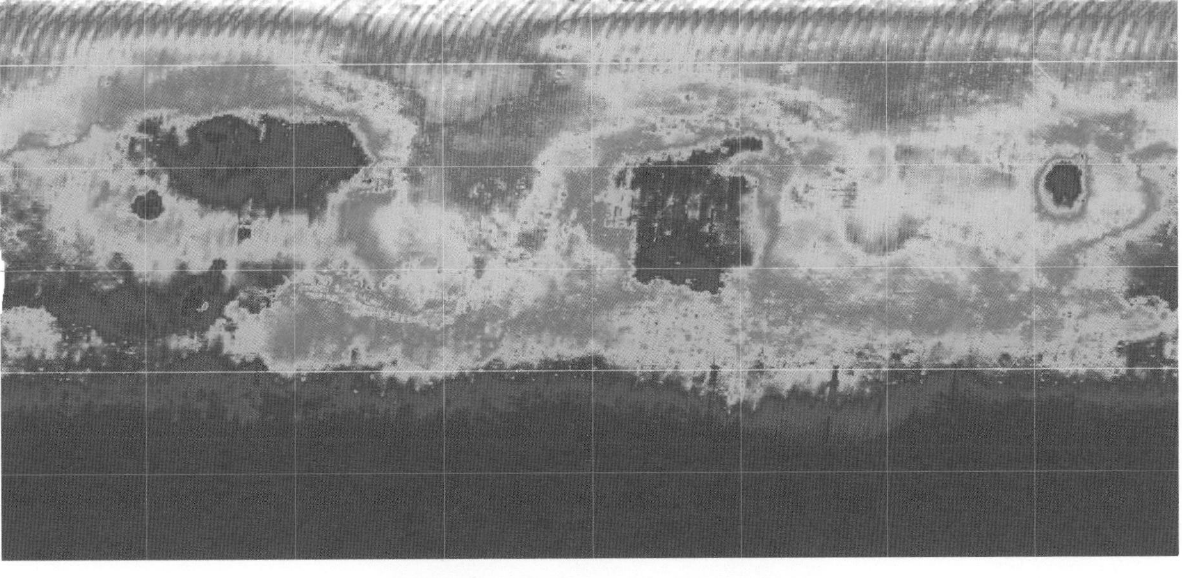

Plate 4 Night-time (02.00 h) temperature of the Martian surface as measured by the thermal emission spectrometer instrument on the Mars Global Surveyor. The data were acquired during the first 500 orbits of the MGS mapping mission. The coldest regions are areas of very fine (dust) grains, whereas the warmest regions are areas of coarse sand, gravel, and rocks. (MGS image PIA02014)

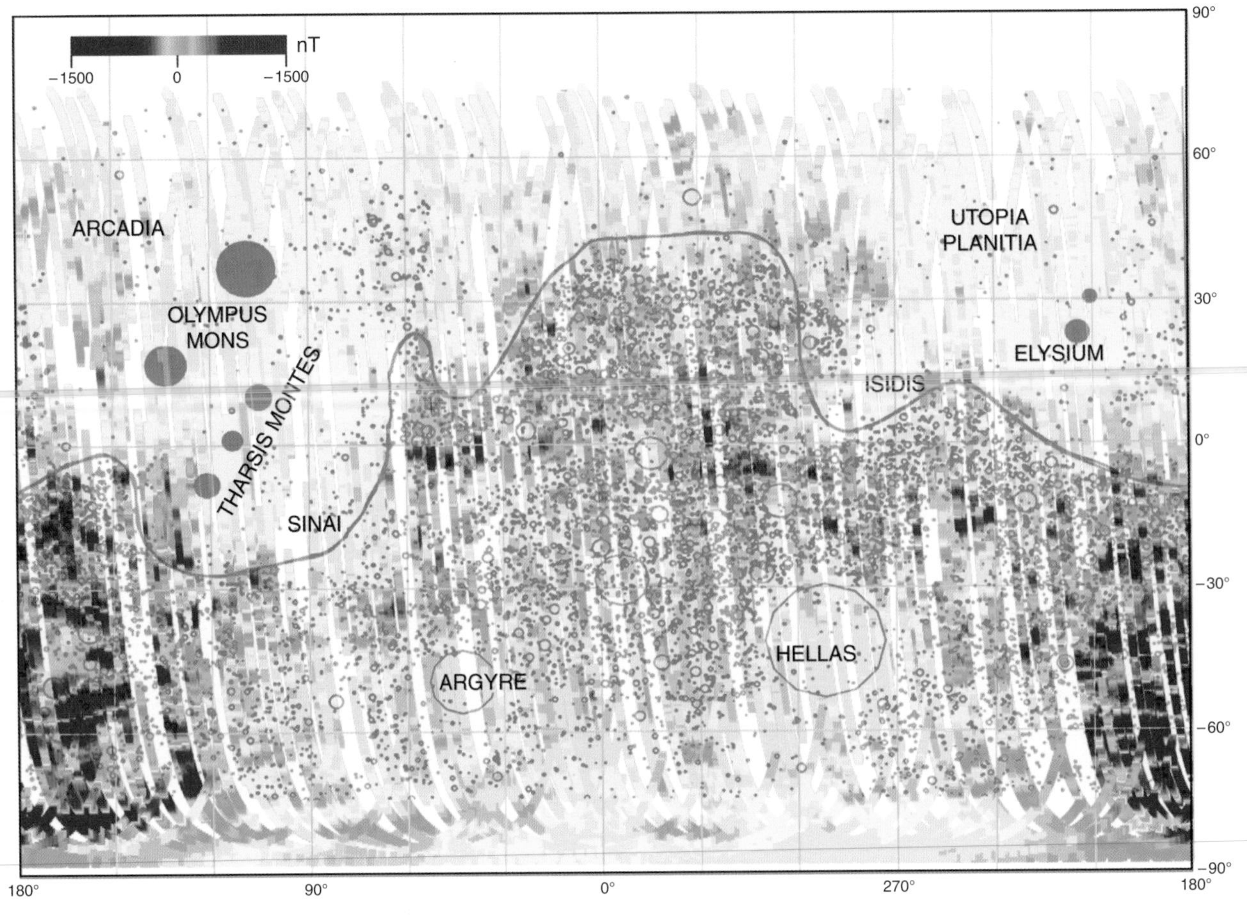

Plate 5 Map of magnetic fields near the Terra Cimmeria and Terra Sirenum regions. The bands are oriented approximately east–west and are about 160 km wide and 960 km long, although the longest band stretches more than 1900 km. The false blue and red colors represent invisible vertical magnetic fields in the Martian crust that point in opposite directions. (Acuña et al. 1999; MGS image PIA02059)

Plate 6 Below and opposite (lower): 360° panorama of the Mars Pathfinder landing site (MPF image PIA01466).

Plate 7 MOLA topographic map of Mars. This represents the first globally distributed high-resolution measurements of the planet's topography.

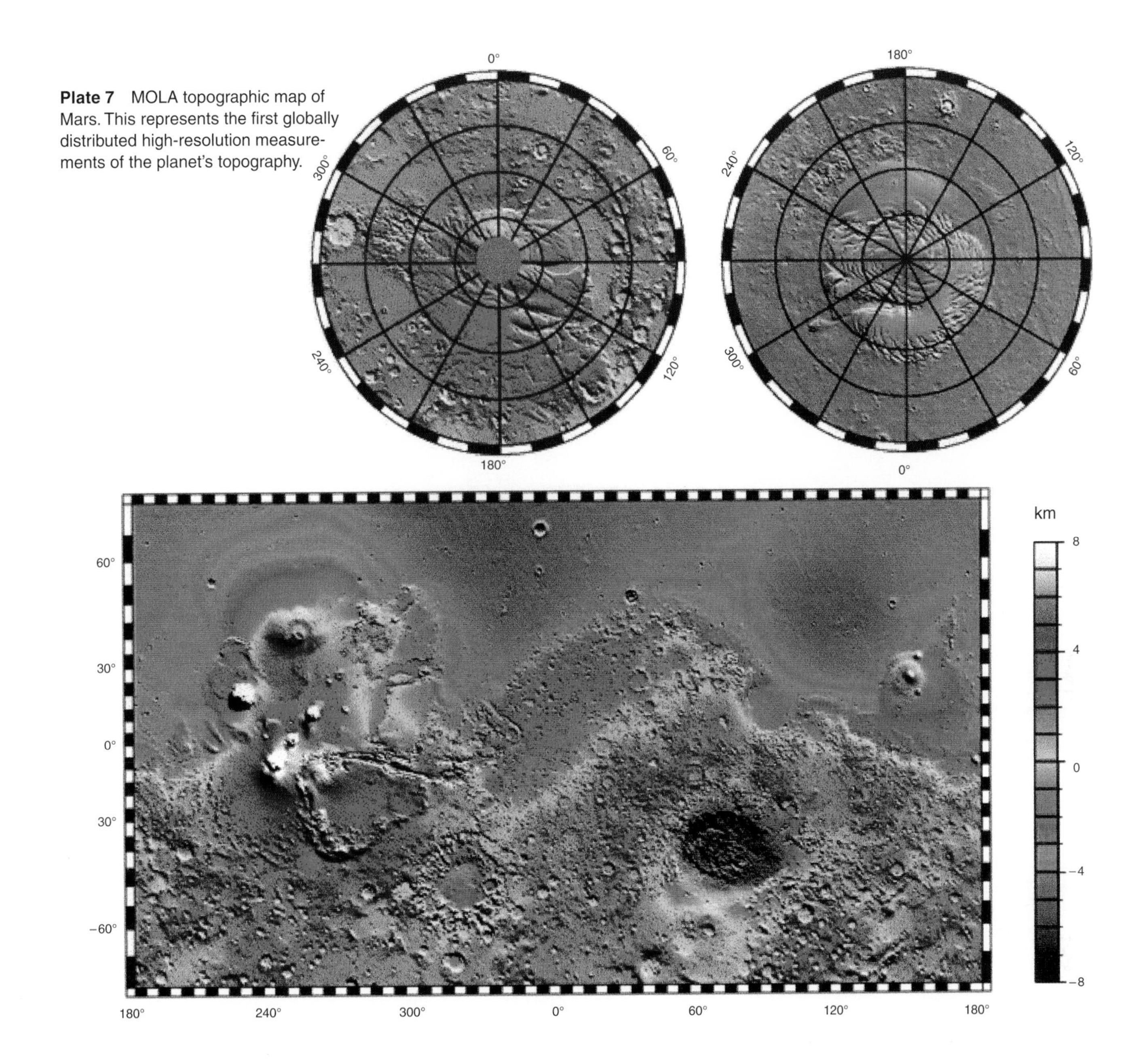

Plate 8 Different classes of rock and soil at Pathfinder landing site. (a) Dark rock type and bright soil type: the rock Barnacle Bill. Reflectance spectra typical of fresh basalt and APXS spectra indicating more silica-rich basaltic andesite compositions. Typified by the small boulders and intermediate-size cobbles at the site. The bright soil type is very common and here comprises Barnacle Bill's wind tail and much of the surrounding soil. The high reflectance and strongly reddened spectrum indicate oxidized ferric minerals. (b) Bright rock type: the rock Wedge. Reflectance spectra typical of weathered basalt and APXS spectra indicating basaltic compositions. Such rocks are typically >1 m across and many show evidence for flood deposition. (c) Pink rock type: the rock Scooby Doo. APXS and reflectance spectra indicate a composition and appearance similar to the drift soil. However, the rock's morphology indicates a cemented or rock-like structure. (This material may be a chemically cemented hardpan that underlies much of the site.) (d) Dark soil type: typically found on the windward sides of rocks or in rock-free areas such as Photometry Flats (shown here), where the bright soil has been stripped away by aeolian action, or in open areas. (e) Disturbed soil type: the darkening of disturbed soil relative to its parent material, bright soil, as a result of changes in soil texture and compaction caused by movement of the rover and retraction of the lander airbag. (f) Lamb-like soil type: shows reflectance and spectral characteristics intermediate between the bright and dark soils. Its distinguishing feature is a weak spectral absorption near 900 nm. (Rover Team 1997)

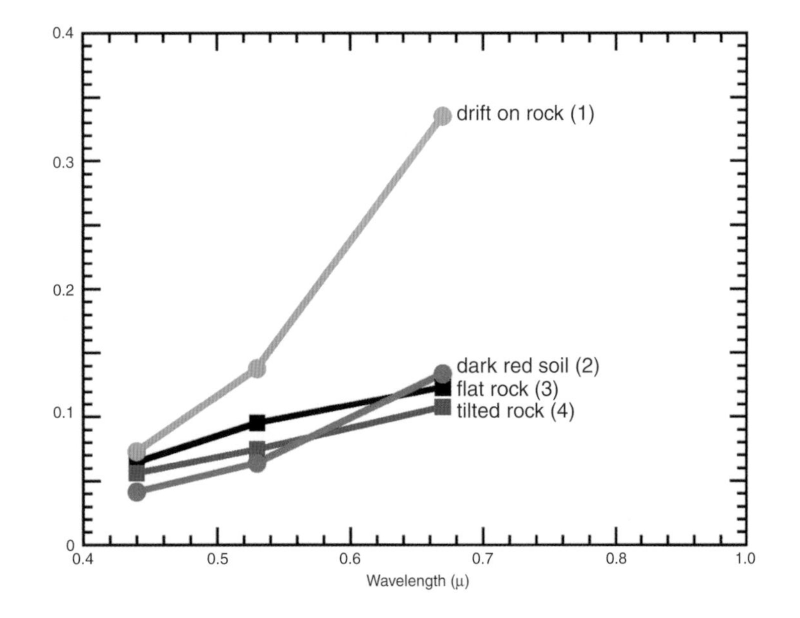

Plate 9 Spectra of Flat Top and adjacent soils at Mars Pathfinder landing site (MPF image PIA00756).

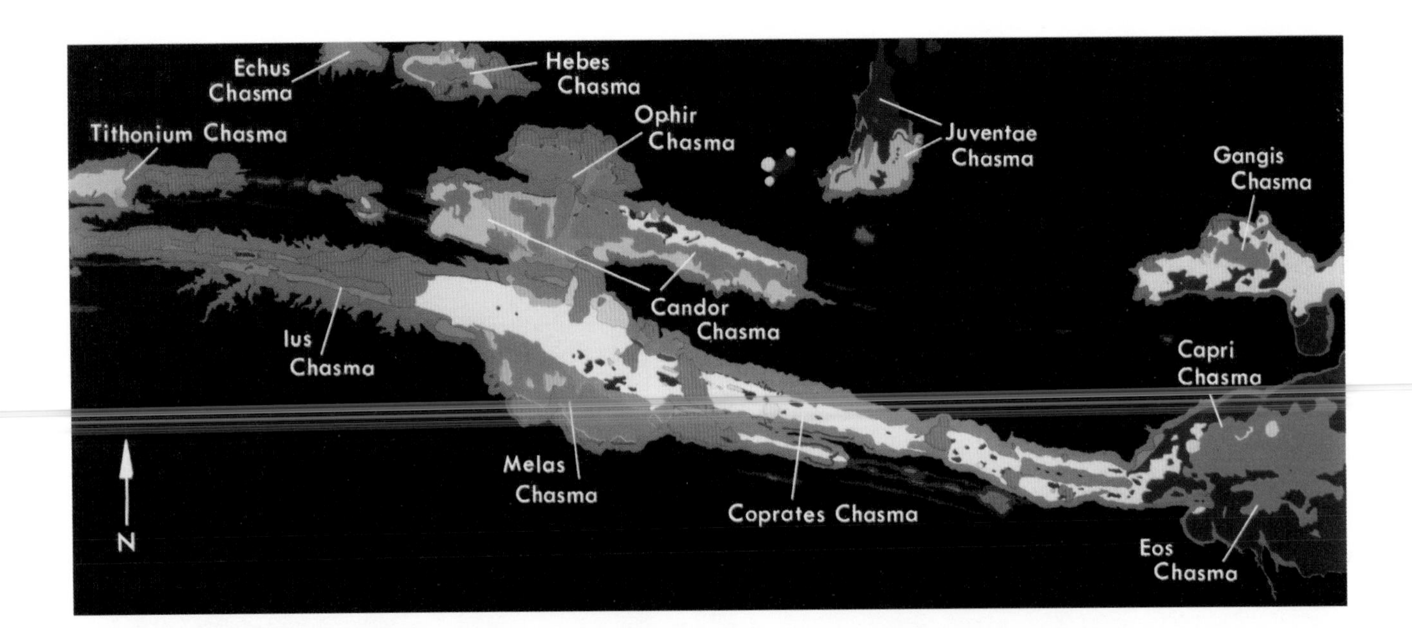

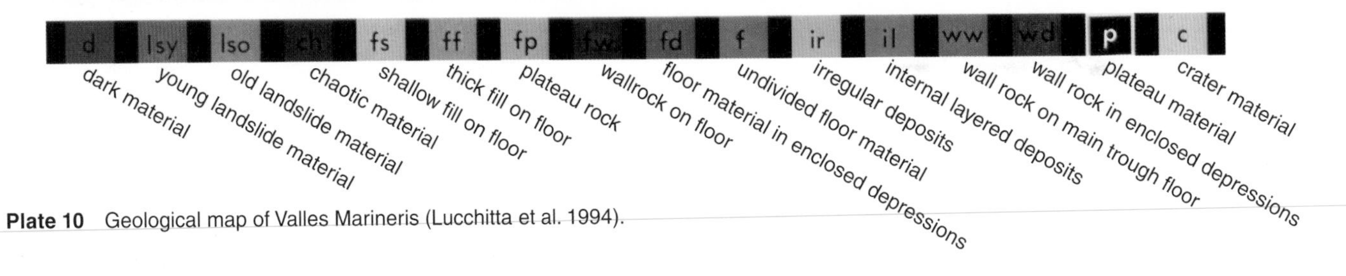

Plate 10 Geological map of Valles Marineris (Lucchitta et al. 1994).

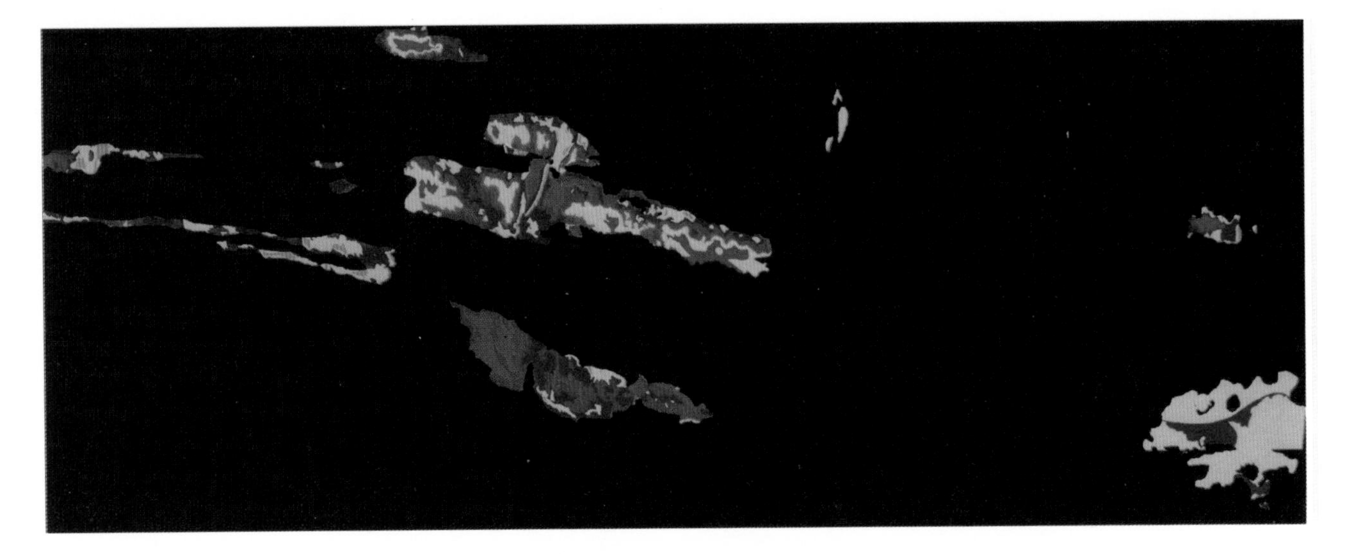

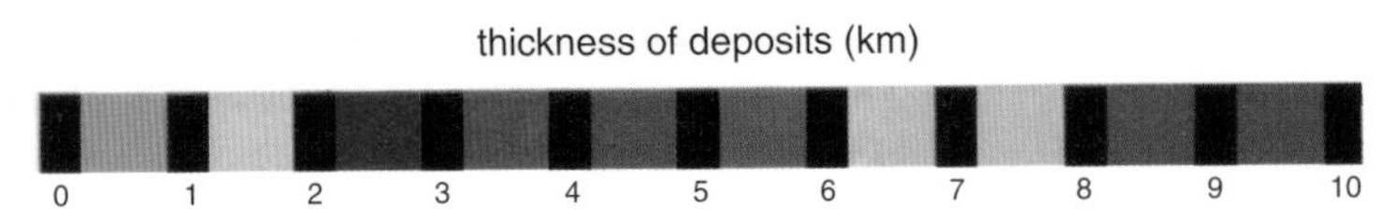

Plate 11 Thickness of deposits inside Valles Marineris (Lucchitta et al. 1994). Contours are in 1 km increments upwards from designated floor. Only areas occupied by deposits are shown.

thought to have been a feature of Mars at that time; hence, significant volumes of volatiles became entrained in magmas rising through the subcrust.

The focusing of volcanic centres around Hellas was doubtless related to the massive impact that produced the basin, which could have generated a major plume beneath the focus of impact (in much the same way as has been proposed for the Moon by Wüllner 1996), with the result that large volcanic centres grew there. Prior to the formation of Hellas, other major impact events may also have determined the position of such plumes within the mantle layer. If the suggestion that the northern "basin" was caused by multiple impacts (Wilhelms & Squyres 1984) proves valid (I personally doubt that this will happen), then early impacts would have been expected to play a major role in determining where plumes were set up prior to that time.

Towards the close of Hesperian times, a further episode of flood volcanism emplaced the volcanic plains of the lowland hemisphere; this was accompanied by continued volcanic centralization in the region of Elysium at Albor Tholus, and in northern Tharsis, where activity became focused on Alba Patera. The vast low-profile Alba Patera was first active during the Early Hesperian epoch, as localized activity became dominant. Its growth involved emplacement of both pyroclastic and effusive rocks. There was also growth of low-profile volcanoes in Syrtis Major Planum and in the Tempe Fossae region. Activity at Alba continued until the Amazonian, a period of around 1.5 billion years. Some explosive activity continued, but there was a gradual change towards more effusive volcanism, with the eruption of very large volumes of low-viscosity magmas such as those found in the Elysium basin area. Later stages in Mars' volcanic evolution saw the growth of huge shields along the crest of the Tharsis Rise and the growth of Elysium Mons.

An analysis of the symmetries in Martian volcano distribution by Matyska et al. (1998) illustrated how the major central vents show an axisymmetry and a reflecting symmetry about a plane whose normal is the axis of symmetry; the latter is skewed by around 30° to the rotational axis and is thought to represent the axis of symmetry for interior dynamics in early Mars. After the formation of Hellas, these authors proposed that an antipodal effect may have focused magmatism in the Tharsis region and at Alba Patera. Because the production of Hellas must have been related to a very large impact, they suggested that mantle circulation was forced into an axisymmetric bipolar pattern. Plume–plume interactions ensued, causing individual plumes to merge, transferring the bipolar plume structure into a single long-lived stable plume, persistently supporting Tharsis until today.

The nature of Martian magmas

The morphology and dimensions of the majority of lava flows discernible on spacecraft imagery are consistent with their being of fluidal basaltic composition. Recent analysis of what were once believed to be flows of extremely low viscosity on Alba Patera (and therefore possibly more mafic than basalt) indicates that many are probably basaltic after all, with viscosities higher than at first believed. The notion that some Martian magmas may have been of komatiitic composition is also a possibility that receives some support from Phobos 2 spectral analysis. Raitala (1990), in an analysis of the evolution of the Tharsis region, cited experimental research into komatiites undertaken by Takahashi & Scarfe (1985). This suggested very deep and very hot early planetary mantle development with which was associated rising diapiric plumes from several hundred kilometres' depth, generating komatiite magmas by partial melting at depths of 150–200 km. In one sense the presence of such a deep source on Mars would rather conveniently account for the crustal doming of Tharsis, the huge volume of extruded material and the peculiarities of certain lavas, for instance those associated with Alba Patera.

Chemical analysis at the Pathfinder landing site in Ares Vallis confirmed the dominance of basaltic rocks in this area, but also established the presence of blocks that were either lavas or composite (volcaniclastic) rocks very similar to andesite in composition. It remains unclear whether such rocks indicate that evolution of Martian magmas went beyond the stage of Al-rich basalt fractionation (presumably by normal fractional crystallization) or imply a mixing of basaltic and other components during erosion and subsequent outflow-channel deposition. However, first indications suggest that the range of igneous rock types on the planet is wider than previously believed.

Spectral analysis of what were identified as relatively pristine volcanic materials in Syrtis Major, Valles Marineris, Ophir Planum and Sinus Meridiani by the Phobos 2 imaging spectrometer (Mustard et al. 1997) established that mafic mineralogical composition in these units is dominated by low- and high-calcium pyroxenes, a two-pyroxene nature similar to that of the SNC meteorites and terrestrial komatiites, and it implies a high degree of partial melting of mantle material depleted in aluminium. The wide range of ages of the analyzed units (estimated at 3 billion to 180 million years) indicates that the depletion in aluminium of the Martian mantle happened early in the planet's history and that little subsequent bulk evolution has occurred. This is perhaps not surprising, since Mars lacks the crustal recycling mechanism of

Earth. The Al depletion that characterizes the mantle may be explicable by the early development of a thick basaltic crust relatively enriched in aluminium.

Volcanoes, loading and lithospheric stresses

The extended cycle of volcanism in the Tharsis region has prompted much research and debate. The geological history of the rise must be closely related to mantle dynamics; study of Tharsis has involved analysis of its volcanic history, tectonics and gravity. At this point the effects of volcanic loading are discussed; tectonics will be discussed in detail in the next chapter.

Quite clearly, the formation of large volcanoes loads the underlying lithosphere and may lead to lithospheric flexure and faulting. In turn, lithospheric deformation affects the stress field beneath and within the volcanic pile, and it can influence the way in which magma is transported towards the surface. McGovern & Solomon (1993), using a finite element code, calculated the stresses and displacements caused by a volcanic load emplaced on an elastic plate overlying a viscoelastic mantle. They established that, as a result of flexure, there are three regions where stresses become large enough to cause failure by faulting:

- at the surface of the lithosphere just beyond the perimeter of the volcano
- near the bottom of the elastic lithosphere beneath the centre of the volcano
- on the upper flanks of the volcano at an early stage in its growth.

The dominant mode of failure predicted for the first region is normal faulting. This is consistent with circumferential graben observed around the Tharsis Montes and with the basal scarp of Olympus Mons, widely interpreted as a large-offset listric[1] normal fault. Normal faulting, mostly radial, is predicted for the second region, whereas failure in the third region is predicted to consist of thrust faulting, circumferential on the upper and middle flanks, and radial on the lower flanks. In growth models, the lower part of the volcano is later covered by younger units, characterized by lower stresses and not predicted to fail. The well defined concentric terraces that have developed on the upper flanks of Olympus Mons may correspond to the predicted circumferential thrust features.

For volcanoes detached from the underlying lithosphere, predicted failure in the structure takes the form of radial normal faulting near the volcano's base. The addition of a local extensional stress arising from the regional topographic slope yields a pattern of predicted faulting that closely matches that observed on the Tharsis Montes, including the development of radial rifts on the lower volcano flanks to the northeast and southwest and the asymmetric formation of circumferential graben along the flanks. This stress state is also consistent with the aureole deposits of Olympus Mons being the result of gravity sliding along a plane of decollement.[2] For a given load increment, the first mode of near-surface failure for most of the area immediately around the load is circumferential normal faulting and graben formation. As the volcano grows and the flexural response to the increasing load develops, the predicted failure mode in a portion of this annular region surrounding the volcano changes to strike-slip faulting. However, because normal faulting has been predicted to have taken place earlier, it is likely that release of later stresses will occur by reactivation and growth of these normal faults and graben, rather than by the formation of new strike-slip faults.

Stress fields within a volcanic structure have an important effect upon the way in which magma is transported and erupted. Where flexure affects a volcano that is attached to the lithosphere below it, horizontal propagation of magma from the summit towards the flanks will be favoured (Crumpler & Aubele 1978). Also, stress evolution may have an effect on the location of magma reservoirs. For a volcano to grow, one important condition has to be met: magma must find a way towards the surface. Maximum compressive stresses will be experienced in the first volcanic products, which will be found near the base of the final load. At these levels, magma may be prevented from rising towards the summit and, in consequence, it may pool at low levels in the edifice, being intruded as dykes and sills. Alternatively, the high pressures may be relieved by fracturing, in which case this would favour magma collection and the formation of magma reservoirs. Such reservoirs might give rise to flank eruptions, stress conditions allowing. This being so, the additional loading of the volcano's lower flanks might relieve compressive stresses near the summit, allowing further eruptions to take place at a high level. In this way, eruptions might well alternate between summit and flanks, at least during a part of the volcano's growth cycle.

1. Normal fault in which the angle of dip increases with depth.

2. A plane of detachment between rock units.

9 Tectonics and the growth of Tharsis

As the Mariner 9 spacecraft orbited Mars, it revealed not only many enormous shield volcanoes and the planet's polar caps, but the enormous equatorial canyon system, named Valles Marineris after the probe that discovered it. Unbeknown to astronomers, it was this enormous feature that had been depicted unwittingly on many early drawings as one of the notorious "canals". Subsequently, spacecraft instruments revealed the presence of a major rise in the Martian crust, centred on the region of Tharsis, from which radiated out an extensive system of fractures; a smaller rise centred on Elysium. Valles Marineris form an integral part of a planetary-scale fracture complex and has its western end firmly within the perimeter of the rise.

One-plate and multi-plate planets

Earth is composed of lithospheric plates that move slowly about on a mantle layer that is ductile and in constant, slow, motion. Convective movements within the upper part of the mantle layer drive plate movements that may be convergent, divergent or passive. Divergent boundaries lie above rising convection cells, and convergent ones above sinking ones. New oceanic crust is generated at convergent boundaries, whereas existing crust is destroyed and recycled along inclined subduction zones located along convergent boundaries. By this process Earth constantly recycles its crust, returning it, along with volatiles and sea-floor sedimentary materials, to the interior, where they play a vital part in the melting and transformation of the subducted materials.

Terrestrial plate tectonics appears to have been in operation for most of Earth's life and it has been instrumental in ensuring that the planet has remained dynamic, with almost continual volcanism, tectonism and sedimentation. It appears unique among the terrestrial planets in this respect. However, it is possible that a form of plate tectonics once operated on Venus, since tectonic features within the ancient crust of the highland massifs attest to lateral movements of individual crustal blocks. It has also been suggested that plate tectonics may have characterized early Mars (Sleep 1994), although this suggestion has not received universal support. Certainly, at the present time, both Venus and Mars are one-plate worlds, and it is for this reason that the distribution of their volcanic and tectonic features is markedly different from those of Earth.

Much discussion has focused on explaining the low elevation and relatively young age of deposits within the northern hemisphere of Mars. One proposal (Wilhelms & Squyres 1984) has it that a massive early impact was responsible for the first-order difference between the two hemispheres. Recent geomagnetic, gravity and topographic data indicate that this is very unlikely to have happened and, in any case, not everyone held this view. In an effort to find another explanation, Sleep (1994) suggested that some form of plate tectonic activity may have been responsible for the subduction of thick ancient highland crust during Late Noachian to Early Hesperian times, with a process akin to sea-floor spreading producing thin oceanic-type crust, which now underlies the northern plains. In his preferred reconstruction, a divergent plate boundary extended from north of Terra Cimmeria between Daedalia Planum and Isidis Planitia, whereas subduction occurred beneath Tharsis Montes and Tempe Terra.

This idea has not met with widespread approval, and recent magnetic data from Mars Global Surveyor

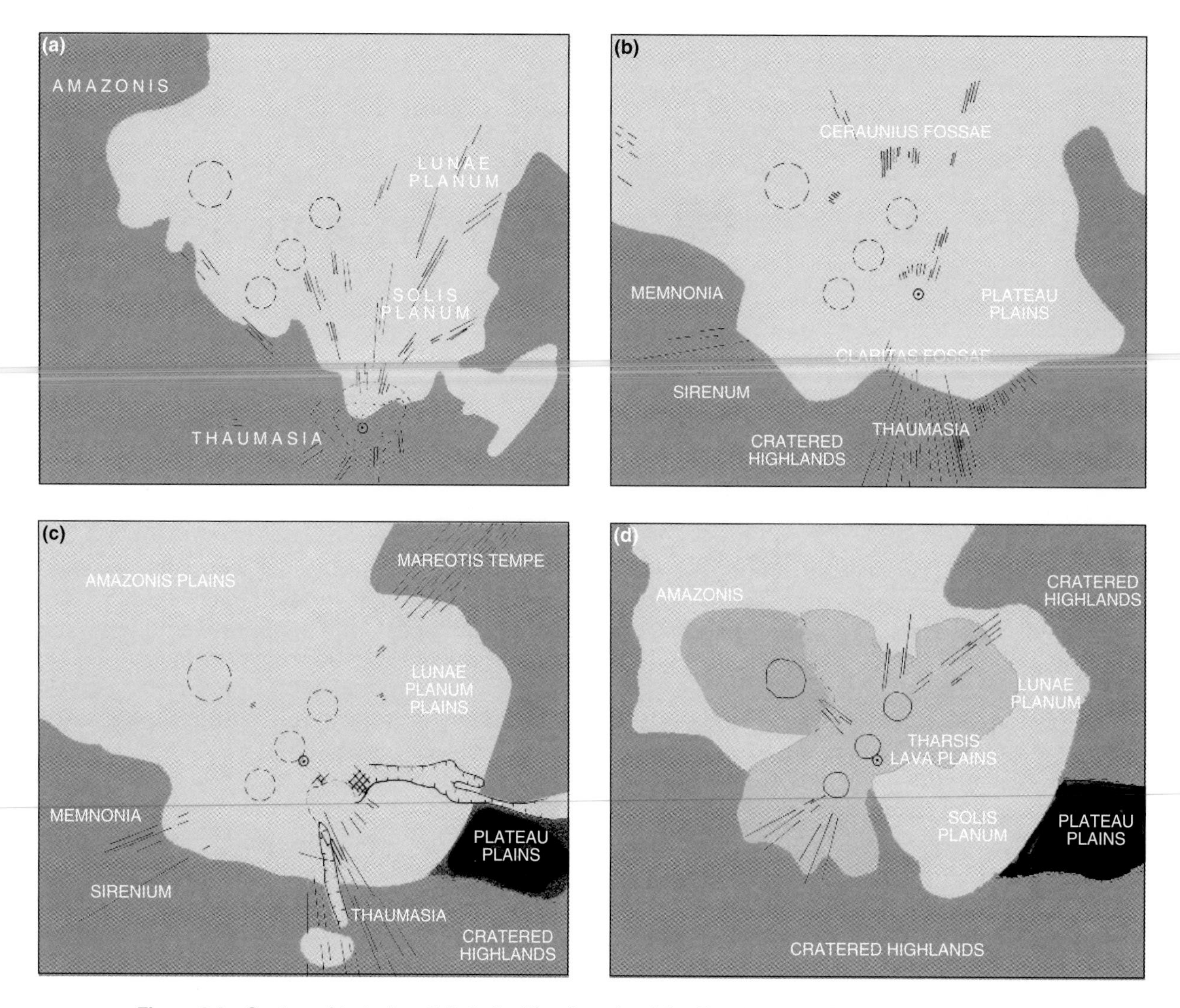

Figure 9.1 Centres of tectonic activity in the Tharsis region (after Frey 1979 and Plescia & Saunders 1982).

has shown no support for this particular lowland hemisphere scenario (Acuña et al. 1999). However, the discovery of regular magnetic striping among the Noachian rocks of the southern highlands, specifically in the region of Terra Sirenum, raises anew the possibility of lateral lithospheric motions during the early part of the planet's history (Kerr 1998b). Therefore, plate tectonics cannot be entirely ruled out, at least on early Mars.

Distribution of tectonic features on Mars

One way of seeking to comprehend the tectonic history is to map faults and ridges that outcrop within geological units of known age. One of the earliest structural analyses was that by Frey (1979), who noted that structures near Valles Marineris and Echus Chasma could not be explained by a single-stage uplift of the Tharsis region and, after studying fracturing in the south Tharsis region, he concluded that there had been at least two major upwarping events, the earlier having been centred on the ancient plateau of Thaumasia (Fig. 9.1a). The generally north–south structural pattern was later overprinted by regional Tharsis-related faults, particularly the west–east Valles Marineris system. Frey envisaged the Thaumasia region as an older Tharsis-type crustal dome.

In the early 1980s a further analysis was completed by Plescia & Saunders (1982), who defined four centres of tectonic activity, the earliest being that defined earlier by Frey. The three subsequent foci were northern Syria Planum (Fig. 9.1b), two centres near Pavonis Mons (Fig. 9.1c,d). The second of the Pavonis centres was characterized by fracturing that postdated growth of the large Tharsis shield volcanoes. Their analysis clearly illustrated that each episode of tectonism was associated with a topographically elevated region and

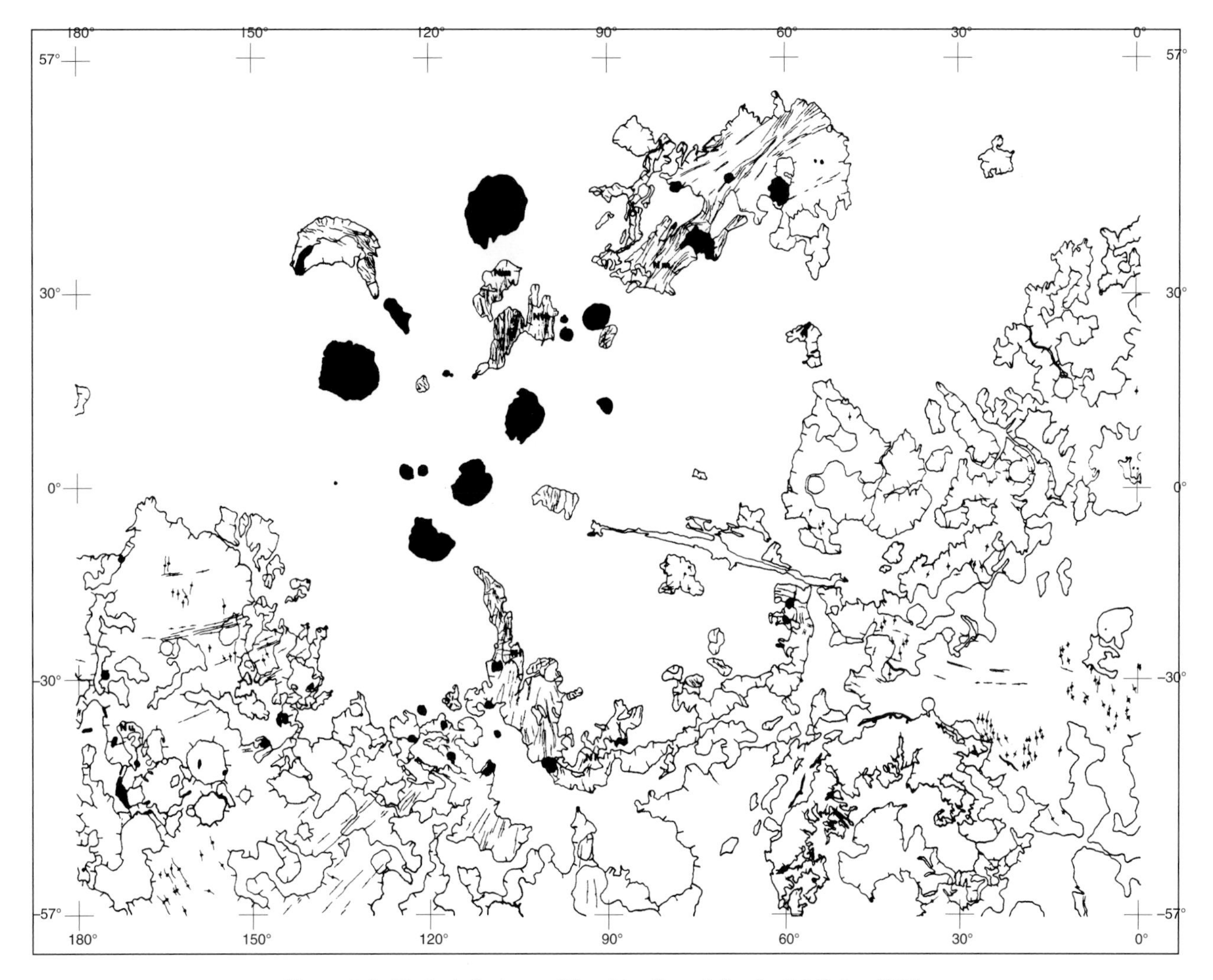

Figure 9.2 Tectonic features of Noachian times (after Scott & Dohm 1990).

that it pre-dated associated volcanism. This implies that fracturing was related to an early stage of the tectonism associated with the development of topographical highs, rather than to loading by extrusive volcanic deposits.

Since the Plescia & Saunders work was published, many new geological maps of Mars have been produced by the USGS. Structural analysis of the western hemisphere of Mars by Scott & Dohm (1990) showed that in Noachian times (Fig. 9.2) the majority of faulting was associated with three foci: the Tharsis Montes with a prominent NE–SW axial trend, the Syria Planum rise (centred at 15°S, 105°W) and the Acheron Fossae structure, which is located north of Olympus Mons. Ridges are widespread but it is not clear whether they are of Noachian age or were emplaced later, during Hesperian times. During the Hesperian period, widespread fracturing continued along the Tharsis Montes axis and in Syria Planum; it was also initiated with a radial pattern at the large volcanic centre of Alba Patera (Fig. 9.3). However, it did not continue at Acheron Fossae. Ridges, most of which pre-date the faults of this age, and which outcrop in an elliptical pattern surrounding major volcanic centres, ceased to form in the Hesperian. There was then a general decline in tectonism during the Amazonian period. However, at Alba Patera some older radial fractures were rejuvenated and concentric faulting was initiated around the summit (Fig. 9.4). Minor faulting also was associated with Olympus Mons and its aureole, some of which postdated several late Amazonian volcanic flows.

The faults so characteristic of Mars are predominantly graben: extensional faults produced by fracturing of the relatively brittle crust. Individual graben vary between 1 km and 5 km wide and may extend, usually as *en echelon*[1] families, for thousands of kilometres. In ancient rock units, several sets of

1. Internally parallel groups arranged in sequence.

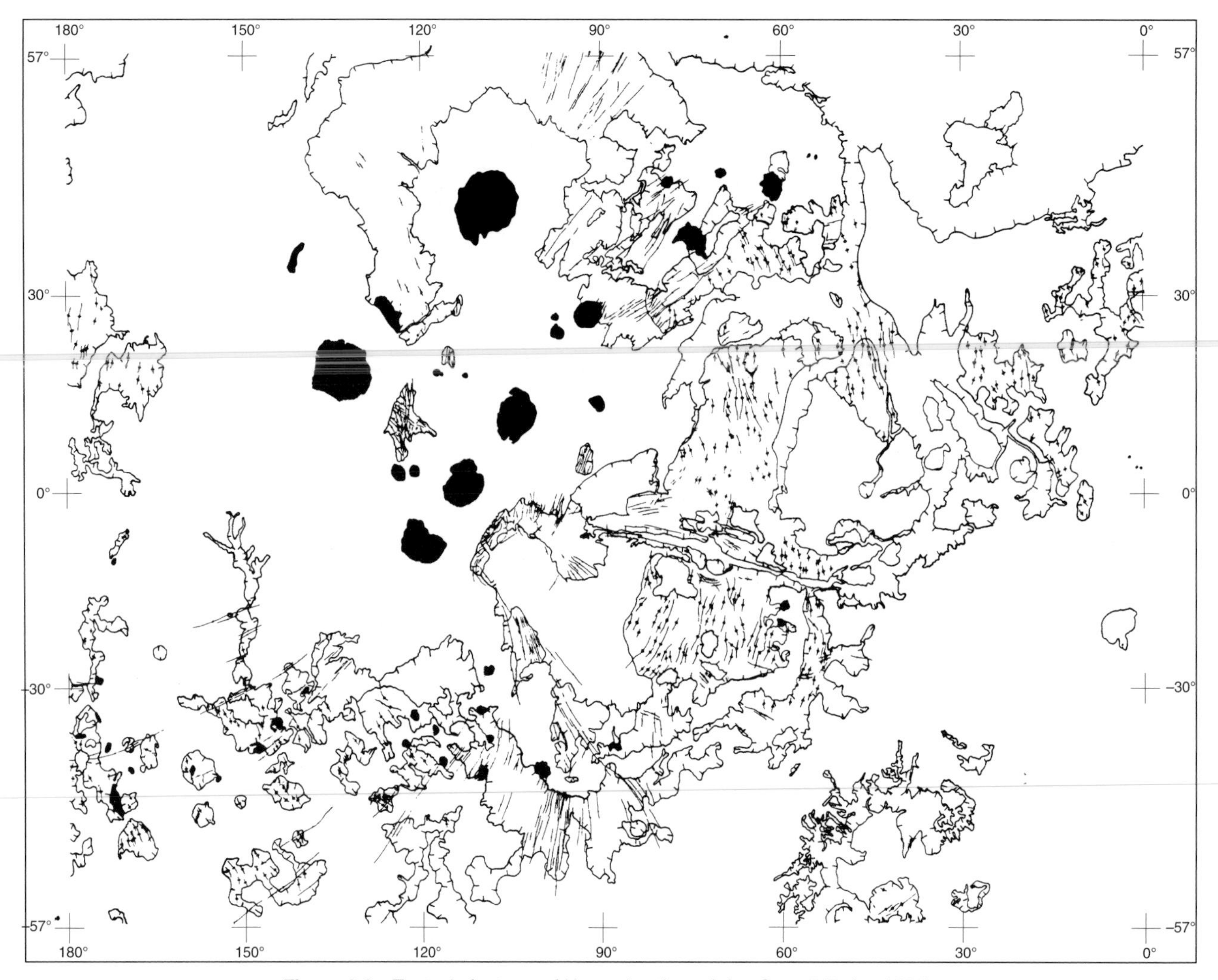

Figure 9.3 Tectonic features of Hesperian times (after Scott & Dohm 1990).

fractures, each of a different age and (sometimes) orientation, break up the surface into complex blocks (Fig. 9.5). The more spectacular of the fracture families are related to major arches in the lithosphere, which, in the case of those associated with Tharsis, spread across a third of the planet.

Faulting in the Elysium region was less intense than in Tharsis. The major episode of fracturing occurred during the Early Amazonian, when Elysium Fossae were produced (Mouginis-Mark et al. 1984). The majority of graben have a pronounced WNW–ESE trend. Subsequently, in the late Amazonian epoch, further fracturing occurred in southern Elysium Planitia, producing long curving WNW–ESE fractures such as Cerberus Rupes. Curving ridges are most prominent in the east of the region, in southern Arcadia; the youngest are of Early Amazonian age.

The extensive graben faults surrounding Alba Patera were analyzed by Turtle & Melosh (1997), who, by mapping the stress field from the fault patterns and using this in tandem with the Goddard Space Center's digital elevation and gravity models, arrived at an estimate of 10–50 km for the thickness of the elastic lithosphere in this region. They also constrained a 1.5–2.0 kbar extensional stress field from the observed strike of the faulting.

The extensive ridge systems, such as those that characterize the ridged plains of Noachian and Hesperian age, are considered to have a compressional origin and to have formed in rocks whose ages range from the Late Noachian to Lower Amazonian. Many geologists consider it likely that they formed preferentially in volcanic flows, as they did on the Moon. They are most readily mapped on the smooth plains of regions such as the Chryse basin, Lunae Planum and Syrtis Major Planitia, but are widespread, if less readily observable, elsewhere, including the northern lowlands. Studies by Chicarro (1989) have shown that the first recorded period of major ridge formation was during the Upper Noachian and that, at this time,

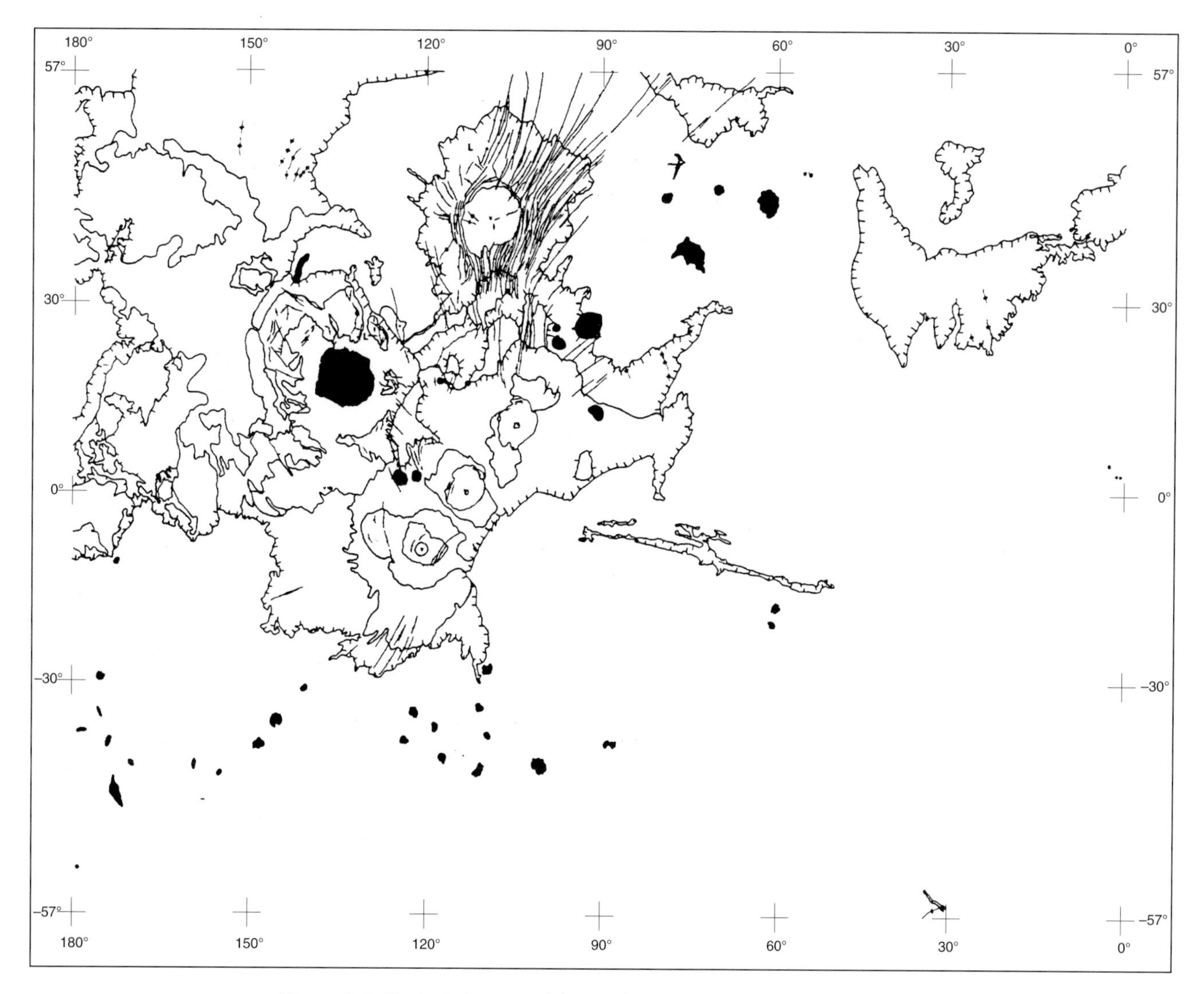

Figure 9.4 Tectonic features of Amazonian times (after Scott & Dohm 1990).

ridges with over 1 km amplitude formed. However, the most prominent ridges are younger and are represented by those that outcrop on Lunae Planum; these are of Lower Hesperian age; indeed, they are of the same age as most of the better-developed ridged plains units in the southern upland hemisphere. This Lunae Planum age is widely used as a basis for comparing crater ages between different rock units on Mars. The development of such ridges appears in some regions to have been related more to localized tectonics than to a global compressive regime; thus, old impact basins and major rises may have dictated ridge orientations at this time. However, mapping by Lucchitta & Klockenbrink (1981) has shown that many ridges are concentric about a focus at 10°N, 112°W, that is, near to the centre of Tharsis (Fig. 9.6). Ridges also formed in the Lower Amazonian but are less prominent than those that were emplaced in the preceding epoch. The number of craters suggests that ridge formation spans the period 3.85–1 billion years ago.

Wrinkle ridges are considered to represent various degrees of crustal shortening. A study of those in Arcadia Planitia by Plescia (1993) indicated that most have a N–S strike but that orientation is strongly controlled by underlying topography, such as isolated knobs and crater walls. In this area the ridges have an average width of 3425 m and accommodate an average crustal shortening of 3 m (caused by folding) and of 55 m (caused by faulting), giving a mean of 58 m. The average total shortening across the three W–E transects studied by Plescia was 0.9 km, representing a regional compressive strain of 0.06 per cent, somewhat less than that observed in Lunae Planum.

Compressional structures also include lobate scarps and high-relief ridges (Watters 1993). Although wrinkle ridges account for about 80 per cent of the total

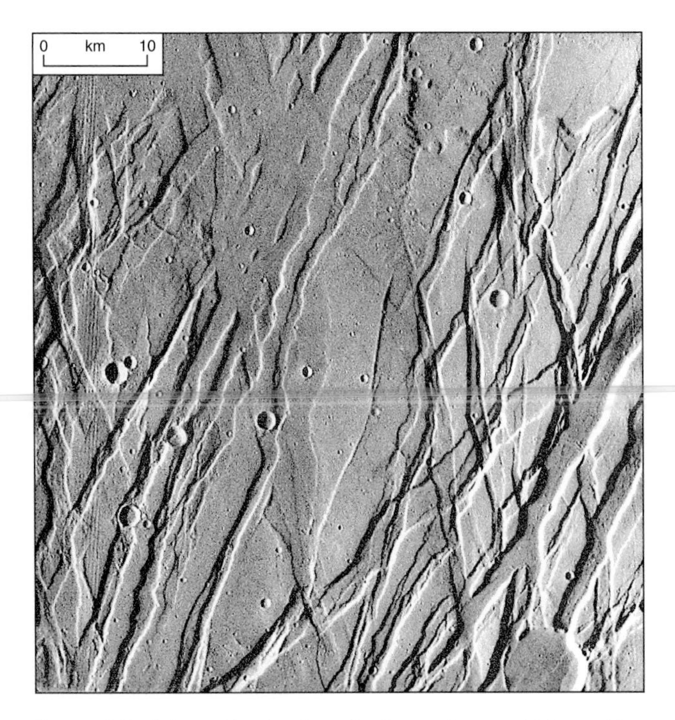

Figure 9.5 Intersecting fractures in Tempe Fossae. Viking orbiter image 627a53; centred at 35.38°N 83.17°W.

cumulative length of recognized contractional landforms, they are confined almost exclusively to smooth plains units. They are generally absent from the highland terrain, where contractional features are either lobate scarps (22%) or high-relief ridges (8%).

The pattern of contractional landforms in the western hemisphere predictably reflects the hemisphere-scale influence of the Tharsis Rise and, although there is no equivalent regional trend in the eastern hemisphere, a prominent regional pattern is found associated with Hesperia Planum. Locally, contractional features parallel the dichotomy boundary in the eastern hemisphere, suggesting the influence of stresses related to the dichotomy itself. Careful study of the ages of the deformed units reveals that the peak of deformation occurred during the Early Hesperian epoch, a time when extensive volcanic resurfacing affected much of the planet. It is therefore possible that the Early Hesperian global volcanism gave rise to a punctuated period of rapid planetary cooling and concomitant global contraction that contributed to compressional tectonism during that period.

Coprates Rise and South Tharsis Ridge Belt

Recent topographic data from the Mars Gobal Surveyor MOLA experiment reveal quite clearly the existence of curving ridge that extends southwards from near Tharsis Montes, passing through Claritas Fossae, bordering Solis Planum on its southern perimeter and then curving northeastwards along the eastern border of Thaumasia (Fig. 9.7). This elevated region on the southern flank of the Tharsis Rise evidently escaped volcanic resurfacing because of its high elevation, and it would appear reasonable to assume that the southern part of the Rise formed, at least in part, by structural uplift. There may also have been some contribution from deformation of the lithosphere beneath this region.

A detailed stratigraphic and structural study made of the Coprates Rise and South Tharsis Ridge Belt indicates that geological units ranging from Middle Noachian to Early Hesperian age were involved in horizontal shortening of both crust and lithosphere. This generated uplift and asymmetric east-vergent[1] folding of a wide region of stratified deposits. This asymmetric east-vergent asymmetric anticline strikes N–S and extends for at least 900 km south of Coprates Chasma (eastern Valles Marineris), comprises the Coprates Rise, which is built from Middle to Upper Noachian intercrater plains that, as we have stated, were deformed during the Late Noachian and probably into the Early Hesperian. The Rise is embayed by ridged plains at its margins and defines the eastern edge of the Thaumasia Plateau. Analysis of structural patterns within this region shows that normal faulting could not have been the prime cause of the Rise; rather it was the folding that was associated with buckling instability, possibly exacerbated by thrust faulting.

The Rise is part of an arcuate belt of ridges and scarps that extends over a region measuring 2500× 6000 km. Within this, ridges are spaced 150–500 km apart and are related to deformation associated with an early stress centre located at 20–25°S, 90°W in Solis Planum. Topographic profiles across the northern part of the Rise indicate that the eastern flank is steeper than the western, and its sequence of tilted strata is thicker, indicating asymmetric folding (Fig. 9.7). The plethora of N–S-striking arches within Thaumasia that parallels the margins of the Rise may have been formed contemporaneously with it. During the deformation the layered rocks were also intensely fractured in places, for instance in the area of Nectaris Fossae. These fractures are normal faults, graben and minor rifts with a mainly W–E trend, and are mostly limited to the Rise, not cutting the adjacent ridged

1. A term from structural geology. In every fold there is a plane bisecting the structure (axial plane). The direction in which the upward-pointing end of this plane points is known as the vergence. Most folds are asymmetric and therefore the vergence is seldom vertical, but usually points in a specific direction.

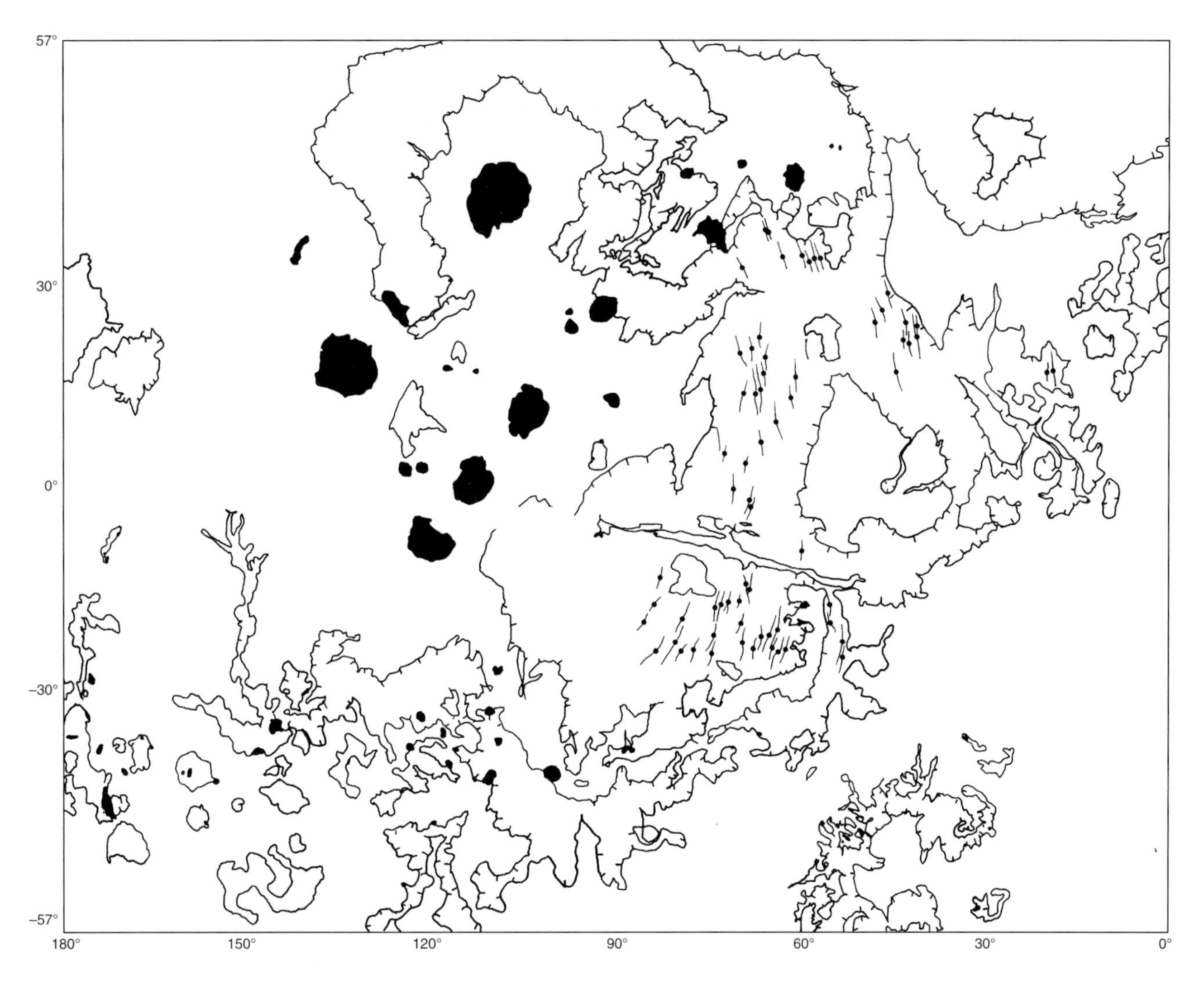

Figure 9.6 Orientation of ridges in Hesperian units surrounding Tharsis. These appear to be roughly concentric about a focus at 10°N 11°W. Central eruptions are shown as solid black areas. (After Scott & Dohm 1990.)

plains. During the early part of the Hesperian epoch, west–east compression generated north–south-striking wrinkle ridges, mainly in ridged plains units. During this period, arching of the Coprates Rise may have continued, elevating and folding what may be the lowermost strata of the ridged plains units.

Schultz & Tanaka concluded that the Rise was formed by 2–4 km uplift of units ranging from Upper Noachian to Lower Hesperian age, giving rise to a broad asymmetric anticlinal fold. By using a dislocation analysis approach, they rule out normal faulting as a means of generating the structure. If thrust faulting had been responsible, it would have required a hypothetical thrust 100 km deep and around 175 km to the west. This would have required an elastic lithosphere at least 100 km thick in South Tharsis during Noachian times. In acknowledging that such a process may have contributed to crustal arching, on the basis of the quasi-periodic spacing of the Coprates Rise and its associated ridges, Schultz & Tanaka prefer the notion that folding was the principal cause.

To the west of the Coprates Rise lies the Thaumasia Plateau and Daedala Planum, the latter being composed largely of young volcanic flows from Arsia Mons; together they comprise a large volcanotectonic province of Noachian to Lower Hesperian age. The Thaumasia province is characterized by normal and graben faulting, wrinkle ridges and broader crustal arches. In a search for broader signs of contractional processes, Schultz & Tanaka discovered 29 large-scale ridges and scarps of possible contractional origin and found them to be generally concentric to south Tharsis. They grouped this family of major structures into what they term a South Tharsis Ridge Belt, within which most of the ridges deform Noachian units and are embayed by Lower Hesperian smooth ridges and

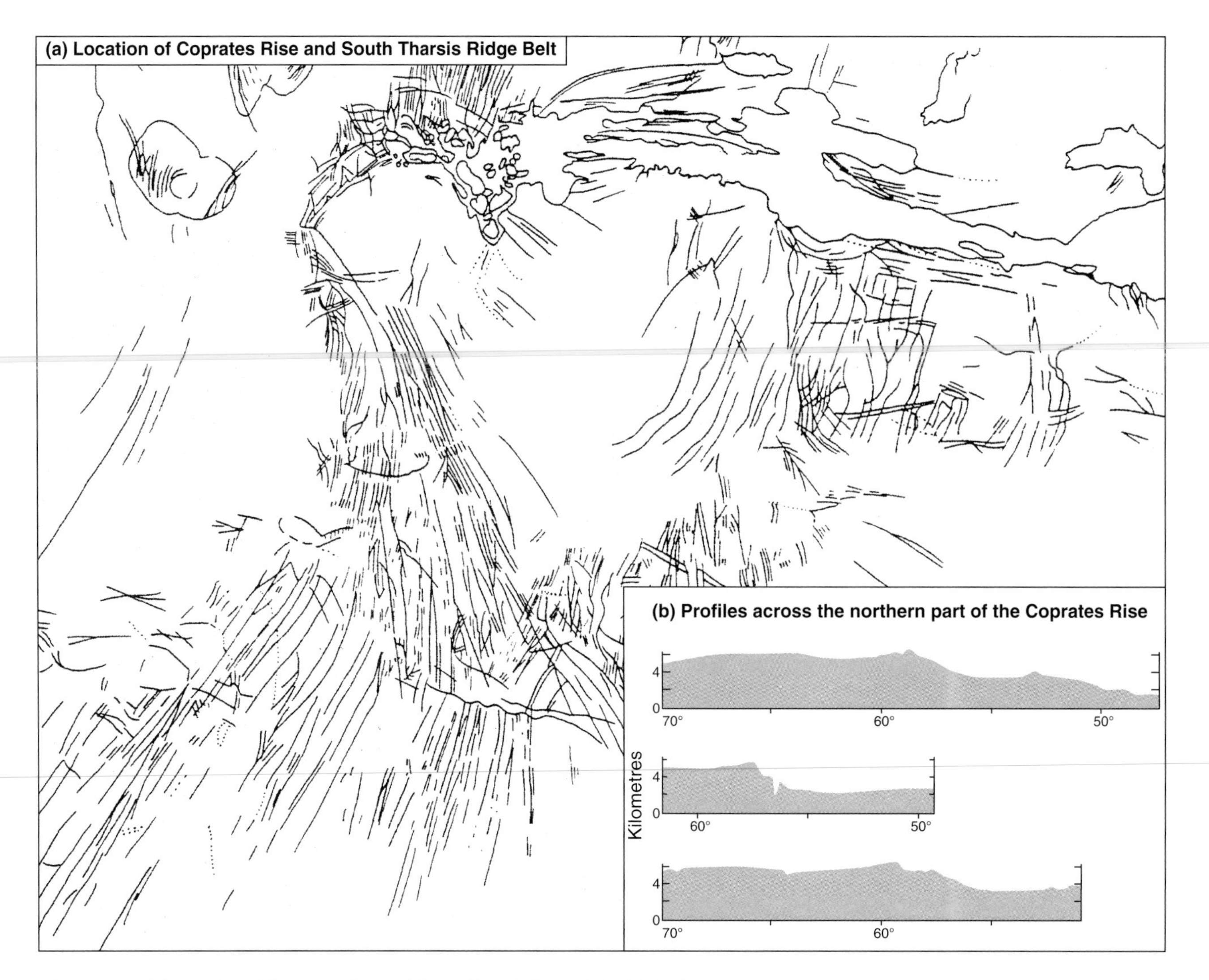

Figure 9.7 **(a)** Location of Coprates Rise and South Tharsis Ridge Belt. **(b)** Profiles across the northern part of the Coprates Rise (after Roth et al. 1980).

fractured plains. In consequence, most of the ridge belt must be of Noachian age. Individual structures are high and tens of kilometres across, and are bordered and crossed by crested ridges 5–10 km wide, also high sinuous scarps. Overall the ridge-type structures are separated by 150–500 km and have heights from 150 m to 2 km.

The large-scale spacing of the ridges is in accord with the 100–1000 km wavelength range predicted by Zuber & Aist (1990) for whole-lithosphere folding on Mars. Because horizontal shortening of the lithosphere as a result of compression would be expected to thicken it to some degree, the average elevation of shortened regions may be expected to have increased. This would apply to Thaumasia (which is isostatically compensated) and South Tharsis generally, as well as regions such as Terra Sirenum. This prediction, together with the ridge morphology and its association with the Coprates Rise and smaller-scale wrinkle ridges, suggests that the South Tharsis Ridge Belt formed by a combination of buckling instability and thrust faulting. Since the Coprates Rise does not continue north of Valles Marineris, it is reasonable to conclude that crustal and lithospheric properties (e.g. stratification and stress state) were very different beneath Lunae Planum and Thaumasia while the ridge belts were being generated. Also, the very different scales of the smaller wrinkle ridges with the South Tharsis Ridge Belt and the more widely spaced major ridges and arches suggest that, whereas the latter are related to whole-lithosphere deformation, the former may have been in part a response to decoupling between shallower and deeper layers, perhaps along a weaker layer such as a brecciated mega-regolith.

What general significance has all this to the evolution of Tharsis? First, the ridge trends within the

South Tharsis Ridge Belt are consistent with their being associated with a volcanotectonic focus located in Solis Planum, as was established by Frey (1979) and Plescia & Saunders (1982). The ridges of Terra Sirenum, on the other hand, appear related to another centre, located closer to Syria Planum. The north-northeast strike of the Coprates Rise and adjacent ridges and valleys implies lithospheric shortening that was orientated in an east-southeast direction. In broad terms, this pattern is consistent with the maximum compressive stress pattern derived by Banerdt et al. (1982) for isostatic adjustment of the lithosphere in response to a Tharsis-centred load.

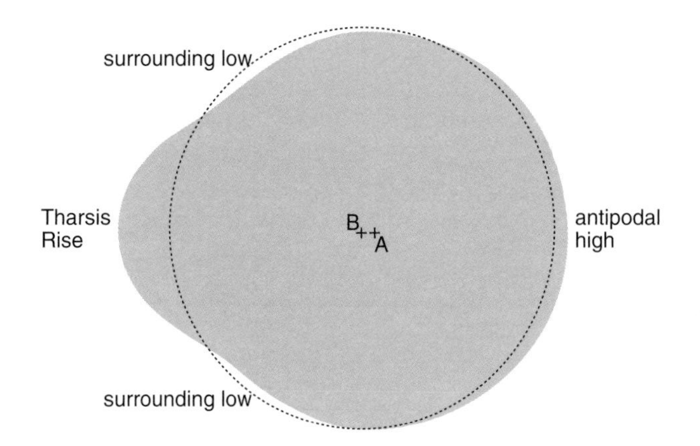

Figure 9.8 Equatorial cross section of the Martian geoid.

The gravity field of Mars

Recently, gravity data have been extended and refined by results from Mars Global Surveyor. Gravity varies from place to place on the Martian surface, as it does on Earth; this reflects the asymmetric distribution of mass within the planet's crust and mantle. One of the best ways of describing the shape is by the geoid; on Earth this is a level surface that approximates closely to the ocean surface.

The Martian geoid

Mariner 9 and Viking gravity data were analyzed by Balmino et al. (1982). Because of an unequal mass distribution within Mars, it was established that the geoid departs from the ideal fluid ellipsoid, in some places rising above it and in others falling beneath it. Thus, there is a marked geoid high associated with Tharsis, which indicates that isostatic compensation at shallow depths is not complete. The fact that the geoid high is shifted somewhat to the north of the equator, whereas the topographic high is to the south, is explicable by noting that the Tharsis Rise straddles the upland/lowland boundary, and so the load is a few kilometres thicker to the north. Another high is antipodal to Tharsis, and a geoid low surrounds it. If we consider an equatorial cross-section of the geoid (the solid line in Fig. 9.8) the explanation for the surrounding low and antipodal high becomes clear: beyond Tharsis, the figure of the geoid approximates to a circle centred at point A. On the other hand, the dotted circle centred at point B represents the equatorial cross section of Mars without the Tharsis Rise. The displacement of the centre of the circle from point B to A results in a geoid low surrounding Tharsis (the part of the section where the dotted line is outside the solid one) and a geoid high opposite to it (where the solid line is exterior to the dotted one).

A more recent study, using Doppler tracking from three orbiting spacecraft, honed the model, showing that there is good accord between the main topographic features of Mars, such as volcanic shields, impact basins and Valles Marineris (Smith et al. 1993). Most of the major topographic features exhibit greater anomalies than in earlier models; in particular it became possible to resolve all three of the Tharsis Montes. Furthermore, more detail was revealed for the equatorial canyon system and for both Isidis and Argyre. However, the hemispheric dichotomy has no discernible gravitational signature. A harmonic analysis of topography by Bills & Nerem (1995) confirmed that these data provided a reasonable approximation to the true topography of the planet.

Mars Global Surveyor gravity data

The radio science investigation undertaken by Mars Global Surveyor provided, for the first time, global high-resolution gravity models for Mars from X-band Doppler tracking data (Smith et al. 1999b). It provided new insights into the way in which Mars has responded to such events as major impacts, volcanism, erosion and internal changes brought about by varying mantle dynamics. Plate 2 shows the new data.

The largest free-air gravity anomalies are associated with Olympus Mons and the three Tharsis Montes; each exceeds 3000 mgal, which is more than an order of magnitude greater than those of Earth for similar wavelengths. This implies that Mars' lithosphere is able to support the large stresses associated with major surface and subsurface loads with relative ease. This is presumably a function of the relatively rapid loss of planetary heat from Mars as compared with Earth, because of its smaller mass. One very striking aspect of the gravity field is the limited range of anomalies over a large portion of the planet: large anomalies are associated only with Tharsis, Elysium and Isidis. A broad gravity high (~100 mgal) centred on Hellas is presumed to be related to the excavation of crustal material during impact (Plate 2a).

The areoid (gravitational potential) of the planet shows two hemisphere-scale features of more than 2 km dynamic range (Plate 2b). The one associated with Tharsis is virtually circular, which contrasts strongly with the complex physiography of this region. The core of the Tharsis areoid includes the Tharsis Montes inside a single major anomaly, but Olympus Mons is isolated to the northwest, whereas Alba Patera has a more subtle signature to the northeast. Such an arrangement is consistent with the topographic isolation of each of the three centres and it implies that each had a separate mantle source region.

Zonal variations in the anomalies also reveal differences between the upland and lowland hemispheres. The absence of gravity anomalies beneath major topographic features in the southern hemisphere contrasts strongly with the presence of substantial anomalies in the north, but without topographic expression. Evidently the southern uplands are isostatically compensated, akin to the condition of the lunar anorthositic[1] crust, having had much more time to equilibrate than the younger northern regions. Recent magnetization data support this notion. There is also an increasing gravity variation from south to north, suggestive of a latitude-dependent variation in isostatic compensation, and probably implying a decrease in crustal thickness from south to north, along with the pole-to-pole topographic slope. The gravity signature differs significantly between the hemispheres, which leads to the conclusion that the lithosphere of the northern hemisphere is thin but strong and that of the south is thick but weak.

A feature of the gravity signatures of major impact basins in the southern hemisphere is that they exhibit primary negative anomalies with an annular form, and small central positive anomalies. This is similar to lunar impact basins that were not infilled by mare basalts. The small central anomaly is presumably associated with mantle uplift, and the annular negative anomaly with a mixture of uncompensated crustal thickening and lithospheric flexure.

A substantial anomaly is also associated with the impact basin of Isidis, which shows a strong positive anomaly surrounded by a negative annulus – in contrast with the southern hemisphere situation – and illustrates a greater flexural response of northern lithosphere to loading. The Elysium rise and Olympus Mons exhibit similar patterns and each deviates significantly from isostatic equilibrium, implying that they are relatively young and continue to be supported by a mechanically strong crust.

The deep canyon system of Valles Marineris is characterized by the largest negative anomaly on the planet; it is flanked on each side by gravity highs (see Ch. 10). Such an arrangement is fully in accordance with the large mass deficit expected beneath this vast depression, and is in accord with the fact that it is not yet isostatically compensated. On this basis, the canyons are seen to be younger those of Argyre and Hellas.

The negative anomaly associated with the equatorial canyon system can be traced into the region of Chryse. Proposed as an ancient impact basin by Schultz & Frey (1990), Chryse lacks any mass anomaly and any hint of a circular topographic low, and is therefore atypical of northern basins. The 200 mgal negative anomaly is taken by Smith et al. (1999b) to indicate the removal of more than 2 km of crustal material by the fluvial processes that carved the huge outflow channels during the early stages of Mars' history.

When the gravity and topographic data from Mars Global Surveyor are combined, they reveal that the flat northern lowlands may well have been a region of elevated heat flow early in the planet's history, which subsequently became an area of rapid water accumulation. The latter was hinted at only as recently as March 2000, when the probe revealed what seem to be huge buried channels that extend well into the northern lowlands from north of the dichotomy boundary. On average, these are about 1190 km wide and 1500 km long, and have long since been buried by thick accumulations of sedimentary rocks. The difference between the two hemispheres is also brought out in a recently compiled crustal thickness map that shows the 75 km-thick southern hemisphere crust thinning regularly towards the northern lowlands, the latter region and Arabia Terra's crust having a thickness of around 35 km.

The evolution of Tharsis

The oldest unambiguous structures identifiable as having a link with the Tharsis Rise are certain circumferential faults that outcrop in the Claritas Fossae region on the southwest side of Syria Planum (Phillips et al. 1990). Since these are Noachian, the rise must be as old as this. However, it should be noted that, although flexural uplift may have been responsible for the Claritas Fossae structures, it does not necessarily imply that the subsequent massive elevation of the Tharsis region was largely a response to such a phenomenon. For instance, it could equally well be that major volcanic construction preceded fracturing.

1. Anorthosite is a crystalline igneous rock composed mainly of calcium-rich plagioclase feldspar.

However, it is clear that uplift of Tharsis continued until at least the Late Hesperian and possibly the Early Amazonian period.

Any successful model for the evolution of Tharsis must incorporate the notion of lithospheric compression occurring beneath the region during Noachian and Early Hesperian times. It must also satisfy the gravity data. In this context, the Coprates Rise may best be interpreted as a Tharsis-related structure that was produced as part of an early stage in the region's deformation. During the same period, the elevated region of Syria Planum – which lies some 2500 km west of the Coprates Rise – experienced local normal faulting, along with extensive magmatism. Volcanism also characterized much of the South Tharsis Ridge Belt at the same time. It is uncertain whether the two regions were at this time linked dynamically, but there can be no doubt that they were regions of enhanced heatflow and active tectonism during this period. The development of ridges in Arcadia Planitia, on the western edge of Tempe Terra, around Hellas and in southeast Isidis Planitia, also took place during Noachian and Early Hesperian times. Evidently this was a period during Mars' early history that was characterized by crustal and lithospheric arching, probably accompanied by thrusting.

One obvious process that could have generated the Rise is some form of mantle convection; in actively uplifting the Martian lithosphere it would also be expected to produce widespread fracturing. Convection could have been a response to radioactive heating or caused by lower-density mantle material beneath Tharsis, produced as a result of inhomogeneities inherent in the accretional process. To what extent such conditions were engendered by the antipodal impact at Hellas is unclear. Kiefer & Hager (1989) suggest that the Tharsis and Elysium rises were supported by internally heated convection, which generated narrow plumes. Thus, there might be one or two major plumes beneath Tharsis, plus a few smaller ones.

Willemann & Turcotte (1982) demonstrated that the lithospheric stresses needed partially to support the mass of Tharsis are reasonable and consistent with the observed radial fracture pattern. Their model predicts that the thickness of the elastic lithosphere ranges from 110 to 260 km, whereas the thickness of the Tharsis load itself is 40–70 km, which is in broad agreement with the figures arrived at by Comer et al. (1985). Their modelling indicates that near the perimeter and beyond Tharsis the stress at the top of the Martian lithosphere indicates a horizontal deviatoric stress (deviating from equilibrium) near the surface, which reaches a maximum between 30° and 60° from the centre of loading, which is therefore where fracturing may be expected. Linear deflection theory predicts that, if Tharsis were purely a dome in the elastic lithosphere, faults in this region should be thrust faults. Since they are not – they are graben – it appears to preclude Tharsis being simply a lithospheric dome. However, this does not preclude the idea that Tharsis can be explained as a crustal dome over a downward deflection of the lithosphere.

Banerdt et al. (1982) ruled out the possibility that Tharsis was supported purely dynamically, since such a hypothesis would require that west–east trending grabens should characterize the highest parts of the rise, whereas the observed faults are radial. Furthermore, in an isostatic configuration, grabens will tend to form on, and be confined to, topographic highs and, where associated with such highs, there are preferred directions of maximum gradient, so grabens will form normal to these directions. On this basis, they should trend either north–south or, in the north of the Tharsis region, NNE–SSW. Since this condition evidently prevailed in the earlier stages of its evolution, an isostatic regime must have been eatablished during the early stage of its development. They propose the following scenario:

- *Stage* 1 Tharsis began to be built by a combination of constructional volcanism and isostatic uplift. Most of the differentiated igneous materials remained above their source regions (Finnerty & Phillips 1981, Finnerty et al. 1988). Because mass was neither added nor removed from Tharsis, an

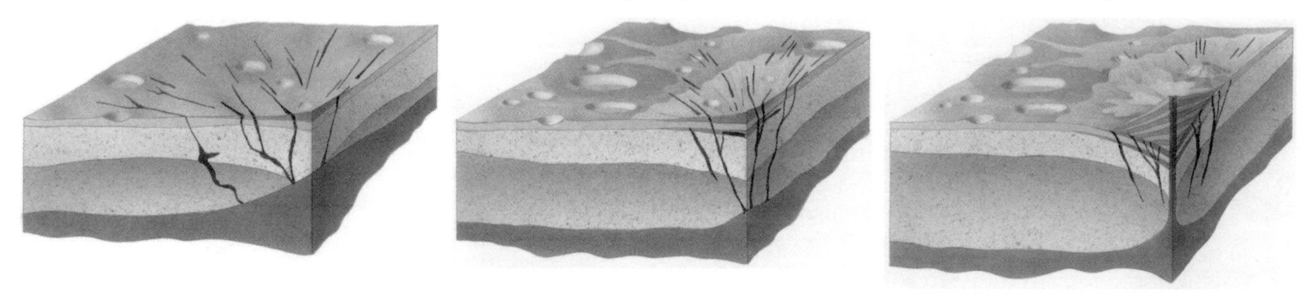

Figure 9.9 Tharsis evolution according to Solomon & Head (1982).

isostatic regime prevailed; however, because isostatic stresses are proportional to topography, at some stage they exceeded the finite strength of the near-surface rocks, with the result that grabens formed in the elevated regions.

- *Stage 2* The lithosphere became thicker and reached the point where complete isostatic compensation of the growing volcanic pile was not achieved. Furthermore, volcanic materials may have been extruded from regions not beneath Tharsis itself, increasing the likelihood that non-isostatic conditions held. At this stage the volcanic load began to cause flexure of the lithosphere. Eventually, the load increased to such a degree that failure took place, and graben faults radial to Tharsis were generated.
- *Stage 3* By this time the lithosphere had become so thick that addition of increased volcanic loading had little if any effect on the stress levels: no further regional failure occurred. This phase is represented by the youngest volcanic flow plains and shield volcanoes. As we have noted earlier, these developed their own local stress regimes.

An alternative is to view the Rise purely as a massive volcanic pile, an hypothesis published by Solomon & Head (1982). On this basis, because the primordial Martian lithosphere was laterally inhomogeneous, global and local stresses would have been caused preferentially where there was thin lithosphere, beneath Tharsis and probably Elysium. Once fractures had been propagated there, magma would be able to reach the surface and, once a high flow of thermal energy was established, it would have maintained relatively thin lithosphere beneath the sites of active volcanism and would have had the additional effect of concentrating fracturing (Fig. 9.9). During the Early Hesperian, with the onset of volcanism in the Tharsis region, the lithosphere probably remained relatively thin, whereupon the response of the lithosphere to volcanic loading would have been indistinguishable from local isostatic compensation by crustal subsidence. However, as Mars cooled so the elastic lithosphere's thickness increased, as did the amount of topographic relief and volcanic loading that could be supported by it. Eventually, a stage was reached when the thickened lithosphere could support at least some of the loading by regional flexure and the increased strength of the lithosphere itself.

This model scores over several of the others since it requires no abnormal chemical or dynamical properties to be sustained for lengthy periods in the Martian mantle. It also predicts that the remnants of ancient cratered terrain that outcrop on the surface of the Tharsis Rise must also be of volcanic origin, since they were an integral part of the growing volcanically generated dome.

Further constraints will doubtless be imposed on many of the scenarios proposed. More and more refined data are arriving all the time, as the current spate of orbiting and landed spacecraft continue to operate. However, one fact that cannot be disputed is that the mechanism that produced this major feature of Mars is of equal importance to the convection pattern that drives Earth's plate tectonics and it played a vital role in determining the development of the planet.

10 Valles Marineris

Mars' equatorial canyon system is over four times deeper, six times wider and at least ten times longer than Earth's Grand Canyon. In scale and form it is more closely comparable to Earth's East African Rift Valley than to the Colorado River's impressive cleft. Complex in form, it is a maze of frequently interconnecting canyons that straddle the Martian equator between longitudes 30°W and 110°W (Plate 14). With a total length of approximately 4500 km, it stretches for nearly a quarter of the circumference of Mars; individual canyons are over 200 km wide. Its western extremity is located among some of the highest plains on Mars, on the flanks of the Tharsis Rise. The deepest sections plunge 11 km below the high plains. At its eastern end it merges into an enormous region of chaotic terrain located between Chryse Planitia and Margaritifer Sinus, and thence into the Simud Vallis outflow channel system. It is without doubt unique in the Solar System, in terms of both scale and physiography.

Early ideas about the origin of the canyon system tended to focus on erosion by water or thermokarst[1] activity (McCauley et al. 1972, Sharp 1973a), but such an origin is now less favoured. Genesis via collapse following withdrawal of subsurface magma was cited by McCauley et al. (1972) and Schonfeld (1979), and into voids created by tensional fracturing by Tanaka & Golombek (1989). Major objections have been raised against both of these ideas. Currently, the most favoured hypothesis is that the chasms were rift faults subsequently enlarged by collapse and erosion (Plescia & Saunders 1982, Sleep & Phillips 1985, Lucchitta et al. 1994).

1. A type of landscape production involving the melting of frozen groundwater in periglacial climates.

A recent study by Anderson & Grimm (1998) of gravity–topography relationships indicated that the thickness of the elastic lithosphere was less than 30 km, corresponding to a heatflow greater than 20 mW m^{-2}, at the time the canyon system was formed. This places some useful constraints on its formation, and the authors conclude that, despite the apparent absence of distinct faulting within the system, a wide-rift (as opposed to narrow-rift) model is better supported by the data, and that early downfaulting localized later erosion and collapse, which modified and enlarged the system.

Physiography and topography

The topography of the Valles Marineris system, together with implications for its erosional and structural history, was discussed by Lucchitta et al. (1994). They combined a simplified geological map with the digital terrain model of Wu et al. (1991), noting depths of the order of 11 km in the central canyons. More recent MOLA topographic data (Smith et al. 1999b) confirms that the deepest section lies in the central part (in Coprates Chasma), plunging to 11 km below the outer plains surface around 300°E. Here also the floor elevation is at its lowest in absolute terms, at ~5 km below datum (Plate 14). This new information also allows accurate determination of the gradients of the floor, which, in the region of Coprates Chasma, slopes gently up hill towards the east at an angle of ~0.03°, a gradient that is nearly constant for a distance of about 1500 km. An interesting implication of this is that the uphill slope would have inhibited the passage of water eastwards towards the chaotic terrain and outflow channels that flowed towards Chryse, unless the water depth had been sufficient to overcome a relief difference of ~1 km.

The general physiography of the canyon system is depicted in Figure 10.1. In the west, near the focus of a plethora of radiating fractures that centre on the Tharsis Rise, and beginning about 600 km east of the summit of Arsia Mons, is the region incised by relatively short and often interconnecting canyons called Noctis Labyrinthus. This extends over an area of at least 400 000 km^2 and borders the northern side of Syria Planum. The much fractured ground into which the labyrinth is incised is embayed and sometimes overrun by Upper Hesperian volcanic flows of the Syria Planum formation and is therefore older than this. The fracturing continues along the western side of Syria Planum and towards Solis Planum as a group of NNW–SSE graben called Claritas Fossae. To the north, fractures trend just east of north but are soon obscured by the volcanic flows associated with the Tharsis Montes.

Noctis Labyrinthus gives way eastwards to the central section of the system. This comprises a series of impressive parallel-sided canyons that extends for over 2400 km and which, in places, is 700 km wide. Recent MOLA data indicate that within this central section lies the deepest part of the system, plunging to 11 km below datum. The general parallelism of the canyon walls and pervasive W–E to WNW–ESE trend of the system as a whole implies a development along major extensional faults. Farther east, the nature of the canyons changes, defining the third geomorphological section. Here the parallel-sided canyons give way to less well defined depressions with increasingly hummocky floors. Typical of these are Eos and Capri Chasmata, which have distinctly scalloped walls and blocky floors, and merge eastwards into chaotic terrain that extends ENE towards Aureum Chaos. North of them, and separated by a remnant of Lower Hesperian plateau sequence 200 km wide, lies Ganges Chasma, on whose floor are several spectacular landslides. This likewise merges eastwards into a large area of chaotic terrain (Fig. 10.2).

Noctis Labyrithus is unlike the remainder of Valles Marineris, having a maze-like appearance and comprising a group of intersecting canyons, the position and orientation of which must have been controlled by sets of intersecting fractures. The latter presumably developed in response to the doming that accompanied the growth of Tharsis, and which elevated the region 10 km above datum. In places the faults that controlled canyon development can be observed intersecting the walls, usually at angles close to vertical. Individual canyons tend to be narrow and rather short. Furthermore, because the controlling faults define an intersecting network, the canyons break up the high plateau surface into a region of massive blocks. Towards the west the labyrinth is composed of rows of coalesced pits, individual depressions often having different floor levels. This suggests that canyon growth was accounted for largely by subsidence as opposed to longitudinal removal of debris by fluvial runoff or ice erosion. Towards the east, pits tend more often to coalesce and ultimately join with canyons of the main section.

East of Noctis Labyrinthus two major canyons bound Sinai Planum on the north side. These are Ius Chasma to the south and Tithonium Chasma to the

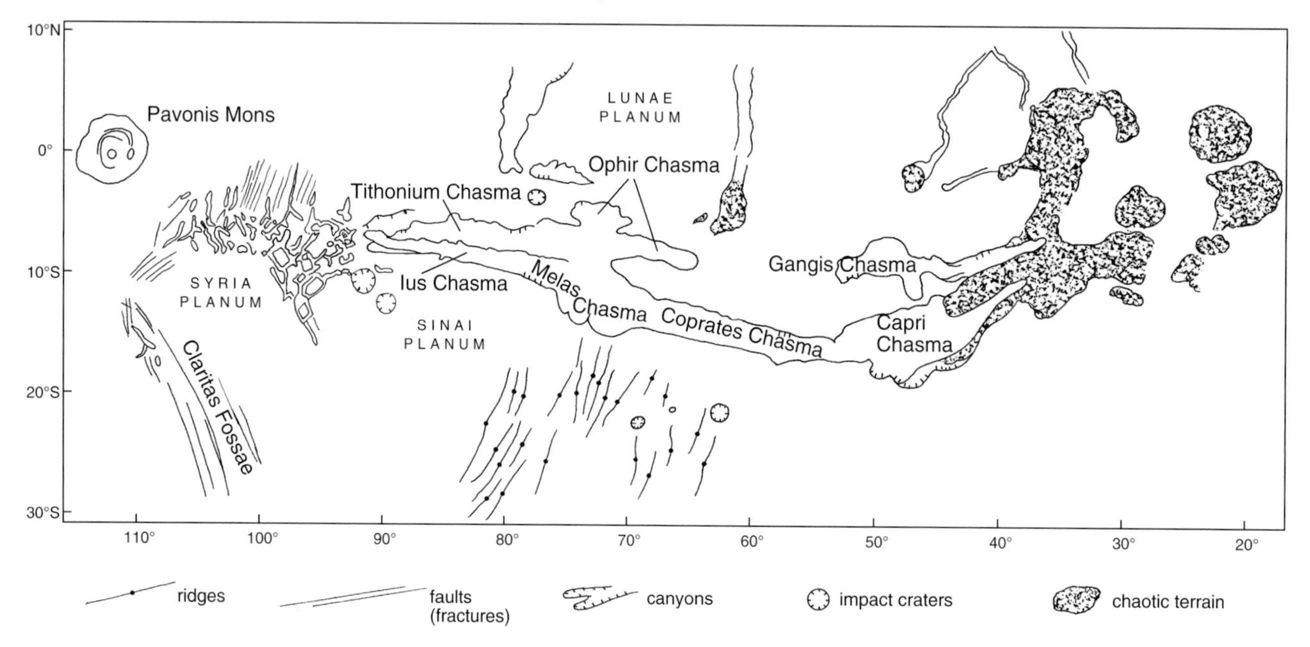

Figure 10.1 Generalized map of Valles Marineris.

north. The Hesperian plateau surface on both sides of each is incised by many graben and pit rows that share the same trend (Fig. 10.2). Sections of these are often offset by transverse faults, some of which are themselves graben. This central 2400 km-long section is characterized by east–west trending straight-sided canyons, and within it a variety of landforms have developed, including flat-floored graben, graben with subsidence pits, chains of coalescent depressions and parallel-walled canyons.

The more northerly of these two canyons forms a single trough and eventually narrows towards the east, until at about 78°W it is separated from the broader central section by a line of large depressions. Its more southerly counterpart comprises two troughs separated by a ridge; towards the east, each of these becomes wider and eventually the two merge into the broad Melas Chasma.

Melas Chasma forms a part of the most continuous section of the system, passing laterally into Coprates Chasma, which in turn passes into Eos Chasma, the last merging with the chaotic terrain that develops from Valles Marineris at their eastern end. Between 65°W and 80°W, Valles Marineris are at their widest (Fig. 10.3). Here, Melas, Candor and Ophir Chasmata run in parallel, forming a broad and deep depression 600 km wide. Candor and Ophir Chasmata are closed at both ends, as is Hebes Chasma, the more northerly depression, which is incised into the south side of Lunae Planum. On the plateau surface beyond their perimeter, the structural trend is continued by lines of partially coalescent pits, which is further evidence of a strong underlying structural control on canyon formation.

Separating Ophir, Candor and Melas Chasmata are steep ridges, which may be partially breached; the inter-canyon rocks appear to be remnants of the ancient plateau surface. A particularly prominent gap in the south wall of Candor Chasma has allowed debris to flow southwards across the canyon divide and onto the floor of Melas Chasma (Fig. 10.3). The divide between Candor and Ophir Chasmata is built from the older plateau rocks, but stacked up against it and partly burying it are younger, finely gullied, relatively smooth deposits. A mesa-like remnant of similar rocks, located west of the former, exposes distinct layering on its western side. These stratified sequences appear to rest discordantly upon the underlying bedrock and were assigned an Amazonian age by Scott & Tanaka (1986).

Great importance has been attached to study of the layered interior deposits that are a feature of parts of the central canyons and the adjacent troughs of Juventae, Gangis, Eos and Capri Chasmata (Fig. 10.3). These units are often found in the form of erosional remnants (i.e. mesas). They appear to hold an important key to the geological history of the region. Such deposits are not found in Ius Chasma or the northern branch of Melas Chasma, or Coprates Chasma; rather, within the last are found remnants of wallrock and plateau materials, some with old impact craters on their surfaces, which must have been downfaulted from their original high level. Different again are the chaotic floor deposits typical of the troughs that merge into outflow channels (e.g. Juventae, Gangis and Coprates–Eos Chasmata). These will be discussed below. The general geology of Valles Marineris are shown in Plate 10.

Form of canyon walls

The steep canyon walls tend to be relatively smooth but deeply gullied, with many rounded embayments etched into the plateau surface. The shoulders between individual gullies are sharply defined, branching and fluted; they resemble terrestrial escarpments formed largely by dry mass wasting under both desert and glacial climates (Lucchitta 1981). In many places the shoulders are truncated by low fault scarps and, where this occurs, as it commonly does in Coprates Chasma, triangular facets develop. This faulting must have followed gully development, suggesting that tectonic activity accompanied both canyon formation and modification. Talus slopes of 30° form large aprons at the foot of the walls.

Stratigraphic considerations suggest that the bedrocks behind the equatorial escarpments are, in the upper regions, resistant volcanic flows and, lower down, less well consolidated ancient breccias. The entrapment of groundwater or ground ice within the relatively porous breccias may have played an important part in the generation of landslides.

Wall retreat and landslide deposits

Several huge embayments give many sections of the canyon walls a markedly scalloped appearance. This is particularly true of the northern walls of Ius and Tithonium Chasmata, and also parts of Ganges Chasma. Such embayments are attributable to wall failure. The scars often have a somewhat rectilinear form, which is another indicator that existing structural controls dictated canyon geometry. Where sections of the steep walls have failed, massive landslides have left vast alcoves and spread debris over the canyon floors. One particular slide in Ius Chasma is 100 km wide and extends from one wall to the other; others are actually banked up against the opposite rampart.

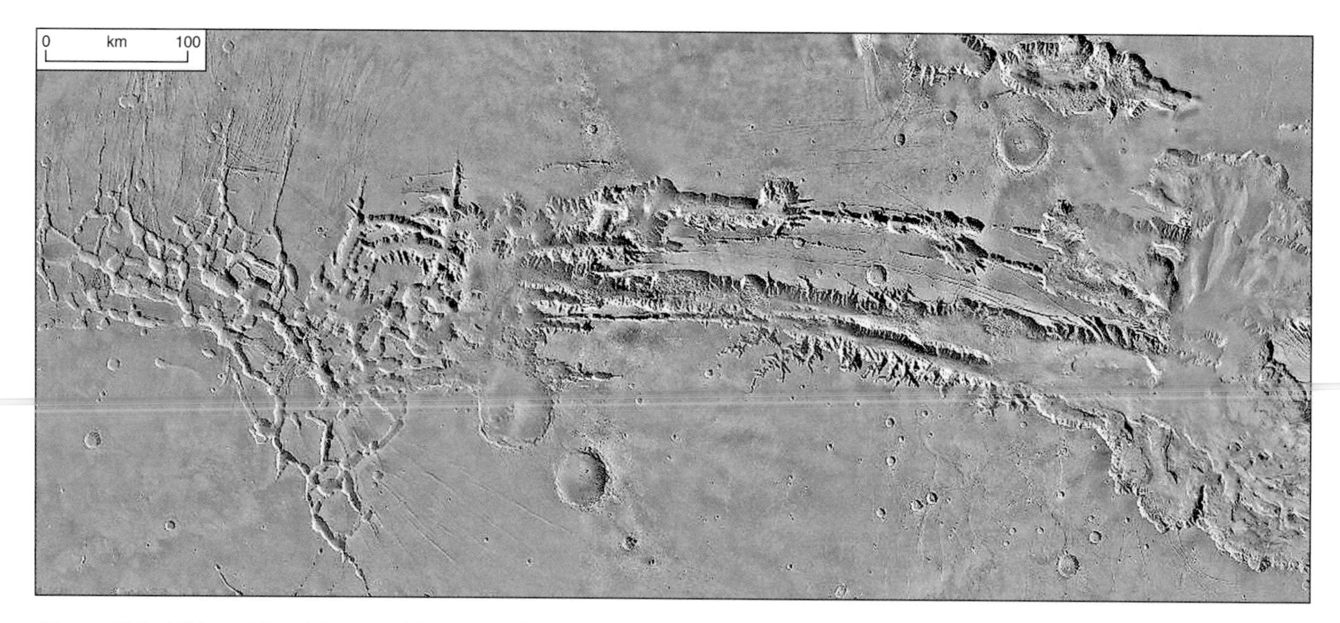

Figure 10.2 Viking orbiter photomap of the western section of Valles Marineris, from Noctis Labyrinthus, through the parallel-sided section of Tithonium and Ius Chasmata, to the irregular Ophir and Melas Chasmata. Note the graben and pit rows, the ridged plateau surface to the north and south, and the short side canyons. Centred at 6.5°S 87.5°W.

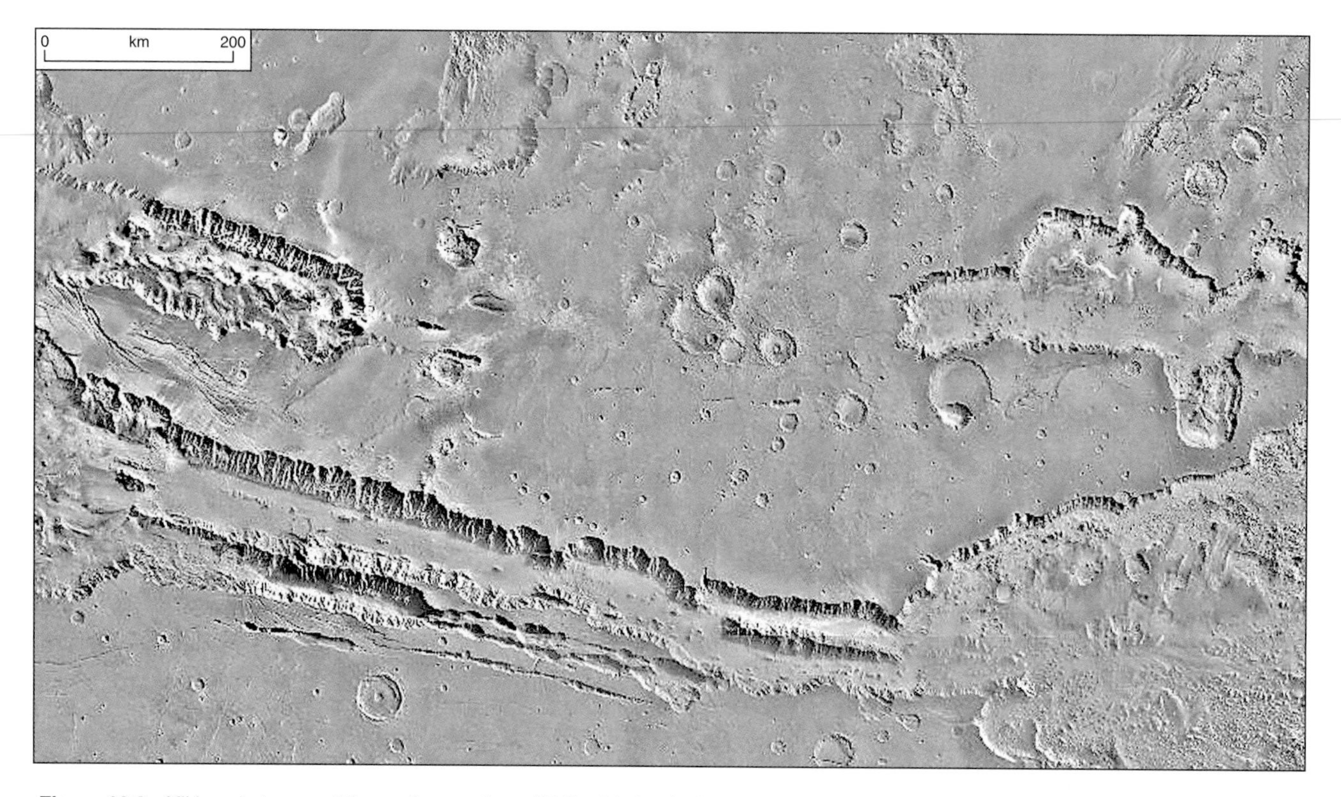

Figure 10.3 Viking photomap of the eastern portion of Valles Marineris, from the irregular canyons of Gangis and Capri Chasmata, through the parallel-sided Coprates Chasma, to the eastern irregular canyons of Gangis, Capri and Eos Chasmata. Centred at 8.5°S 58.8°W.

Landslide morphology is variable. The overall form appears to be dependent upon whether or not slippage was constrained by existing topography (Lucchitta 1979). Slides that were largely unconfined usually have large slump blocks at their heads and vast grooved aprons. On the other hand, where the flow of debris was confined between canyon walls, as in narrow troughs, debris aprons usually have transverse ridges, particularly where the slide materials became piled up against obstacles near the toe (Fig. 10.4).

By measuring the dimensions of obstacles overridden by certain slides, Lucchitta (1978) estimated transport velocities to have been 100–140 km s^{-1}. Such high velocities imply that the material must have had a very low coefficient of friction, which could be accounted for by assuming the flow to have been lubricated by a cushion of air, as is suggested to have occurred in certain terrestrial slides (Shreve 1966). However, this seems unlikely to have occurred on Mars and either water or ice (or both) are more likely to have helped both trigger and fluidize slippage. Certainly, the depth of canyon walls is greater than the calculated depth to the base of a permafrost layer (Fanale 1976); thus, water-saturated materials with low shear strengths could well have existed behind canyon walls, making them susceptible to failure.

The longitudinal striations seen on some Martian debris aprons are very similar to those observed in parts of Alaska, where slides have developed above glacier ice (Lucchitta 1981). The best known of Martian striated landslides evidently emerged from a 2 km-high backwall and moved at least 60 km from their source across the floor of Ganges Chasma (Fig. 10.4). Although Shreve (1966) explains the features of the Alaskan Sherman Landslide as being attributable to air cushioning, it seems that such a catalyst is not necessary to produce the features of striated slides in Valles Marineris. Rather, the frictional resistance of the debris would have been low in any case, because of the enormous energy and velocity attained by the material during a descent of several kilometres. Slides may therefore have originated as massive mudflows with large slumped blocks at their heads. Precisely what triggered collapse is speculative, but seismic tremors associated with faulting or volcanism (or both) could have liberated water trapped in aquifers behind a mantle of ice coating the canyon walls. Most appear to be relatively thin, as is discussed on p. 112.

Many of the landslides appear to be of roughly similar age (Lucchitta 1979). They were produced after the major episode of faulting that generated escarpments of great relief in the equatorial regions. Indeed, most slides were emplaced after the escarpments had been gullied and dissected by tributary canyons. This episode seems to have been approximately contemporaneous with the major volcanism of Tharsis Montes.

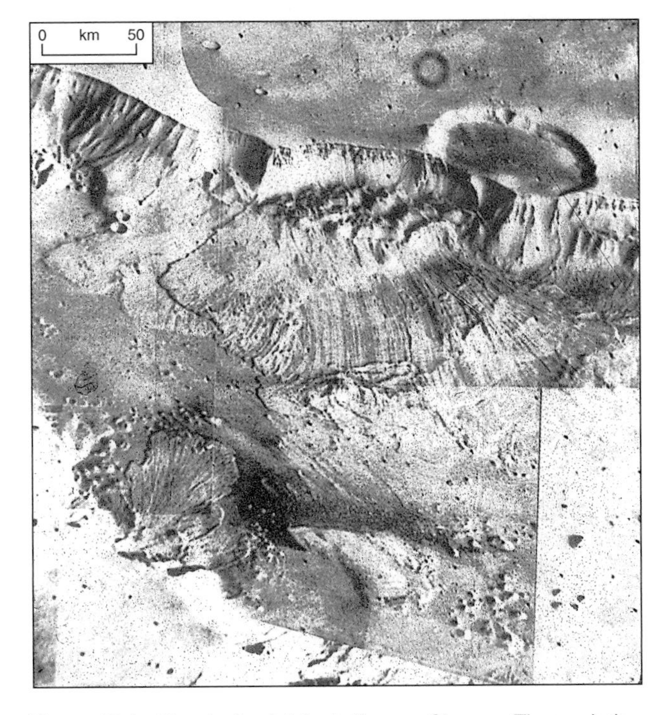

Figure 10.4 Massive landslides in Ganges Chasma. The one below the north wall has an upper blocky layer (probably disrupted caprock) and a striated lower layer. Viking mosaic P-16952; centered at 7.6°S 48.2°W.

Tributary canyon development

Although scalloping is common on north canyon walls, the development of tributary canyons is characteristic of several sections of south walls, particularly on the south side of Ius Chasma, where the canyon backwall is deeply dissected (Fig. 10.2). The canyons tend to be relatively short, with V-shape cross sections and rather rounded headward terminations; also, they have developed along orthogonal fractures.

Sharp (1973a) has suggested that side canyons were generated by a sapping process involving groundwater seepage – probably as a result of the sublimation of buried ground ice – which would have weakened the side walls. The preferential development of such canyons on the south side of Ius Chasma was likened by Sharp & Malin (1975) to the situation at the Grand Canyon, where a regional southerly dip (which controlled the southward migration of groundwater down the bedding planes) has meant that there is a better development of tributary canyons up dip on the north rim. Should this in any way be analogous, then, by implication, the situation at Ius Chasma would be the reverse of this.

Canyon interior deposits

Within the walls of the canyon system are widespread sedimentary and volcanic deposits generally considered younger than both the plateau surface and the canyon floor. These have been studied by various groups, including Nedell et al. (1987), Geissler et al. (1990), Komatsu et al. (1993) and Lucchitta et al. (1994). Layering is present on quite a small scale, certainly down to the resolution limit of the available imagery. At the junction of Ophir and Candor Chasmata, the horizontally bedded deposits discordantly overlie older rocks (Komatsu et al. 1993). Evidently they formed after the canyons had developed and it is also apparent that they themselves have subsequently been eroded. McCauley (1978) suggested that, since such finely layered deposits imply deposition under quiescent conditions, such that they could well represent lake sediments.

The interior layered deposits are particularly well seen in, but not confined to, the central section of Valles Marineris and there is ample evidence that strongly suggests that these were deposited after canyon formation and are not simply remnants of the plateau into which the canyons were incised. The materials can be subdivided into:

- thick floor-filling materials
- dark materials that tend to line the bottom of canyon walls
- irregular materials with moderate to high albedo that unconformably overlap more disturbed and eroded units, and are therefore relatively young
- interior layered deposits, largely forming mesas or underlying benches abutting against wallrocks (Lucchitta et al. 1994).

Very recently Malin & Edgett (2000) reported instances within western Candor Chasma where light-coloured stratified units emerge from beneath darker mantling materials. They consider that the former "continue into and under the walls". If this is so, then it would imply that most, if not all, of the interior layered deposits would be much older than Valles Marineris, turning the traditional view on its head. Having studied the MGS subframes presented as evidence, I have to say that I am unable to convince myself that one can prove this and, until a wider debate of these new images has been completed, have presented the pre-2000 view of canyon development. By the time this book is in print, a clearer picture may have emerged.

The layered units that outcrop widely within Juventae, Gangis, Eos and Capri Chasmata – where they often outcrop as mesas – are evident on south-side benches in Ophir, east Candor and Melas Chasmata, but are noticeably absent from Ius, the northern component of Melas and Coprates Chasmata. The highest terrain in Figure 10.5 is the relatively smooth plateau near the centre of Coprates, where slopes descend to the north and south (upper and lower part of image, respectively) in broad debris-filled gullies with intervening rocky spurs. Layering is clearly visible in the steep slopes on the spurs and gullies.

The laminated deposits, which are generally rather thin, may reach a thickness of 3 km in Candor Chasma. Figure 10.6 shows a thick layered sequence exposed near the middle of the western part of this canyon. The rocks outcrop in two buttes (upper and lower right) and a stepped mesa (centre right). The only younger sediments that can be identified are thin deposits of dust. These may simply represent dust fallout from the atmosphere or be wind drifted. Either way, a veneer of this material coats most mesa tops within the canyon system.

The lighter-coloured material that outcrops on mesa tops absorbs strongly in the red part of the spectrum and has very low thermal inertia (Lucchitta 1987). This could conceivably represent palagonite tuff. Where the covering is relatively thin, streaky light and dark deposits outcrop; these are spectrally blue and probably represent volcanic materials. If this is so, they could be among the youngest volcanic rocks on the planet.

A spectrum obtained of low-albedo deposits in Hebes Chasma by the MGS thermal emission spectrometer (TES) was similar to laboratory spectra of augite in the 800–1350 cm^{-1} region; this is consistent with earlier visible–near-infrared Earth-based and spacecraft data (Christensen et al. 1998). However, all the features of the spectrum cannot be explained by the presence of this species alone, suggesting that a significant amount of feldspar must also be present. A 4:1 mixture of pyroxene to plagioclase appears to fit the data best. The same data were also used to estimate upper limits for carbonate (<10%), olivine (<10%), clays (<20%) and quartz (<5%). These low-albedo materials are also relatively free of weathering products such as clays and haematite; hence, that they represent relatively pristine igneous rocks.

Volume and thickness of canyon deposits

The thickest sequences of interior deposits give rise to flat-topped mesas, which at first sight resemble the old plateau surface; however, they do not show the fracturing or pitting so typical of this and they have a smoother appearance. The mesa sides are also finely fluted, largely uncratered and quite different from

Figure 10.5 Coprates Chasma. Multiple pre-canyon rock layers, varying from a few metres to a few tens of metres thick, are visible in the steep slopes on the spurs and gullies and are representative of the volcanic and sedimentary sequence into which the canyon system was cut. MGS image PIA01168; centred at 14.7°S 55.8°W.

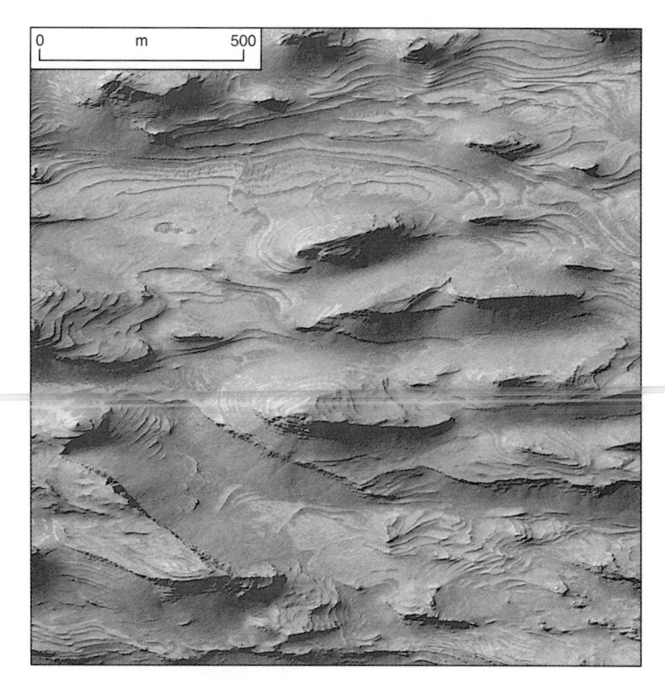

Figure 10.6 Stratified rocks in SW Candor Chasma. The image measures 1.5×2.9 km and is Individual units are 10 m thick. MGS image PIA02839; centered at 7°S, 77°W.

canyon walls. The thicknesses of interior deposits identified within Valles Marineris are shown in Plate 11.

Volumetric analysis by Lucchitta et al. (1994) indicates that, areally, the most abundant rock type within the perimeter of the Valles Marineris system is wallrock (32%). Interior layered deposits occupy 17 per cent, thin floor deposits 16 per cent, other floor units 15 per cent, landslides 10 per cent, and crater material 0.5 per cent. Deepest of the troughs are the northern branches of Ophir, Melas and Candor Chasmata, central Candor, and a trough connecting Ophir with Candor Chasmata. The western part of Coprates and the eastern portion of Ius are also deep. Coprates Chasma becomes shallower towards the east, merging with shallower peripheral canyons that eventually merge with outflow channels. The depth of Ophir, Candor and Hebes Chasmata is much reduced where interior layered deposits outcrop in high mesas and wall benches; in these localities the deposits attain thicknesses of 9 km. Within Hebes and Gangis Chasma, the upper surface of interior layered deposits rises to within 1 km of the exterior plains surface.

The same analysis illustrates that the volume of the interior stratified deposits must be about 60 per cent of the total, vastly exceeding the volumes of all other deposits. Assuming that the irregular deposits and the dark materials are both relatively thin veneers, then the total volume of the interior layered units rises to 80 per cent. The only other areally significant sedimentary units are landslides, and these are generally considered to be quite thin, a conclusion based on the fact that landslides are present in Ius Chasma but absent from Candor, yet both are equally deep.

Layered deposits within peripheral troughs

The peripheral troughs are generally less well delineated by straight wall sections than the central canyons are. They tend to merge with outflow channels, are less irregular in outline than the central chasms, and are shallower, being generally only 2–5 km deep. The stratigraphy of the sequence reveals (Komatsu et al. 1993) diversity of origin, age and post-depositional erosional history for individual canyon sequences. Thus, there is poor correlation between sequences in isolated troughs such as Hebes, Juventae and Gangis Chasmata, suggesting that each represented a discrete depositional environment. Elsewhere, for instance in Ophir, Candor and Melas Chasmata, there are similarities in the succession, which suggests they may have been part of a single continuous deposit. Where the layered units outcrop in such troughs, the stratified units always unconformably overlie the chaotic materials.

The interior layered deposits in Gangis, Capri–Eos, and Juventae Chasmata are 1–4 km thick and overlie chaotic materials (Witbeek et al. 1991, Lucchitta et al. 1992, Komatsu et al. 1993). This means that the chaos regions must have formed prior to the lakes (if lakes indeed formed) that eventually received the layered sediments. Chaotic terrain was presumably formed by transport of rock, water and ice out of the area, giving rise to outflow channels, as suggested by Carr (1979). In order to fill such lakes, the channels must have been blocked by dams, evidence for which is absent. If the dam walls were of ice, its presence could only have been temporary and it is difficult to see how a 4 km-thick layer of sediment could have accumulated to such great depth behind such temporary structures.

Alternatively, the layered deposits could have formed in isolated lakes in regions of chaotic terrain (Komatsu et al. 1993), in which the deep outflow channels currently observed must have been predated by the lakes, which would have drained only after deposition of the layered deposits. Howard (1991) envisaged the existence of such lakes and suggested that chaotic terrain may have formed where confined aquifers formed bodies of ice or subsurface sheets (sills) of liquid water. The latter could have caused collapse of the uplifted roof to form chaotic terrain and lakes, and any ice bodies could have melted or sublimed. The observed superposition of layered sediments on chaotic material suggests that the peripheral troughs also formed first as isolated ancestral basins and may have been at least partly occupied by lakes for parts of their lives.

On the other hand, the layered deposits in the peripheral troughs may be volcanic (Komatsu et al. 1993), as is perhaps true for at least the upper layers of the central troughs. A volcanic composition is suggested by the shape of the free-standing mesas of stratified deposits in Gangis and Juventae Chasmata. The mesa in Gangis Chasma has steep sides, on which light and dark layers are exposed, and a relatively flat top surmounted by knobs. Mesas in Juventae Chasma are similar. Such configurations are like those of table mountains, which are mounds formed of volcanic material erupted beneath ice.

Alternatively, the volcanic material could have been intruded into relatively shallow but completely frozen lakes. The steep sides on the mesa in Gangis Chasma are about 2 km high, consistent with the depth to which water may have been frozen in the equatorial area (Rossbacher & Judson 1981). If the isolated mesas in the peripheral troughs are indeed table mountains, the light and dark layers could be palagonitic tuffs or mafic lava flows (or both), compositions common in table mountains in Iceland (Van Bemmelen & Rutten 1955). A table mountain origin has also been proposed by Croft (1990) for the layered mesa in Hebes Chasma. Certainly, spectral data confirm this suspicion.

Origin of the layered interior deposits

The stratified deposits discovered within the Valles Marineris system are generally believed to hold vital clues in the quest to understand the evolution of the equatorial canyons and areas of chaotic terrain.

The origin of interior layered deposits as fluvial sediments has been questioned mainly because no major channels debouch into the central troughs. It may well be, therefore, that the sedimentary sequence was derived from eroded wall rock (Nedell et al. 1987, Lucchitta et al. 1992). In an effort to shed some light on this possibility, Lucchitta et al. (1994) compared the entire void space within Valles Marineris with the volume of all interior deposits, including landslides, and showed that the void volume of canyons is about six times the volume of the deposits, indicating that, overall, the interior deposits are only a minor feature of the canyons.

Lucchitta et al. also compared the volume of material eroded from the walls with the volume of deposits inside the troughs, assuming vertical fault planes, concluding that the volume of eroded material is larger than that of the deposits (Table 10.1). If the initial faults dipped at only 60°, then the volume of eroded wall material would be less and would be approximately the same as that of the deposits. Both are consistent with the hypothesis that the material removed from the walls could form the interior layered deposits, without any outside contribution. However, because the true depth of the floor beneath the interior deposits is unknown, their estimates probably represent minimum values for the deposits, whose volume may exceed that eroded from the walls. Furthermore, derivation of interior deposits entirely from eroded wall rock would be possible only if all the troughs were interconnected and, as was suggested above, the Coprates–North-Melas–Ius system may not have been in existence when the layered deposits were emplaced. If so, the amount of material eroded from trough walls at that time would have been only about half as much as the estimated total.

Table 10.1 Volume of wall material eroded from individual troughs; percentages are of total eroded wall material (after Lucchitta et al. 1994).

Chasma	Volume km^3	%
Echus	6800	0.93
Hebes	29 730	4.05
Juventae	7340	1.00
Tithonium	55 760	7.60
Ius	167 100	22.79
Ophir	25 070	3.42
West Candor	29 100	3.97
Central Candor	31 980	4.36
East Candor	58 870	8.03
Melas	90 630	12.36
Coprates	173 280	23.63
Gangis	32 680	4.46
Capri/Eos	24 920	3.40
Total	733 260	

Even if Ius and Coprates Chasmata had already existed, it would have been difficult to transport eroded wall material from these troughs through the low area in Melas Chasma and then up hill towards Ophir and Candor Chasmata to form the thick high-level deposits there. Instead, the deposits should have come to rest in low-lying central Melas Chasma instead; however, no major stratified deposits are found there. Furthermore, all of this wall material would have to be transported not only towards the central troughs, but through the relatively narrow isthmus connecting Melas and Candor Chasmata. The inevitable conclusion, therefore, is that the interior layered deposits were not built entirely from mass-wasted wall rock.

If the ancestral central canyons were isolated basins containing segregated ice or water, and if, like the peripheral troughs, they formed mostly from erosional collapse, much of the material in the interior layered deposits could be disaggregated in-situ material reworked by wave or current action. The necessary assumption is that lakes were present and the climate was warm enough for the water not be frozen. In favour of this argument is the observation that the interior deposits occupy only one sixth of the space in the troughs; the material that occupied the space now void could have been condensed into deposits, and the rest could have evaporated or sublimated.

On the other hand, the central troughs in general are much deeper than the peripheral troughs and are bounded by straight walls, suggesting that they were largely controlled by tectonism and that their floors were downfaulted. Another problem is that local reworking of pre-existing material does not explain isolated mesas such as that in Hebes Chasma. In addition, in the peripheral troughs, where collapse processes were common and the existence of former lakes is more likely, we see little evidence that chaotic material, a disintegration product, was reworked into stratified deposits; indeed, on the contrary, stratified deposits always bury chaotic material. Therefore, it is more likely that the ancestral central troughs were largely formed by faulting, that erosional collapse was only a subsidiary process, and that most of the interior layered deposits are not composed of disintegration products reworked in situ.

If the ancestral central troughs were formed mostly by faulting, then the interior void is largely attributable to the downfaulting of the floor, and there could not have been enough material from the eroded walls to build the layered deposits. As no major channels empty into the troughs, the suggestion has been made that additional material may have been derived from subterranean piping (Croft 1989), but the required underground flow would produce carbonates, not detritus, and no spectral evidence for carbonates was found in earlier work, nor has it emerged from MGS data (Christensen et al. 1999). Winds probably also blew in material; however, the diversity of the layers in most places suggests that aeolian material is not dominant. The above arguments lead to the conclusion that another mechanism is needed to supply additional material.

Viking and MGS high-resolution images illustrate how the lower strata building the mesa within Candor Chasma are massive, but the upper beds are more diverse and include finely layered and resistant units. This configuration also seems to be present elsewhere in the central troughs (Nedell et al. 1987, Lucchitta et al. 1992, Komatsu et al. 1993). It is likely that the older massive beds are lacustrine and were built largely from mass-wasted material, an idea supported by spectral investigations (Geissler et al. 1990). The appearance of the upper strata is more compatible with that of volcanic rock, which could have topped the mesas in Ophir, Candor and Hebes Chasmata. As time passes, greater support for this view accrues.

Chaotic material

Chaotic material occurs only in peripheral troughs. It is broadly agreed that the chaotic materials found in Juventae, Gangis and Capri–Eos Chasmata are genetically linked to the outflow channels, withdrawal of water and ground having led to collapse and to the floods that formed these channels (Sharp 1973a, Baker & Milton 1974, Masursky et al. 1977). Evidence supporting a collapse origin is illustrated in Figure 10.3, where the large depression on the south side of Gangis Chasma is seen to be full of slumped and tilted plateau material. This merges northwards into typical chaotic terrain of the main canyon. It seems logical to conclude, therefore, that the peripheral troughs were formed largely by collapse and less from faulting, although the latter undoubtedly augmented their depth.

The chaotic materials can be explained further by the observation that they outcrop on the trough floors near the 4 km contour on the surrounding plateaux. The similarity in elevation supports the idea (Carr 1979) that the outflow channels broke out from confined aquifers that had attained the requisite head of pressure at that height. A postulated 1–2 km-thick layer of ground ice in the equatorial area (Fanale 1976, Rossbacher & Judson 1981) is also consistent with the depth below plateau level of about 2 km, at which chaotic material appears on the floors in the upper reaches of Juventae and Gangis Chasmata. This depth may be the depth of the postulated aquifers.

History of the canyon system

The notion that the canyon system is simply analogous to Earth's East African Rift now appears untenable, since simple extensional tectonics cannot account for the observed configuration of canyons and the pattern of faulting (Wise et al. 1979). Furthermore, the Valles Marineris lack prominent and narrow gravity highs characteristic of the flanks on either side of the East African Rift system.

Canyon inception can be traced to the initial opening of deep troughs parallel to families of shallow graben, which developed radially to the Tharsis Rise, possibly in Late Noachian but more probably in Early Hesperian times. Dating this first phase of canyon-forming activity is facilitated by noting that none of the Late Noachian to Early Hesperian plateau lavas spilled over into the canyons, implying that they were erupted prior to tectonic disruption (Lucchitta et al. 1989). Fracturing widely disrupted the plateau surfaces of Lunae and Sinai Planae, cracking the lava caprock and exposing the underlying brecciated mega-regolith.

That faulting dominated canyon development is clear from the exceptionally straight wall segments in central Valles Marineris. However, exactly how the canyons attained their great depth (up to 11 km) is not entirely clear. Doubtless the negative gravity anomaly known to exist over the central section holds some clue in this respect, but its significance currently remains an enigma. Deepening and lateral growth may have proceeded by a combination of fracturing, crustal spreading, collapse following the withdrawal of magma to lateral sites of effusion, or removal of underlying bedrock by solution or sapping.

The actual volume of removed material was estimated to be 1 million km^3 by Sharp (1973a) and 0.73 million km^3 by Lucchitta et al. (1994). When a comparison is made between the amount eroded and the volume of the interior layered deposits, the latter show a shortfall of one third. Therefore, there is no mathematical problem in deriving the latter from erosion of the original canyon wallrocks. However, there is little direct evidence to suggest that fluvial processes could have achieved this. True, the fluvial landforms that emerge from the regions of chaos into which the eastern canyons merge suggest that removal of large volumes of debris may have occurred there. However, it is difficult to envisage that fluvial activity has removed much debris from the western sections. One possibility is that, as subsidence and wall retreat progressed, downfaulting continued to lower the canyon floors, and interior deposits continued to accumulate. To what extent fluvial, glacial and aeolian processes moved or removed such internal materials remains unknown. However, the presence of the extensive laminated interior deposits opens up the possibility that lakes once existed within the confines of the canyon walls. Draining of these hypothetical lakes eastwards may have given rise to the Simud and Tiu Valles channel systems. Younger materials also accumulated and these now lie unconformably upon the laminated sequence; the rather low albedo and spectral characteristics strongly suggest they are volcanic in origin.

However, in Ophir and Candor Chasmata, lakes could not have been sustained if all the ancestral canyons were interconnected; lake waters in these troughs would have to reach 8 km elevations to deposit the uppermost layered deposits. Such high-standing lakes would have spilled out of Coprates Chasma onto the 4–5 km-high surrounding plateaus. If the lakes were interconnected, their levels inside the central canyons could not have stood higher than 4 km. This line of argument shows how the younger interior layered deposits cannot be lacustrine; they are probably volcanic. On the other hand, the canyons may always have formed isolated basins when the interior layered deposits were laid down. In this case, the northern sections of these canyons and the entire Coprates–north Melas–Ius graben system may have formed later, after deposition of the interior layered deposits.

The peripheral troughs Juventae, Gangis and Capri–Eos Chasmata, reaching depths of 2–5 km, are shallower than the central troughs. Lucchitta et al. (1994) showed that they probably formed from a combination of erosional collapse and structural activity. Furthermore, the presence of chaotic material in these troughs at similar elevations supports the idea that the chaotic material may have formed from release of confined artesian water. In the peripheral troughs, the interior layered deposits postdate chaotic material, indicating that, if the layered deposits are lake sediments, lakes formed after the chaotic collapse of the surface. The lakes were apparently breached only later to form the outflow channels. However, the layered deposits in these troughs may not be lacustrine but volcanic, and this may also be true for those in the central troughs. The volcanic rocks may have erupted below segregated ice masses (Howard 1991), or could have formed in shallow and completely frozen lakes.

Later, the central canyons were evidently widened and deepened with the formation of further graben faults on the north side. The Coprates–north Melas–Ius graben system then cut the entire region, including the ancestral Melas Chasma. These new grabens then probably connected the troughs with one another and with the peripheral troughs in the east. When all the troughs merged, a major flood may have emptied the central troughs. However, it may be more reasonable to conclude that no flood ensued, because the lakes that once existed there had already emptied, and the troughs became filled with sedimentary and volcanic materials. Alternatively, it is still possible that the components of the system were interconnected early in their development, no deep lakes formed, and thus most of the layered sequence is of volcanic origin.

Both of the above scenarios are consistent with recent topographical analysis, but it seems most reasonable to assume that a combination of the two contributed to the development of Valles Marineris. Certainly, lakes were more likely to have existed in peripheral troughs than in the central canyons, and volcanic activity was more important when the younger laminated deposits were produced.

11 Fluvial activity and channels

The major channels that emerge from areas of chaotic terrain, which are located largely in equatorial regions, tend to lack tributaries, emerge fully developed from chaotic regions and become wider and shallower down stream. They have been termed outflow channels (Sharp & Malin 1975). In contrast, channel landforms that tend to be concentrated along the line of dichotomy have flat floors, are characterized by widespread debris aprons and flows, and show strong evidence for an origin in collapse and mass wasting. These are termed fretted channels. Of smaller proportions but extremely widespread within the heavily cratered southern hemisphere are rather different valley networks. These runoff channels have well developed dendritic tributary networks and they may show evidence for more than one period of downcutting.

The principal question arising from the existence of any channels is whether they were generated by running water; if available evidence is found to indicate a fluvial origin, the connotations for the planet's climatic evolution are profound. At present, liquid water is unstable at the surface; the current temperature range of −143°C to +13°C and the surface pressure of 5–10 mbar dictate that liquid water on Mars will either evaporate or freeze, depending upon the latitude, altitude, time of day and season. In view of this difficulty, several investigators have pursued alternatives to water as an erosive agent, including wind, ice, lava and faulting. Although some of these may have played a part in channel formation, the consensus is that running water played the dominant role in generating at least two of the channel types. Very recently, Mars Pathfinder landed at the mouth of Ares Vallis and provided unambiguous first-hand evidence that fluvial activity distributed debris along this major channel.

Distribution of valley and channel systems

Of all the valley systems, the networks that characterize the ancient cratered hemisphere are most like terrestrial river systems, in that they have developed dendritic patterns of tributaries. First recognized on Mariner 9 images, these typically have V-shape profiles and widen down stream. Too numerous to show individually on a general map, they occur widely in the heavily cratered upland southern hemisphere and seem to have been formed very early in Martian history.

More restricted areally, but more impressive geomorphologically, outflow channels generally flow northwards towards the lower northern hemisphere (Fig. 11.1). Major channel systems emanate from regions of collapsed terrain in proximity to Valles Marineris. Thus, Kasei Vallis runs along the western side of Lunae Planum, then turns east-northeast and enters the Chryse plains. Farther east, with a source in Juventae Chasma, is Maja Valles; east again are Nanedi, Shalbatana, Ravi, Simud, Tiu and Ares Vallis, all originating in an area of chaotic terrain in the region of Xanthe Terra and eventually running into Chryse Planitia. Southeast of the Valles Marineris system, and emanating from collapsed ground near Bond and Holden craters, are Nirgal and Uzboi Valles. Mangala Valles is different, appearing to originate in a point source within Memnonia and flow towards Amazonis.

Major channels are also found around the Hellas basin. On the eastern side, Dao, Niger, Harmakhis and Reull Valles flow northeastwards towards Hesperia Planum. Mamers, Aquakuh and Huo Hsing Valles are cut into the upland hemisphere north of Deuteronilus and Nilosyrtis Mensae, and flow northwards towards the lowlands of Vastitas Borealis. On the north flank of the Elysium rise, Hrad, Tinjar and Granicus Valles

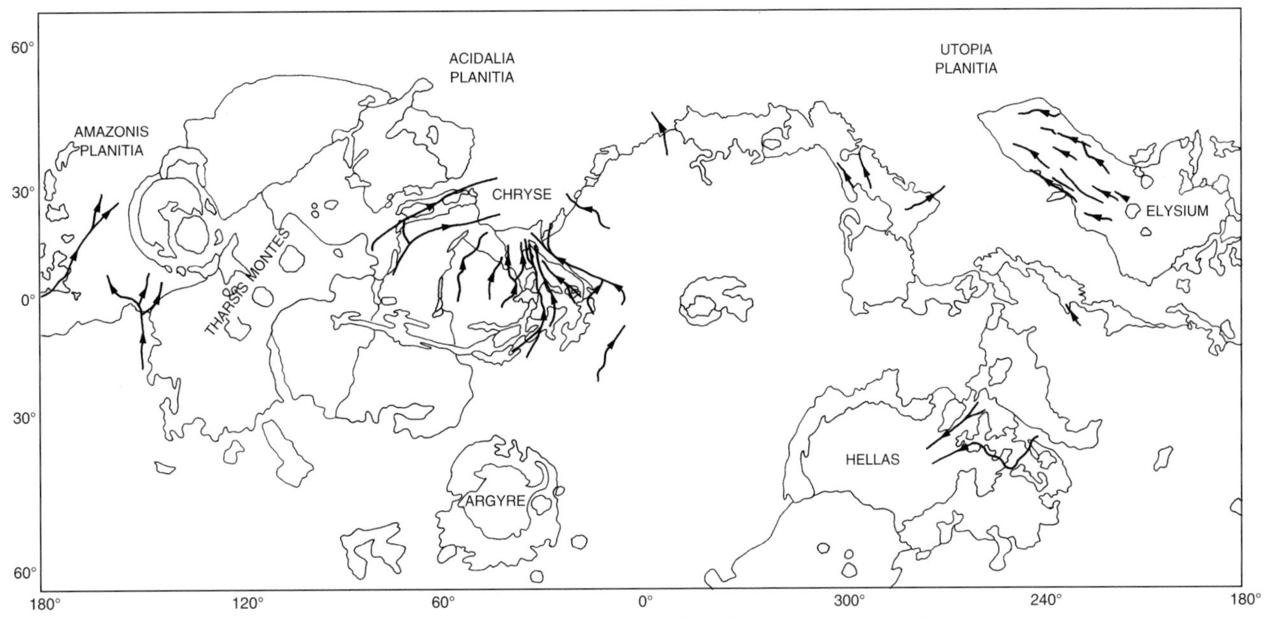

Figure 11.1 Global distribution of outflow channel systems on Mars.

flow towards Utopia Planitia, while Licus, A-Qahira and Ma'adim Valles enter Elysium Planitia from the upland hemisphere on Terra Cimmeria. Finally, more localized channels also found on the flanks of both Hadriaca and Tyrrhena paterae, and on Alba Patera, on the north flank of the Tharsis rise.

The major developments of fretted channels are found along the dichotomy boundary, in particular at Nilosyrtis and Protonilus Mensae. However, smaller areas of similar landforms are also found within some large impact rings close to the line of dichotomy.

Fluvial activity and Mars' volatile history

There is now a consensus that fluvial activity was widespread on Mars during its early history, with major channel-cutting activity during Late Hesperian times. The clearest evidence is provided by the outflow channels. Channel ages, inferred by crater dating, indicate several episodes of flooding and that the most recent may have occurred during the Late Amazonian (~1 billion years ago). Carr (1987) estimated a minimum channel discharge of 7.5 million km^3 of water (equivalent to a global ocean 50 m deep). Baker et al. (1991) argued that, since the channels were produced by multiple flooding episodes, the total discharge may have been much more, perhaps as much as 65 million km^3 of water (global layer 450 m deep). Either way, there was significant water discharge.

A growing body of evidence supports the view that large lakes occupied troughs within Valles Marineris in Hesperian times (Komatsu et al. 1993). Furthermore, theory predicts and geology shows that major impacts released substantial volumes of water; for instance, the Hellas impact alone is estimated to have released 0.28 million km^3 of water (Carr 1984). Finally, major periods of volcanism also contributed significant volumes of volatiles to the atmosphere. Plescia & Crisp (1991) estimated that, during emplacement of the volcanic plains of the Elysium basin, as much as 10 000 km^3 of water may have been exsolved: Mars once had a significant hydrosphere. At present, the volatiles must be concentrated in a cryosphere.

The most obvious reservoirs for volatiles at the present time are the polar caps; it has been widely accepted that exchange of volatiles between the crust and the polar regions takes place only via the atmosphere. However, Clifford (1993) pointed out that, if the planet's inventory of outgassed water exceeds the pore volume of the cryosphere, even by only a few per cent, then a subpermafrost groundwater system of global extent will result, meaning that subsurface transportation of volatiles may supplement long-term crust–atmosphere exchange.

For most of Mars' history, prevailing climatic conditions have ensured that ground ice in low- to mid-latitudes has remained unstable, with the result that there has been a net movement from "warm" latitudes towards the poles (Fanale et al. 1986). Other lines of evidence, including geomorphology, suggest that this has been supplemented by periodic release of volatiles as a result of major volcanism, impacts and catastrophic flooding. The net effect of all of this would be to establish a situation wherein the local equilibrium depth to the melting isotherm was exceeded, at which point the melting of ice at the bottom of the cryosphere would re-establish thermodynamic equilibrium. Percolation of this meltwater downwards into the global aquifer would then cause the water table to rise, forming what Clifford described as a groundwater "mound". Making the reasonable assumption that the subcrust is relatively permeable, this means that the hydraulic head created by the mound would drive an equatorward flow of groundwater (perhaps as much as 100 million km^3) during the course of Mars' history.

In lower latitudes, the existence of a geothermal gradient would result in a net discharge of the groundwater system, as water vapour becomes pumped from the warmer depths into the cooler near-surface crust. Clifford (1993) estimated that a gradient as small as $15\,K\,km^{-1}$ could lead to the vertical transport of the equivalent of 100–1000 km^3 of water over the course of Martian history. Such a process would allow replenishment of much of the volatile material lost through sublimation of equatorial ground ice, major impacts and catastrophic flooding.

The physiography of the early valley networks is quite unlike that of the younger outflow channels, indicating that erosional style has changed with time. Postawko & Fanale (1993) investigated the relative importance of higher surface temperatures on early Mars, as opposed to higher deep-regolith temperatures, in generating less depth of liquid water in the regolith. They note that around 3.8 billion years ago, despite a weaker early Sun, if there had been sufficient CO_2 in the atmosphere the surface temperature could have approached the freezing point for water. On the other hand, if there had been a much higher heatflow at that time – and remember the timing of major volcanism on Mars – then higher regolith temperatures would have prevailed and the depth at which groundwater was encountered would have been shallower. They observe that if the total available CO_2 rose above ~4 bar at this early stage, the greenhouse effect could have played an important part in raising the liquid water level nearer to the surface, at least in equatorial latitudes. Then again, heatflows greater than $100\,mW\,m^{-2}$ could raise the liquid water level from 1 km to less than 350 m. Both (perhaps simultaneously) would have contributed to a very different style of erosional activity than has characterized later Mars.

The widespread collapse landforms in both the equatorial regions and at higher latitudes are powerful evidence that undermining of the crust occurred via sapping processes. Indeed, this appears to have been much more common on Mars than on Earth. Removal of large volumes of ground ice known to exist in the mega-regolith, which is likely to be relatively porous and rather weak over large areas, would clearly lead to subsidence of the surface layer and, if sufficient volumes of volatiles were released, could instigate catastrophic flooding. One obvious means whereby such melting might be catalyzed is by volcanism; the enhanced heatflow accompanying major volcanic episodes (e.g. intrusion of large sills, extrusion of massive volcanic flows) providing the energy necessary to instigate the process. Major impacts would achieve much the same end, and sapping can occur in response to a release in pressure instigated by tectonic re-adjustments such as faulting. It is therefore not surprising that fluvial activity, sapping and catastrophic flooding are intimately connected, in both time and space. The change in style of fluvial activity is probably a function of differences in heatflow and atmospheric composition that typified early Mars.

Valley networks

The dendritic valleys have been termed runoff channels (Sharp & Malin 1975). Typically the smaller ones are 1–2 km wide and may form systems several hundreds of kilometres in length. With the exception of some tributary networks associated with Ius Chasma and some with the volcano, Alba Patera, nearly all are located in the heavily cratered regions between 30°N and 40°S. This can hardly be a coincidence and it seems to imply that they are ancient landforms (Carr & Clow 1981). Indeed, crater counts suggest that most of the dendritic valley networks were formed between 4.0 and 3.8 billion years ago, during the period of heavy bombardment. Although most are relatively restricted areally, others, such as Nirgal and Ma'adim Valles, are significantly more extensive, sharing some of the features of the smaller networks, but tending to have much blunter and less well developed tributaries. Baker (1982), noting the apparent great age of both of these larger systems, suggested that they may have started as runoff channels but were slowly modified in their lower courses by wall retreat.

The drainage patterns are either rectangular or parallel. In the former case, the patterns have developed in areas of fracturing; in the latter, the pattern often results from drainage off the flanks of large impact craters and they have relatively small drainage areas compared with the rectangular ones. Both clearly show a considerable measure of structural control.

The networks are more complex than at first sight. The most widespread valleys have scalloped or runnelled and rather indistinct walls, that have generally been degraded to rather a low slope angle (Fig. 11.2). The larger networks are commonly broken up by areas of cratered terrain, giving them a poorly integrated

Figure 11.2 Channel networks in Terra Tyrrhena, showing parallel drainage patterns and initiation of valleys on outer flank of large impact craters. Viking orbiter image 625a08; centred at 21.7°S 273.8°W.

appearance. Baker & Partridge (1986) have termed these degraded valleys and it is these that exhibit the greater degree of plateau dissection.

More local in their development than the above are deeply incised valleys with steep or even vertical walls and U- or box-shape cross sections (Fig. 11.3). These have been called pristine valleys by Baker & Partridge (1986) and, although such valleys are more local in their development than degraded ones, they may still attain lengths of 600 km. Compared to the degraded valley systems, pristine valleys generally have rather poorly developed tributary networks and often show amphitheatre-like headwalls. Furthermore, the plateau surface in the interfluves is relatively undissected. Pieri (1980) suggested that such characteristics indicate the possible importance of groundwater sapping in their growth. This tends to be supported by high-resolution MGS images, which reveal linear dune forms on the valley interior, straight wall sections with scalloped edges, and stratified rocks of low albedo in the valley sides.

The broad pattern observed is of systems that consist of both degraded and pristine segments. The pristine segments tend to lie in downstream locations, whereas degraded ones occupy upstream positions. Although the former extend into heavily cratered terrain, they tend to have been incised more widely into intercrater areas, which is a natural consequence of their occupation of downstream locations. Degraded valleys also may be observed in downstream positions, but, where they do, they are often obscured either by lava sheets or sedimentary units.

Several features indicate that the degraded networks were incised before the pristine ones (Fig. 11.4). Thus, where degraded tributaries enter pristine valleys, they usually do so as hanging valleys; then again, knickpoints mark positions where pristine valleys end up stream within degraded valleys, giving rise to a cirque-like head. Baker & Partridge (1986) attempted to derive crater ages for the two types and, acknowledging errors inherent in their method, obtained ages of approximately 4 billion years ago for the degraded networks and 3.8–3.9 billion years ago for the pristine valleys.

The geomorphological features of the networks

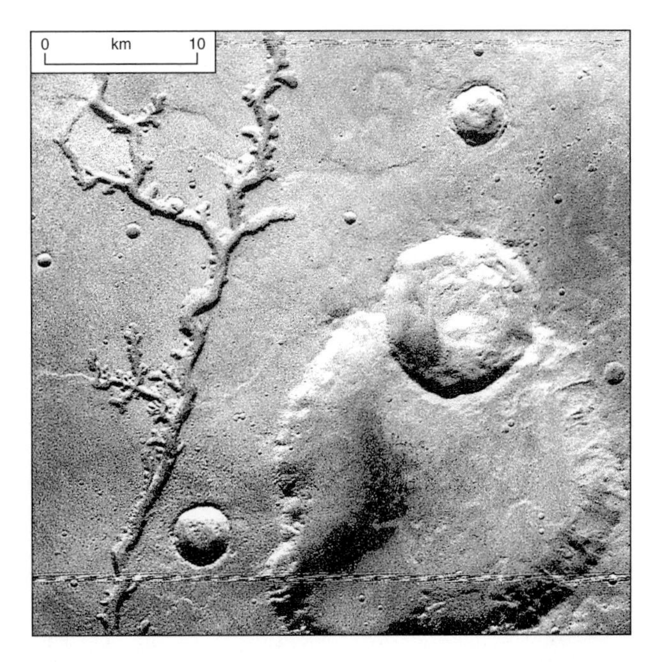

Figure 11.3 Dendritic channel development at Nirgal Vallis. Viking orbiter image 466a52; centred at 27.66°S 45.31°W.

suggest the following scenario. During the period of major bombardment, valley networks were cut into the rugged and primitive cratered crust. The parallel drainage patterns of many principal valleys indicate that their development was strongly dictated by the positions of large impact craters, which provided local gradients down which drainage occurred. A later period of resurfacing produced the intercrater plains, which locally buried sections of the existing channels. Subsequently, after the main phase of bombardment was complete, a second stage of valley formation was initiated. This second generation of (pristine) valleys gradually encroached up into the headward regions of the older networks, occasionally extending into the heavily cratered regions of the plateau.

Even though it is difficult to map the exact extent of the degraded valleys (because of their low slope angles), their very wide extent and high density of tributary development have obvious implications for Martian palaeoclimate. Their close similarity with terrestrial fluvial systems leaves little doubt that they were produced by fluvial action. None of the alternatives so far proposed can account for their production.

Figure 11.4 Pristine and degraded channel development in Terra Tyrrhena. The more northerly (downstream) sections are better defined and more deeply incised than the older (upstream) sections to the south. Viking orbiter image 525a02; centred at 12.24°S 279.90°W.

The more restricted distribution and less well developed tributary systems of the younger valleys have led to several workers attributing their development to groundwater sapping rather than fluvial runoff (Pieri 1980, Baker & Partridge 1986). If this is so, then the implications are that an active hydrological cycle must have prevailed, which could have recharged the high water tables necessary to sustain spring discharge at valley heads. The compelling conclusion is that a period of sustained erosion must have occurred to incise the networks in the first place. It is difficult to imagine any other means for provision of a rechargeable hydrological system without surface runoff. Under present climatic conditions how could runoff be returned to subsurface reservoirs, since the ground is permanently frozen? However, Clifford (1981) has suggested a cycle involving basal melting of polar ice. The most attractive explanation, therefore, is that, when the valleys were cut, climatic conditions were more temperate than they are today, allowing the groundwater system to be replenished from precipitation (Carr 1984). It then becomes possible to envisage valleys being cut by a combination of fluvial action and groundwater sapping.

Recent MGS images tend to support this contention. High-resolution views of Terra Meridiani show valleys cut into low-albedo rocks that are probably basaltic lavas, clearly seen in backwall alcoves (Fig. 11.5). The valley walls are intensely scalloped, and old impact crater rings are visible inside valley confines, having been left by the valley-forming process. Bright material on the valley floor and inside impact craters is evidently aeolian sand or dust. The overall morphology suggests that, whatever process formed the valleys, it was gentle, and sapping seems probable.

Tanaka et al. (1998) showed that the incidence of erosional valleys in the Thaumasia region is greatest on the slopes of Noachian to Early Hesperian volcanoes, close to rift systems and within 100 km of Noachian impact craters greater than 50 km in diameter. Such features occur in regions of steep topography or high relief – quite the reverse of terrestrial occurrences – and it appears to rule out widespread precipitation as the main source of the valleys. Instead it seems more credible that shallow intrusions of hot magma (expected to be present beneath both rifts and volcanoes) and impact melt produced at large impact craters might have led to local hydrothermal circulation and the consequent cutting of valley networks.

Younger valleys (Late Hesperian to Amazonian) originate within 200 km of three large young impact

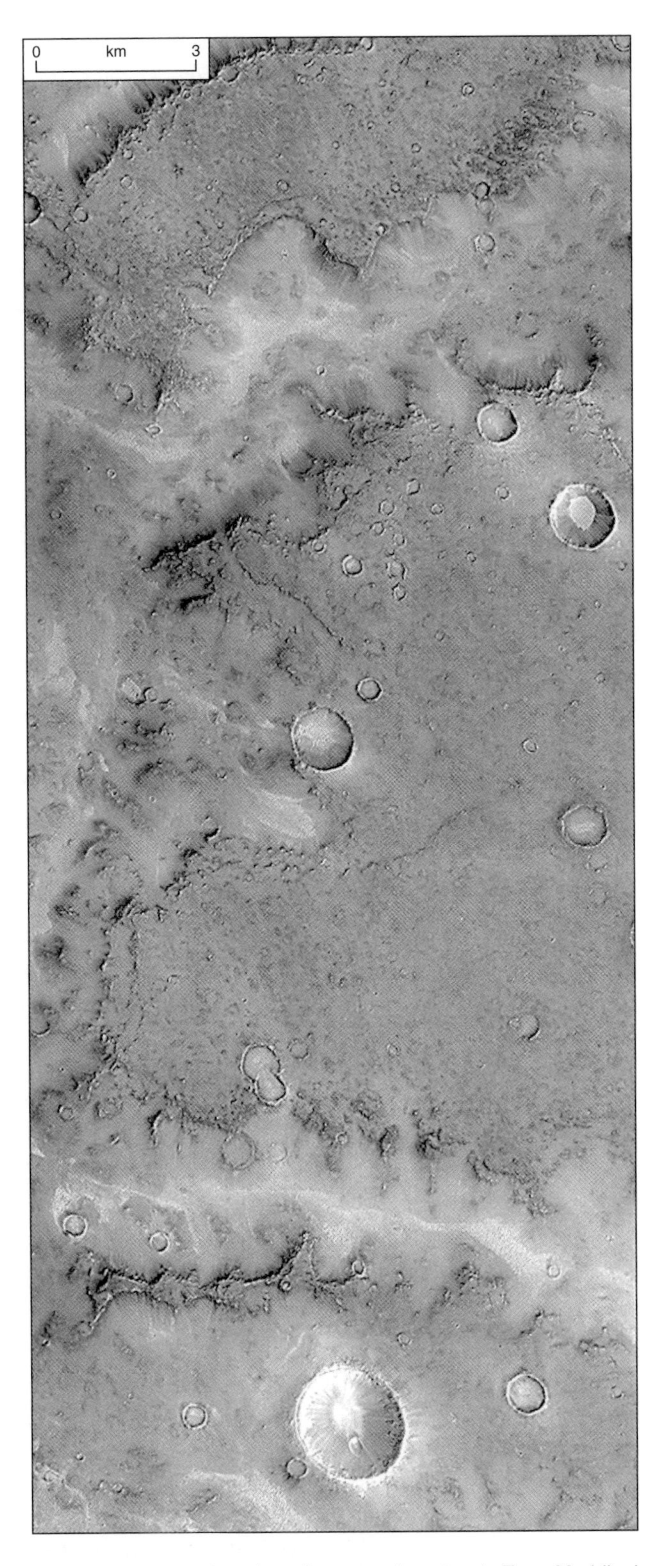

Figure 11.5 A section of a valley network system in Terra Meridiani. MGS image PIA01499; centred at 6.2°S 357.4°N.

craters and near rift systems at Warrego Valles. Tanaka et al. suggested that these may have formed when the cryosphere was at least 2 km thick, thereby inhibiting valley incision by hydrothermal circulation. However, surface eruption of groundwater may have been instigated by impact shock waves and associated fracturing, via seismic pumping. In the case of the crater Lowell, which is 200 km in diameter, the cryosphere must have lain at a depth of at least several kilometres.

The development of short channel systems also characterizes many large (>40 km diameter) impact craters within the Noachian uplands. Forsythe & Blackwelder (1998) identified 144 crater basins with closed drainage systems and no outward-flowing channels. The influent channels are rather short, extending perhaps 10 km from the degraded crater rim, and channel gradients are steep (averaging 13°). They argued that such channels operated as evaporation cells that drew down the water table in their vicinity, evaporation rates of 10–20 $cm\,yr^{-1}$ being estimated. They calculated that, without recharge, one crater basin would effectively drain the Martian mega-regolith of groundwater out to a 100 km radius in about 40 000 years. The many examples of such channels being associated with large craters within such a radius of each other seems to imply that recharge did occur and was maintained at rates of 1–2 $cm\,yr^{-1}$ during Noachian times.

Outflow channels

Outflow channels are spectacular and have provoked much discussion. With the exception of those on the east rim of Hellas and on the flanks of the Elysium volcanoes, they emerge from the cratered plateau, usually from areas of chaotic terrain, and terminate after transgressing the boundary with the northern lowlands. In 1997, Mars Pathfinder landed near the mouth of Ares Vallis, one of the circum-Chryse channels, and provided scientists with first-hand data concerning the geomorphology, rock- and soil chemistry, and spectral characteristics of materials at this site.

Morphology of outflow channels

Outflow channels typically arise fully developed from chaotic regions and are widest and deepest at their upstream ends. Scour marks, although typical of channel floors, are not confined to them, indicating that sheetflow often extended beyond the main channel limits. Scouring is particularly evident where obstacles such as impact craters have caused flow to diverge around them, where ridges, subsequently breached by the flow, have confined flow, and where channels are forced to turn corners and change direction (Fig. 11.6).

Two of the most extensive channel systems emerge from north of the central section of Valles Marineris. On the west side of Lunae Planum, the broad Echus Chasma is the source of Kasei Vallis and, on the opposite side, Juventae Chasma sources Maja, Vedra and Bahram Valles. Both Kasei and Maja Valles exceed 1200 km in length, and sections of Kasei Vallis are over 200 km wide; like Maja Valles, their features are typical of their kind. On entering Chryse Planitia, the flowlines diverge and give rise to a broad complex of shallow channels, streamlined islands and longitudinal grooves.

The source of Simud–Tiu and Ares Vallis systems is in regions of chaotic terrain within Xanthe Terra and enter Chryse on its southeast margin (Fig. 11.7). Typical of such systems, Tiu Vallis is over 600 km long and in places over 25 km wide; Ares Vallis is at least 1200 km long and 100 km wide. Both flow northwards, debouching onto the southern floor of Chryse Planitia and neither has any significant tributaries. Floor features include groups of interlacing channels and streamlined erosional remnants similar to those found in terrestrial braided streams. The teardrop-shape mesas, often topped by impact craters, exhibit stratified scarp faces, indicating either bedrock revealed by erosional stripping or terracing produced by sediment deposition during channel forming. Smooth deposits associated with the Tiu–Simid channel system extend for more than 2000 km from Hydraotes Chaos and in places are 1000 km wide (Tanaka 1988).

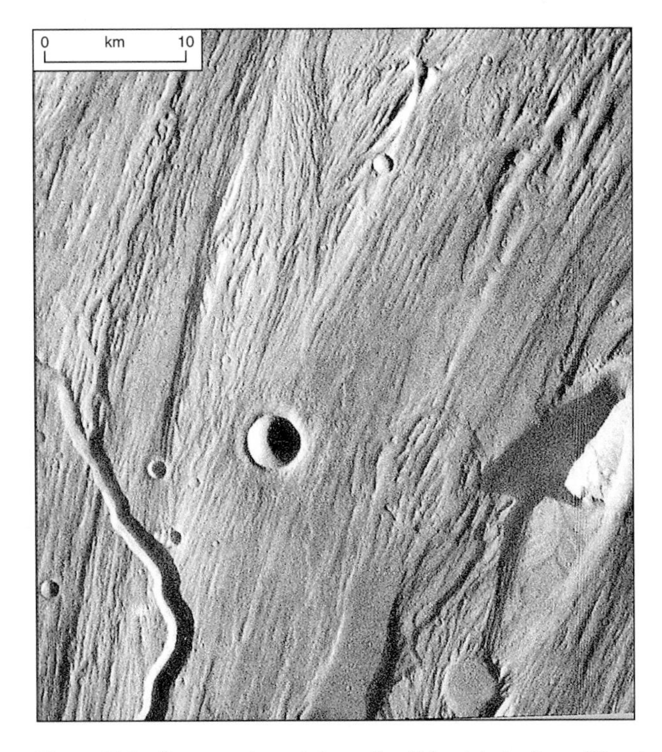

Figure 11.6 Scour marks and streamlined islands in the floor of Kasei Vallis. Viking orbiter image 665a43; centred at 27.37°N 61.38°W.

The region was mapped by Tanaka (1997), who correlated the geological history of Chryse Planitia. The oldest deposits are ridged plains units of Late Noachian to Early Hesperian age that outcrop in western Chryse. Then, in upward sequence come a series of Hesperian channel system materials that end with deposition of Chryse units 1–3 in the region of Ares Vallis. Chryse unit 1 is featureless except for scattered wrinkle ridges and knolls. Chryse unit 2 occurs as remnants that overlie bedforms in lower Ares Vallis and also forms degraded bars in east-central Chryse, north of the MPF landing site. The streamlined bars north of the landing site indicate that erosional agents emanating from the Tiu–Simud system dissected the material (Ward et al. 1999). Chryse unit 3 is younger and comprises thin plains that buried the lower parts of Chryse Planitia, particularly where Chryse 2 had been removed by erosion. Occasional development of boundary scarps and small mud volcanoes is taken by Tanaka (1999) to indicate viscous flow, probably as a massive debris flow. This same sheet may well cover large areas of northwest Chryse Planitia. The youngest unit – the Acidalia unit – buries the above and represents some kind of flow deposit with a source north of the MPF landing site. It is possible that it represents a form of turbidity current set up when massive debris flows entered standing water within the Chryse basin.

Nelson & Greeley (1999) revealed the progression of events that generated the observed channel network and its products. The oldest deposit is a broad smooth-textured unit representative of sheetwash; this blanketed older units such as ridged plains and Noachian cratered terrain. Subsequent fluvial activity incised channels into this, largely in the Late Hesperian. Ares Vallis, relatively narrow (~25–50 km) near its source in Iani Chaos, widens to 100 km down stream and becomes shallower. The initial flooding overflowed the channel banks and extends up to 120 km beyond the main channel. Simud and Tiu Valles originate in Hydraotes Chaos (the latter also in Hydaspis Chaos), whose floods excavated Xanthe Terra highland rocks, carrying debris over 1500 km into Chryse. The sediments from all of these outflow channels were spread out as channelled plains deposits in Chryse Planitia.

Channelling occurred in several stages, the final episode at Ares Vallis leaving various etched units suggestive of a complex fluvial regime (Rotto & Tanaka 1995). Tanaka (1997) suggested the etched deposits may have originated as fan deposits from Maja Valles, because material found in Chryse Planitia resembles the Ares etched deposit. Nelson & Greeley (1999) suggested that the striated units were deposited as a result of flooding from Hydaspis Chaos. Either way, Ares

Figure 11.7 Outflow channels emerging from Aureum Chaos. Viking orbiter image; centred at 7.0°S 32.5°W.

Vallis was re-eroded as floods emanated from Iani Chaos and Aram Chaos, deepening the channel, eroding sediments in central Ares, and etching and streamlining them. It is possible that this outwash originally formed a debris fan in Chryse Planitia, but this must have been smoothed subsequently by flooding episodes or removed except for isolated knobs and mounds. Longitudinal grooves and tear-drop islands characteristic of the channel floor unit imply that flow continued northwards across Chryse. The very late-stage unit crosses the Pathfinder landing site. Tanaka (1999) suggested that the morphology of features mapped within this deposit – flow lobes, internal depressions and longitudinal ridges – is more consistent with deposition by massive debris flows than by simple flooding. He suggested that dewatering of the flow gave rise to mud eruptions that may have formed low pancake-like shields and a couple of rille-like channels. If this is the origin of the deposit, then its features are those of the Pathfinder landing site, which was probably produced by catastrophic flooding.

Not all channels originate in collapsed regions. For instance, the Mangala Valles outflow system is incised into both Noachian cratered highlands and Hesperian lobate plains units of Terra Sirenum. The latter are scoured and truncated by younger channel materials, which also occupy the lowest ground within the Amazonian volcanic plains of Daedala Planum. It differs from other outflow systems: it originates at a point where a major N–S fault intersects one of a series of WSW–ENE-striking grabens (Memnonia Fossae). Here a 5 km-wide breach in a graben wall appears to be the source of a major flood (Tanaka & Chapman 1990).

Zimbelman et al. (1997) calculated that the intensity of flooding was far less here than in Kasei Valles, arriving at a discharge rate of 10 million $m^3 s^{-1}$ for the former, a figure two orders of magnitude less than that of the latter. Whereas the Kasei floods can be traced to the collapsed terrain of Echus Chasma, the source of the Mangala flooding is rather more obscure and, as Zimbelman and his colleagues rightly pointed out, the permeability of the rocks in the source region is of considerable relevance. They suggested that these were basalts on the flanks of the Tharsis Rise (permeability of 1000–100 darcies) and that meltwater flowed off the Rise beneath a cover of ground ice that acted as an aquaclude. The water then found its way into the Memnonia Fossae graben system, following this until it intersected the N–S fault already identified. At this point the meltwater found an easy conduit towards the surface, emerging under artesian pressure generated during its descent from Tharsis. Once it emerged, it flowed northwards across the ancient cratered terrain.

Initially, no clearly defined channel was cut; rather the flow was chaotic. Later, however, ice crusts may have formed, giving rise to local ice dams along the course. The pitted and scoured terrain seen in the distal regions of Mangala Valles may have resulted from early-stage sheetflow: a mixture of meltwater and slush. It was undoubtedly the increasing volcanic activity of Tharsis at this time that generated the meltwater in the first place, perhaps giving rise to a twofold increase in the local heatflow. This would easily have accounted for a substantial local lowering of the permafrost table, possibly by a factor of two. Zimbelman et al. estimated that 0.5 km of lava and its associated thermal energy could have led to melting of a 1.5–2.0 km-thick layer of permafrost, thus making 20 000–200 000 m^3 of water available, quite sufficient to fuel the estimated 5000 billion $m^3 s^{-1}$ minimum discharge rate identified (Tanaka & Chapman 1990). The flow eventually gave rise to an integrated channel system, during whose history there were frequent surges in the intensity of volcanic activity in Tharsis. It was this that caused periodic damming of parts of the channel by ice, which alternated with periods of intense downcutting that left terraces along the channel sides.

As we have seen, outflow channels were generally not cut in one massive flooding episode. This is certainly true of the 1600 km-long channel system Maja Valles that emerges from Juventae Chasma, traverses Lunae Planum, then turns east across Xanthe Terra before entering Chryse Planitia. The initial course of the flood followed a natural valley formed by the eastward slope of Lunae Planum and the Xanthe Terra uplands. The broad (50–150 km) and shallow valley is characterized by scour marks and teardrop islands. It appears that the initial flow ponded about 1100 km from its source, on the northeastern surface of Lunae Planum (De Hon & Pani 1993), before bursting through the crest of the Xanthe highlands and cutting the deep gorges of Bahram, Vedra, Maumee and Maja Valles, travelling a further 75 km into Chryse Planitia. Smaller ponds also existed in the upland terrain of both Xanthe Terra and western Chryse. In its lower course the channels have ill defined boundaries and eventually merge into one broad channel that merges with the plains deposits of Chryse. Eventually the Maja canyon captured most of the flow and became dominant.

De Hon & Pani estimated that, if a single release event cut the outflow system, it could have traversed the entire system in about 40 h. However, as they rightly noted, the valley did not exist at the time of the first release event. They calculated that the first batch of water reached northern Lunae Planum within 19 h and the north-central part of Chryse in 13 days. They suggested that during this time and for the ensuing 4.3 months water was continually discharged from the source region, the Lunae Planum pond emptying at gradually decreasing rates for a further 10 months.

The Pathfinder landing site (Ares Vallis)

Although catastrophic flooding is favoured as the dominant process forming the outflow channels and therefore the unit for the Pathfinder landing site (see Golombek et al. 1997, Komatsu & Baker 1997, Rice & Edgett 1997, Smith et al. 1997), Tanaka (1999) promoted the idea that the relatively smooth plains unit associated with the Tiu–Simud and Ares Vallis systems was actually formed by a debris flow (Tanaka 1999). Nelson & Greeley (1999) and Lucchitta (1998) have offered some support for this idea. It is therefore time to consider the general geology of the landing site in Ares Vallis and see what the new data suggest.

Mars Pathfinder landed near the mouth of Ares and Tiu Valles, on the south side of Chryse Planitia. This site was selected for the widest possible range of rock types eroded from the highland areas by fluvial activity in the Hesperian. As noted above, the deposit at the landing site must represent sediment formed at a relatively late stage and probably from the Tiu–Simud system. This unit (Chryse 3 of Rotto & Tanaka 1995) buries the longitudinal ridges and grooves of lower Ares Vallis. Fluvial, glacial and debris-flow origins have been suggested for it (Fig. 11.8).

The landscape revealed by Mars Pathfinder is complex. Ridges and troughs, larger hills, crater rims, boulders of various shapes and sizes, surface drift deposits and dunes are all found. It is generally assumed that after deposition of a bedrock stratum or strata, subsequent fluvial and aeolian activity (both erosional and depositional) modified this (Ward et al. 1999). Rock types found at the site include basaltic and andesitic lavas, and composite rocks that may be conglomeratic. If the latter proves to be the case, then fluvial activity may have been much more important on Mars than was previously suspected. Etching of vesicular lavas and other rock types, and construction of longitudinal and barchan dunes and of drift deposits attest to recent aeolian activity, which must be the dominant surface process at the present time.

Although some smaller rocks are subrounded,[1] the majority of the boulders and cobbles at the site tend to be subangular, indicating a relatively short period of transport. The more rounded clasts are probably

1. A term used to denote sedimentary clasts that have more rounded corners than angular ones.

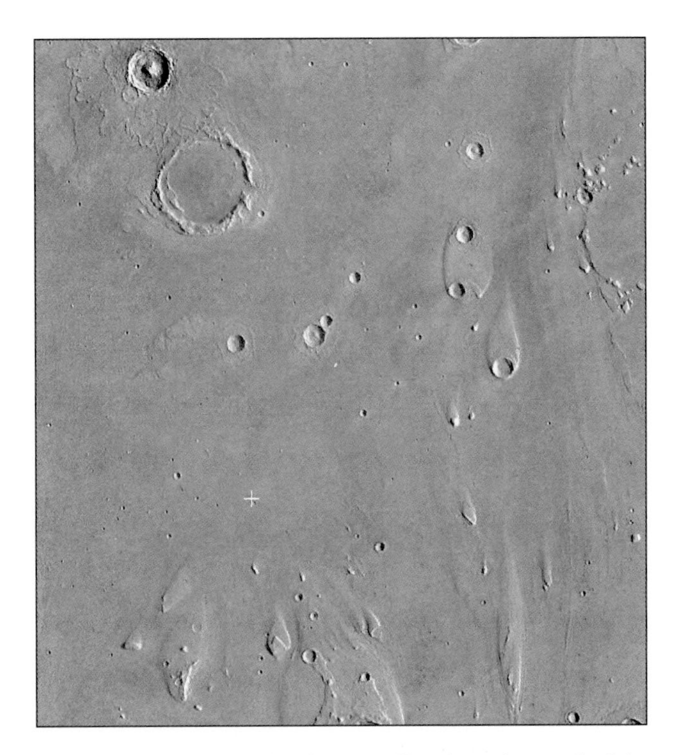

Figure 11.8 The Mars Pathfinder landing site at the mouth of the Ares Vallis outflow system, marked by the white cross. Viking orbiter image M120n032; centred at 19.4°N 33.1°W.

derived from nearby conglomeratic blocks, whose source region is unknown. They were probably subject to a relatively long period of transportation prior to incorporation in these rudaceous[1] deposits, but when and where this occurred remains a mystery.

From the surface, the landing site is seen to be characterized by ridge and trough topography, with ridges being as high as 5 m and spaced 15–25 m apart (Plate 6). The largest of these is about 300 m in length and strikes NE–SW. This direction is the same as several tear-drop islands discernible on Viking imagery. Most of the ridges have boulders and rocks along their crest lines; however, some lack such large clasts and are significantly lighter in hue. The latter probably represent ridges draped by relatively young aeolian deposits.

Many of the medium-size to larger rocks (e.g. Yogi) appear to be perched, suggesting that deflation has occurred. Indeed, Greeley et al. (1999) identified half a dozen such blocks, including Flat Top, which are characterized by a soil-coloured coating along their sides, reaching from the surface to a height of around 7 cm. They suggest that this may be indicative of rocks sitting on a deflated surface, a situation that occurs on Earth where rocks have been more deeply buried in the past, fine soil adhering to their surfaces to a level above the present surface. When deflation subsequently strips off some of the soil, a "tide mark" is left along the side of boulders, its height above the present surface indicating the amount of material stripped off.

It is doubtful that this evidence pins down the precise process responsible for the observed deposits and landforms. As Chapman & Kargel (1999) rightly concluded, despite the availability of detailed geomorphological information from the Mars Pathfinder landing site and high-resolution Viking imagery, no tight constraints are imposed by the data, and any or all of the proposed processes may have contributed to the outflow channels and their associated deposits. When high-resolution imagery is obtained by Mars Global Surveyor, further clues may be provided.

Ages of the channels

Unlike the runoff channels, which show a narrow range of ages, the outflow channels appear to have been formed over a lengthy period (Masursky et al. 1977). For instance, some large discontinuous channels (e.g. Ladon Vallis), which are emplaced in the cratered uplands, pre-date the decline in impact rates that marked the end of the Great Bombardment, around 3.9 billion years ago. On the other hand, Mangala Vallis, which debauches into Amazonis Planitia, is very sparsely cratered and has been assigned an Upper Amazonian age (Scott & Tanaka 1986).

The plains and channels in the Lunae Planum–Chryse Planitia region were studied by Theilig & Greeley (1979), who defined at least four phases of channel development, the first two of which affected the cratered terrain and which may have resulted in the formation of a groundwater system that provided a source of ground ice. These both pre-dated the formation of the extensive volcanic plains of Lunae Planum and Chryse Planitia. They concluded that most channels began downcutting in later Hesperian times.

Moore et al. (1995) concluded that the various circum-Chryse outflow channels must have formed over most of Martian history, in general, becoming younger the closer they get to the Chryse lowlands. They also propose that the highlands peripheral to the Chryse basin may have acted as a recyclable aquifer via what they term groundwater seep. Thus, if a large ice-covered lake once existed within the basin, then it might continually seep out through the mega-regolith, the strata of which, assuming Chryse was formed by impact, would be tilted radially away from the centre of the basin. They estimated that a lake 200 m deep and 200 km across could empty and thereby recharge a source region near the chaotic terrain, say 1000 km

1. Sedimentary rocks with an average grain size greater than 2 mm.

from Chryse, in less than 1 million years. The viability of this process is uncertain; there is currently no proof that a lake ever existed, nor evidence of seepage. Recent MGS gravity data also throws doubt on the impact origin of Chryse, further detracting from this idea. More recent mapping attributed the main circum-Chryse channel activity to most of Hesperian time (Rotto & Tanaka 1995, Tanaka 1997, Nelson & Greeley 1999).

Terrestrial analogs of outflow channels

Martian outflow channels have been likened to the channelled scablands of eastern Washington state, USA (Baker & Milton 1974, Baker & Kochel 1978, Baker 1982). They represent some of the most widespread terrestrial catastrophic flood features and are the only ones that approach the dimensions of Martian outflow channels. In a comparison between the scablands and the Martian outflow channels, Baker (1982) and Baker & Milton (1974) pointed out that teardrop islands, longitudinal grooves, angular channel-floor depressions and terraced margins are common to both. The scablands were produced when one of the containing walls of Lake Missoula (an ice dam) was breached during the Late Pleistocene, causing a catastrophic flood that inundated the surrounding region. The enormous lake covered much of Idaho and western Montana.

Figure 11.9 Fretted channel development in the Protonilus Mensae region, showing emergence from large degraded impact crater acid lineated debris flows on channel floors. Viking orbiter image 268s71; centred at 41.6°N 312.4°W.

Fretted channels

Flat-floored and somewhat sinuous fretted channels penetrate deeply into the upland hemisphere. Anything up to 40 km across, they frequently merge with large impact craters, which are usually highly modified and degraded (Fig. 11.9). Almost everywhere, talus aprons may be seen at the base of escarpments or isolated mesas. In the latter case these may extend up to 20 km across the surrounding plains. The surfaces of such aprons are finely striated, the striations being arranged normal to the scarp walls. Where the debris is confined, as it is between channel walls, the striae are longitudinal; they diverge and converge around obstacles, and, where two valleys meet, the striations merge, rather like those in modern terrestrial glaciers. The striae evidently represent flowlines. The scarcity of impact craters indicates that they are very young.

The observed relationships imply that the ancient cratered plateau has been – and probably still is being – eroded by a process of scarp retreat, the eroded material being slowly moved towards lower ground by gravity (i.e. by mass wasting). Squyres (1978) showed that the most prominent debris flows outcropped in two 25°-wide latitude bands, centred on 40°N and 40°S. Currently, in these zones ice precipitates out from the atmosphere during mid-winter and accumulates on the ground. He postulates that this ice mixes with debris eroded from the scarp faces, and forms a mixture that shares many of the features of a terrestrial rock glacier. Downslope flow is largely via slow movement of the interstitial ice, the debris glacier creeping towards the northern lowlands. The continual removal of the debris from the scarp faces allows erosion to proceed, inexorably adding to the degradation of the cratered plateau (see Ch. 13 for further discussion).

12 The polar regions, ice and wind

Mars' polar caps are large enough to be visible from Earth, even with modest telescopes. For many years prior to spacecraft exploration, observers noted seasonal changes in the size and shape of the caps, together with albedo changes in the adjacent regions. These are a function of the Martian seasons, which, because the planet's axial inclination is closely similar to Earth's, lead to alternate melting and refreezing of the ices present, rather as happens on Earth. They are of particular interest for two reasons: first, they are major reservoirs of volatiles and, secondly, they are regions of present-day geological activity. The cyclical movement of volatiles into and out of the caps is an important part of the planet's hydrological cycle.

Extensive layered deposits surround the polar ice and pass beneath it. These are incised by long sinuous valleys and canyons that extend towards latitude 80°. Surrounding these is a broad zone of dunes, which are a manifestation of the activity of global winds and local storms. Great importance attaches to Martian winds, since they are responsible for the movement of not only sedimentary materials but also volatiles, which may become entrained in the dusty atmosphere. There is a clear link between the seasonal cycle of carbon dioxide and those for dust and water.

There is little doubt that, once plans had been laid to send Mars Polar orbiter and Mars Polar lander, intense interest was aroused in the poles, in particular the proposed landing site. Regrettably, neither spacecraft achieved its goal, but despite this failure much has been learned from Mars Global Surveyor. Its high-resolution imaging, laser altimetry and the thermal-emission spectrometry experiments have all played a part in enhancing our knowledge of the high latitudes.

The polar caps and seasonal variations

Like the polar caps of Earth, those on Mars wax and wane with the seasons. At the present time, southern summers are shorter but warmer than those in the north, and winters are longer and colder. This reflects the fact that Mars reaches perihelion during southern summer, as does Earth, with the result that the climate experienced by the southern high latitudes is much more extreme than that of the north. Because of this, the south cap shows the greater variation in size, at its maximum extending as far as 50°S, around 15° closer to the equator than its northern counterpart. The general physiography of both polar regions is beautifully depicted in Figures 12.1 and 12.2.

Physiography of the polar regions

The deposits associated with the polar regions consist of layered terrain and residual ice, the latter constituting the perennial ice deposits that are distinguished from seasonal frosts. Circumpolar deposits consist of a variety of types of mantled plains, isolated outliers of residual ice and extensive dunefields. Results based on Mariner 9 data indicated the northern cap to rise 4–6 km above its surroundings, with two elevation maxima at ~88°N, 120°E, and 83°N, 10°E (Dzurisin & Blasius 1975). The new MOLA data give a very different picture. In fact, the highest point of the northern cap is within a few kilometres of the rotational pole and has an elevation of –1950 m with respect to the average elevation at the equator. The surrounding terrain ranges from –4800 m to –5200 m, giving a cap relief of about 2950 m. (Zuber et al. 1998).

Layered ice deposits represent the most distinctive

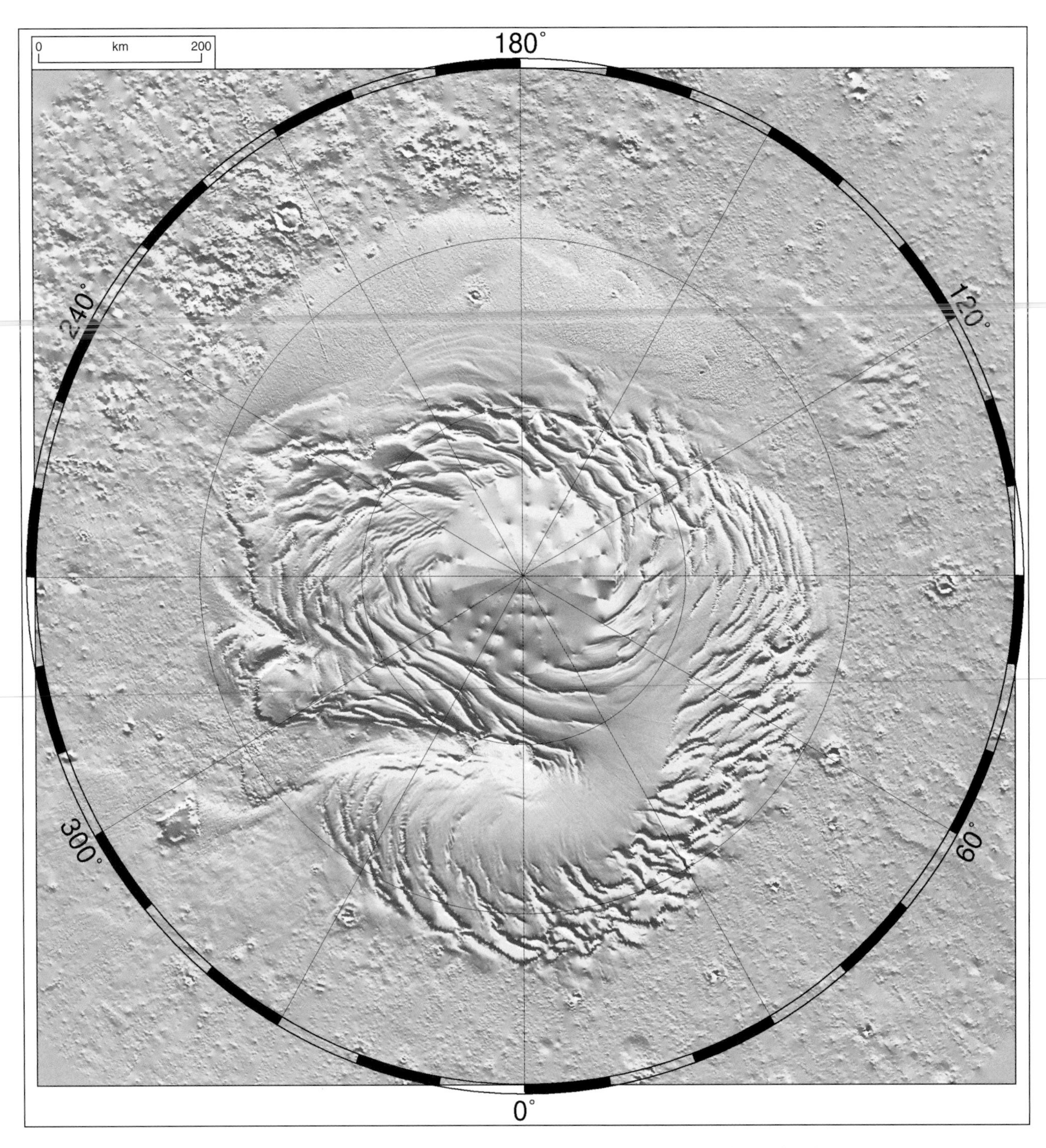

Figure 12.1 Polar projection of MOLA topography between 75°N and the north pole.

element of the northern cap. These sit on top of plains units and form the permanent ice cap. Topographic profiles across the ice cap (Fig. 12.3) show that the surface rises sharply by about 1000 m at the edge of the layered deposits, in contrast to the 2000 m rise estimated previously.

The surface beyond the cap slopes gently downwards towards the pole at all longitudes in the northern hemisphere, and most steeply close to the Tharsis Rise. This is very significant as it means that there will be a tendency for any liquid water to flow towards high northern latitudes, either at or below the surface.

MOLA quite unequivocally showed that the northern hemisphere is both low and smooth, with a gentle poleward decrease in slope (Smith et al. 1999a). The difference in elevation between the southern and

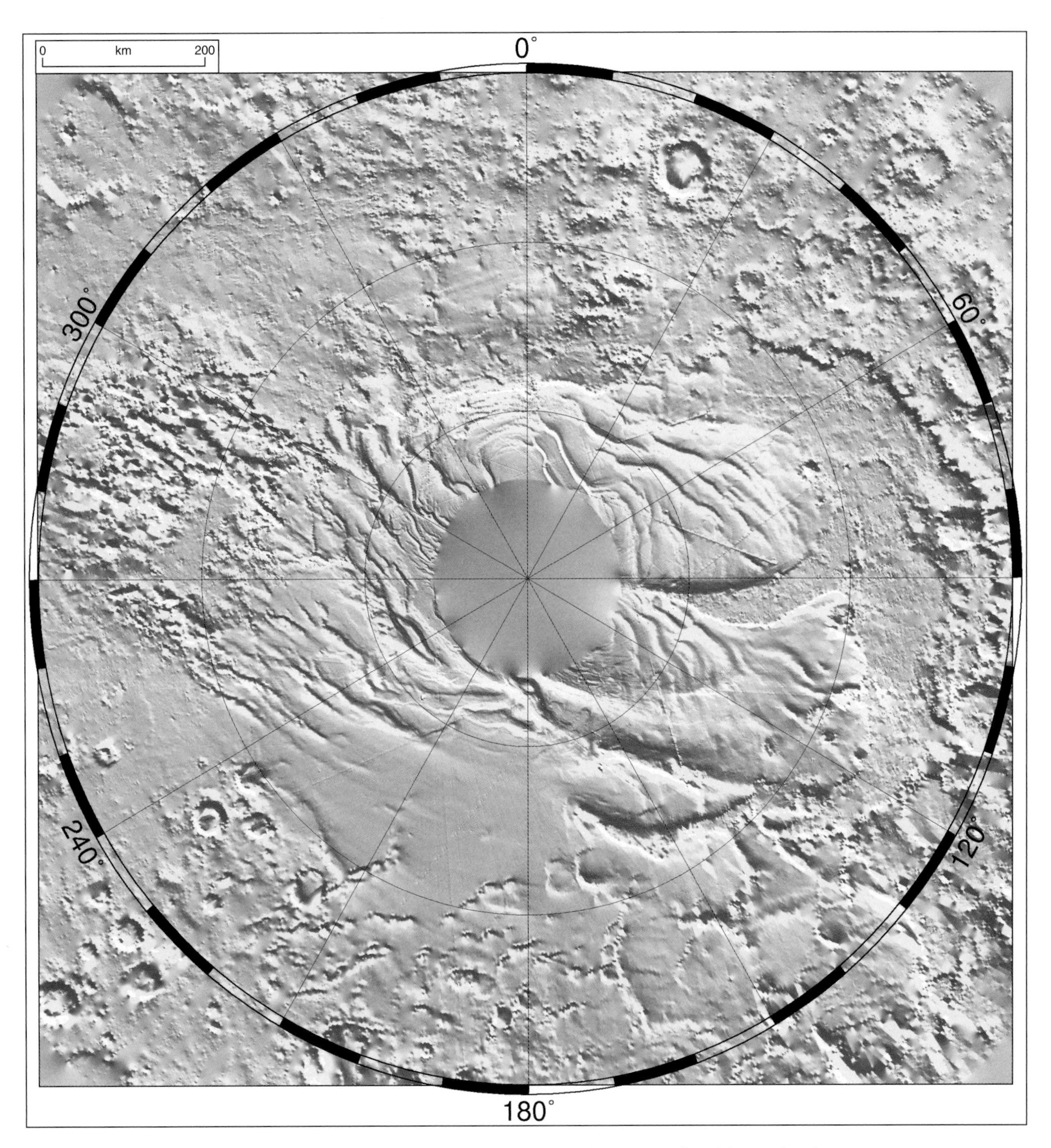

Figure 12.2 Polar projection of MOLA topography between 75°S and the south pole.

northern hemispheres – known to have been in place very early in Mars' history – must have dominated the transfer of volatiles throughout the planet's history. This hemispheric elevation contrast must have played a major role in the present distribution of surface volatiles. MOLA data, including measurements of elevation, reflectivity and pulse width, have also provided new insights into the physiography of the caps and surrounding layered terrain. The ice cover and swirling canyons are clearly seen in the above figures. Earlier work led to the suggestion by several groups of planetary scientists, including Thomas et al. (1992), that the layered terrains are ice rich; however, prior to the arrival of MOLA, the amount of ice present was poorly known. Things have now changed.

MGS pass 21 made a 75 km-long traverse of the

northern layered deposits at longitude 358°, between latitudes 78.8°N and 80°N (Smith et al. 1998). It was found that elevation increased to 1210 m above the dune-covered surface to the south, the spacecraft overflying two ridges and troughs during the pass (Fig. 12.3). Since the MOLA backscattered energy from the ridges exceeded that of the adjacent terrain by a factor of three, it is reasonable to assume that the former are made from or covered in ice. The measured high reflectivity correlates well with the extent of the residual ice cap observed by Viking orbiter during Martian summer. Lower reflectivities were detected in the troughs, implying that the ice here has entrained dust.

The two other MOLA passes of similar polar terrain reveal that steep slopes (> 15°) characterize the edge of the layered terrain. These are unlikely to be representative of any permanent feature and the break in slope may well be attributable to ablation. The troughs may also be ablation features, for both low elevations and reflectivities were recorded here too. Reflectivity also decreases monotonically away from the perimeter of the layered terrain, meaning that frost probably extends southwards to at least 76.5°N. Taken together, the MOLA data are consistent with the view that the layered deposits have a thick ice cover. A 1 km grid of polar topography shows that large areas of the top of the permanent ice cap are smooth, regional slopes over many tens of kilometres being approximately 0.2° (Fig. 12.3b). The combination of imagery and MOLA data has enabled production of the first 3-D view of the northern pole (Fig. 12.4).

High-resolution images and MOLA data obtained during March 2000 elicited compare-and-contrast

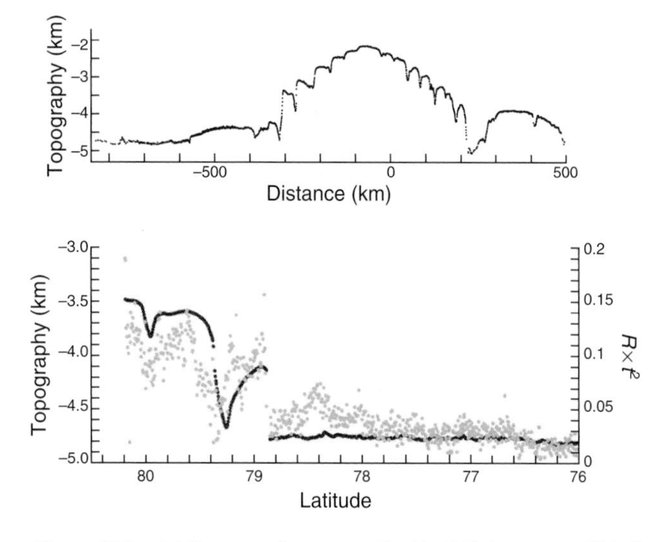

Figure 12.3 **(a)** Cross section across the North Pole on pass 404 of Mars Global Surveyor; the cap has a maximum elevation of 3 km above the surrounding terrain; vertical exaggeration 100:1. **(b)** Topography of north polar layered terrain; vertical exaggeration 50:1. (After Smith et al. 1997.)

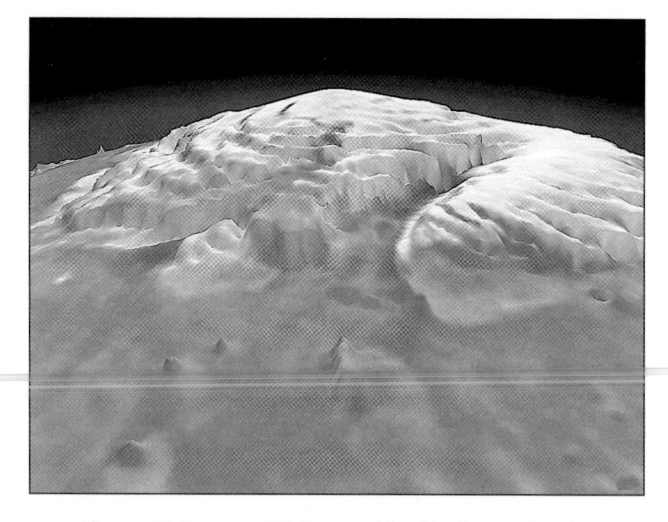

Figure 12.4 MOLA 3-D image of the Martian north pole.

statements from NASA's imaging team, likening the two caps to varieties of cheese. Thus, the north cap is viewed as resembling cottage cheese, having a relatively flat pitted surface, whereas its southern counterpart, with its larger pits, troughs and flat-topped mesas, is likened to Swiss cheese. The differences in morphology appear to indicate that the polar regions have experienced different climates and geological histories for at least several millions of years.

Seasonal behaviour of the polar regions

Each cap consists of two components: a seasonal one and a permanent or residual one. The seasonal caps consist largely of carbon dioxide, which condenses out of the thin atmosphere during autumn and then dissipates in spring. As it condenses, clouds of carbon dioxide accumulate, hovering over the poles and obscuring the process of cap growth. After studying the seasonal variations in atmospheric pressure recorded by the Viking spacecraft, Hess et al. (1979) estimated that the seasonal cap may be only a few tens of centimetres thick. Recent MGS images have recorded the sublimation of seasonal frost around the northern cap (Fig. 12.5). During August 1998, and within about one month, dark spots appeared within the shrinking frost cover. The latter, representing frost-free patches of dark dune material, were subsequently attacked by the wind, which formed small streaks across the adjacent frost surface.

As each hemisphere experiences spring, its cap retreats, leaving behind temporary outliers of ice, which do not last long, generally disappearing within days. Eventually a residual cap is left, this being incised by many deep valleys that gradually emerge from the

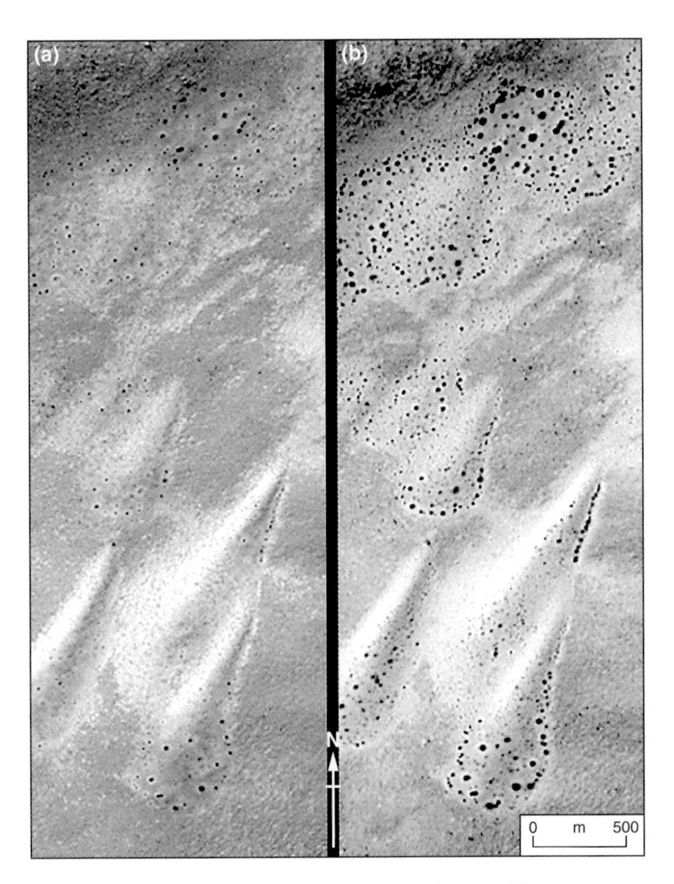

Figure 12.5 **(a)** As the north polar seasonal cap sublimes away, a portion of the huge circumpolar dunefield is covered by winter frost (23 August 1998). **(b)** By September 1998, dark spots have appeared on the northern dunes, the dark streaks emanating from them sharing a common orientation indicative of removal by prevailing winds. (PIA02302)

frost cover. Their orientation gives to each residual cap a distinctive swirl pattern (seen in Figs 12.1, 12.2). The northern residual cap is about 1000 km across, but the southern shrinks to a mere 350 km.

The northern residual cap is almost certainly water ice. Temperatures of –68°C have been measured above it, which is well above the frost point of CO_2 but close to the frost point of water in an atmosphere holding little precipitable H_2O. The southern residual cap is a mixture of water ice and CO_2 ice, and much lower temperatures were recorded (around –113°C); little if any water was detected.

Between 1990 and 1997 the Hubble space telescope imaged Mars and monitored parts of four consecutive north polar cap retreats, spanning $L_S = 335.65°$ to $L_S = 144.56°$ (Cantor et al. 1998). Observations in 1991 confirmed a standstill in retreat of the cap at $L_S = 32.25°$ and $L_S = 59.88°$ in the longitude range 270–280° at latitude 70°N. A similar standstill had been reported at latitude 67° in previous years; this was not observed between 1992 and 1997. Changes of this kind are generally considered to reflect variable activity of dust storms.

During 1996 and 1997 the HST also covered the entire spring season in the northern hemisphere. It observed that local dust activity occurred close to the edge of the retreating cap during early spring. James et al. (1999) ventured the suggestion that such activity may be the result of winds associated with fronts moving across the seasonal cap. These could lift dust exposed on the cap during CO_2 sublimation.

The HST was also used in the near-infrared (1042 nm) between February 1995 and June 1997 to obtain measurements of the planet at the highest spatial resolution made to date (Bell et al. 1999). They reveal that there have been substantial changes in the surface albedo patterns over the past two decades and also show that the atmosphere has a diffusely scattering nature, even during times of low dust opacity.

Wind and dust storms in cap behaviour

The differences between the two residual caps are intimately linked with the movement of dust and other atmospheric activities, particularly that of CO_2 gas, and ices and clouds of both CO_2 and H_2O. Dust storms are a feature of the Martian seasonal cycle; thus, Viking recorded over 35 during 1977 and two of these gradually developed into global events. The Hubble space telescope has recorded many more, often following their development in considerable detail. The global storms tend to coincide with the perihelion retreat of the southern polar cap. Because of the eccentricity of the planet's orbit, insulation is 40 per cent greater than at aphelion and leads to increased wind activity; however, it is doubtful that it could raise enough dust to generate the massive storms observed. More importantly, near perihelion a large temperature gradient exists between the newly exposed circumpolar surface and the residual cap. Where considerable topographic features exist, this gradient is believed to be sufficient to generate winds fierce enough to raise large amounts of dust and inject them into the atmosphere (Kieffer & Palluconi 1979).

Another dust-raising phenomenon is found in tidal circulation on a global scale. The wind strength largely depends upon the degree of atmospheric heating, which in turn is influenced by the amount of dust in the air. If the air is heavily dust laden, winds will tend to become more intense until they are strong enough to raise more dust. This regenerative process begins on a local scale, when tidal winds and disturbances arising from local topography produce small-scale dust storms, which are believed to be fairly widespread. Support for this idea also comes from the generally

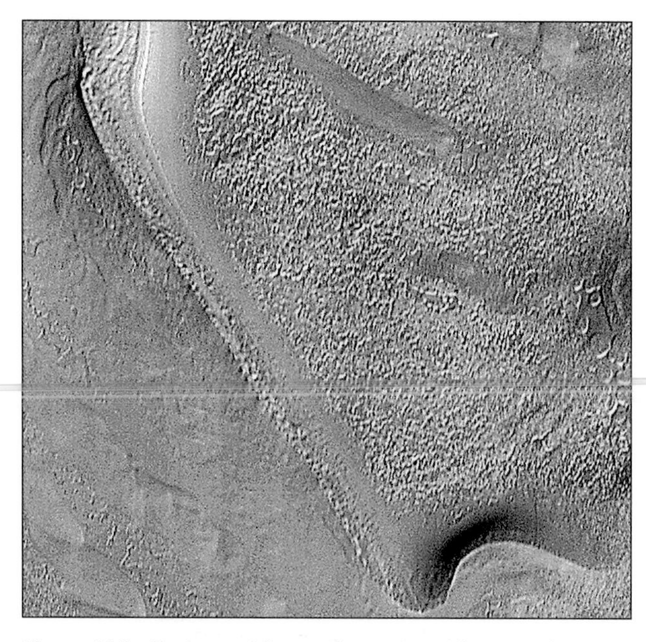

Figure 12.6 Textures of the south permanent ice cap, showing a 30 km^2 area at latitude 87°S 77°E. MGS image PIA00809

pinkish hue of the Martian sky as imaged by both the Viking and Pathfinder landers, suggesting that the air may be more or less permanently dust laden.

Returning to the differences between the polar caps: the northern seasonal cap grows during perihelion, when large amounts of dust are held in the atmosphere; it is likely therefore that much dust is entrained in the ice that is precipitated. Thus, during northern spring and summer, dirty CO_2 ice is exposed to solar radiation through a relatively transparent atmosphere; thus, all of the CO_2 dissipates. This leaves only a water-ice residual cap. In contrast, when the southern seasonal cap forms, the air is relatively clear. The clean CO_2 ice cap is then protected from the Sun's rays by a dirt-laden atmosphere, with the result that much of the CO_2 ice remains in the residual cap.

The MGS orbiting camera observed the recession of the southern cap during spring 1997, during which the Martian southern hemisphere remained fairly clear of dust. The extent of the cap, including the peninsula of ice known as the Mountains of Mitchel (typically observed during spring retreat), was almost identical with that observed by Viking during 1977, when severe dust storms affected the planet. It seems, therefore, that the annual sublimation of CO_2 remains unaffected by variations in annual dust-storm activity.

The MGS camera also recorded the textural features of the permanent cap surface, revealing a variety of striations apparently indicative of materials of varying resilience. Figure 12.6 shows both the residual ice cap and adjacent layered units (Fig. 12.6).

The new TES data also confirmed earlier Mariner 9 and Viking observations that cap retreat is asymmetric (Christensen et al. 1998); indeed, it became markedly more asymmetric as the spring season progressed. Atmospheric particulates were pervasive in spectra of the south polar region, and cloud spectral features are strong near the edge of the retreating cap. Observations with a resolution of 45 km revealed that 200 km-wide dust clouds formed over the cap, suggesting that small dust storms started near the cap edge and possibly within the seasonal polar cap. Of course, this is the present-day behaviour of the polar caps. The pattern must change with time because of the processional cycle; thus, 25 000 years from now, the southern pole should have the water-ice residual cap.

Laminated polar deposits

Close to both poles, the thick sequences of layered deposits obscure the older cratered plains down to about 80° latitude. These extensive blankets of sediment are each broken by a series of swirling valleys, in the walls of which a prominent light/dark stratification is revealed (Fig. 12.7). Within the valleys, whose width may be as much as 100 km, the stratification is predominantly horizontal. The surface of the layered units is everywhere smooth and lacks impact craters, and implies they must be very young. Estimates suggest this age to be about 100 million years.

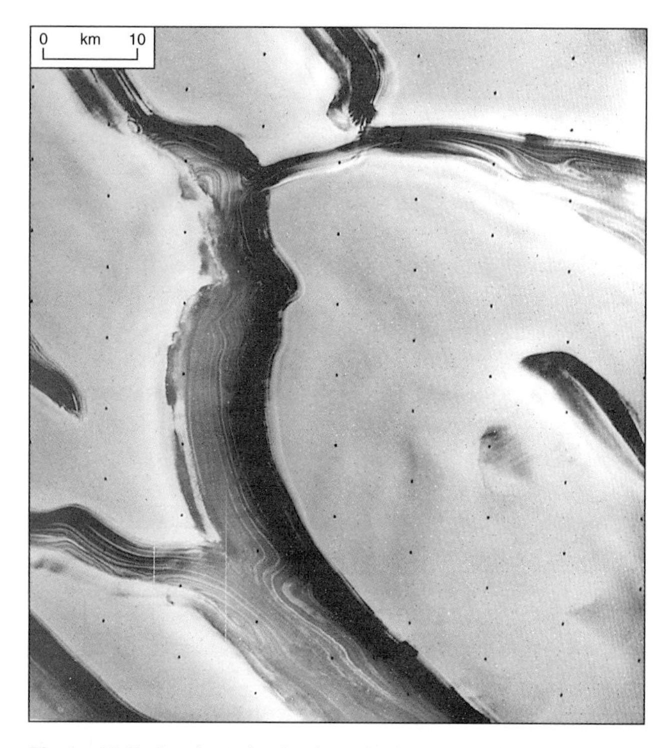

Figure 12.7 Laminated polar deposits exposed in the walls of polar valleys. Viking orbiter image 065b74; centred at 82°N 338°W.

However, it is likely that only the visible deposits are this young, and that they obscure older yet similar deposits beneath. Similar older deposits have been located elsewhere on the planet, and although these now appear to be devoid of ice, Schultz & Lutz (1988) have suggested that they may be evidence for polar wandering on the planet, as a result of changes in Mars' moment of inertia.

Extent and thickness of layered deposits

The deposits extend between 1000 km and 2000 km from the poles. The alternating sequence of light and dark strata, each of which is typically tens of metres thick, outcrop often as a series of steps with cliffs and terraces. Seasonal frost (largely CO_2 with traces of water) forms each winter and extends down to latitudes ~60°. This deposit then sublimes each spring, leaving the permanent residual caps. The summer equilibrium temperature of the north cap suggests that it is composed of water ice, whereas the temperature regime of the southern pole implies CO_2 ice, although water ice may exist at greater depths. In the southern hemisphere, layered deposits overlie older cratered terrain, whereas in the north they overlie cratered plains deposits (Malin et al. 1998). Some surfaces within the area of the southern layered deposits are smooth and almost featureless, but sinuous scarps revealed by MGS high-resolution images indicate that materials of differing resilience are present (Fig. 12.6).

Various estimates of the thickness of the sequence have been made. The general consensus is that they are 1–2 km thick in the south and 4–6 km thick in the north (Dzurisin & Blasius 1975). Around the northern pole, the valleys spiral outwards in a counterclockwise direction; around the southern pole, the reverse happens. Cutting across this general trend are two very prominent and much larger valleys, one in the north (Chasma Boreale) and the other in the south (Chasma Australe).

Thin alternating light and dark bands probably preserve the record of seasonal and climatic deposition and, indeed, erosion of dust and ice. Topographic profiles do not show steps that correlate with individual layers at the edge of the cap, which is not surprising given the intricate multi-scale nature of the layering revealed by the MGS camera (Fig. 12.8).

Figure 12.8 MGS high-resolution image of an area of polar layered deposits. Note the angular unconformity with horizontal strata overlying inclined ones. PIA02071; centred at 78.5°N 120°W.

Composition, volume and volatile content of the layered deposits

As we have seen, the polar deposits are accumulations of ice and dust. The preferential accumulation of such deposits at the north pole may reflect the fact that the atmosphere is dust laden when the northern seasonal cap forms, the carbon dioxide scavenging dust from the atmosphere as it condenses. As the precessional cycle takes its course, presumably the southern pole becomes the site of preferred dust and ice deposition.

The two polar caps are the largest permanent reservoirs of surface volatiles on Mars, the southern being apparently smaller. However, the layered deposits associated with the southern cap are more extensive and less symmetrical than those in the north. Residual ice, which persists throughout the seasonal cycle, is much more limited in extent at the southern pole and is offset from the present rotational pole towards 37°E. Furthermore, the south polar region is more elevated within the residual deposits (87°S, 10°E), where a broad dome is present with more than 3 km of relief at one end of the cap.

What appears clear, now that better altimetric data is to hand, is that the area of probable ice-rich material greatly exceeds the region of residual ice that is seen in imagery of the southern cap. Support for this idea comes from the existence of distinctive plateaux regions that correlate with layered terrain units, as would be expected if the layers were deposited on cratered terrain. In addition, impact craters within the plateaux share unusual geometric properties with their counterparts in the northern polar region, which are seen to have formed in an ice-rich substrate. This similarity suggests that significant portions of the south polar ice cap may be buried beneath mantling dust deposits.

MOLA profiles across both poles show a striking correspondence in shape that argues for similarity in composition and suggests that the southern cap may have a significant water-ice component. The surface exposure of the residual southern polar cap has been observed to display a CO_2 composition that led to the idea that CO_2 is the dominant volatile of the southern cap. However, recent experiments on the rheology of solid CO_2, coupled with relative elevation measurements, suggest that water is the more likely dominant volatile of the southern cap.

Smith et al. (1999a) calculated that the volume of the southern ice cap is 2–3 million km^3 and that it extends for 1.44 million km^2. The average relief above the surroundings is 815 m. The volume of the southern polar cap (ice and layered deposits) is greater than the northern, the volume of which is more difficult to establish because of the presence of a considerable dust load. For an average thickness of 1030 m and an area of 1.04 million km^2, Zuber et al. (1998) estimate a volume of 1.2 million km^3. Together, these estimates give a total polar surface volatile inventory of 3.2–4.7 million km^3, equivalent to a global layer 22–33 m deep.

Spectral data for the cap regions

The polar caps are characterized by highly variable low-emission zones that exhibit anomalously low brightness temperatures and erratic behaviour. It has been suggested that their occurrence is in some way related to the condensation of CO_2 in the Martian atmosphere (Forget et al. 1995), and that precipitating CO_2 clouds with particles larger than 10 μm radius and CO_2 snow deposits with millimetre-size grains could account for the observed features.

Analysis of Viking IRTM data allowed Forget & Pollack (1996) to determine the true temperatures of the polar caps by removing the effects of the low-emission events. They found that there was a systematic temperature decrease towards the pole in both hemispheres, and that the southern cap is about 5 K colder than its northern counterpart. Temperatures in the north were calculated to be about 145.5 K and at the south they average 141.1 K. The low-emission events appear to occur much more frequently in the northern hemisphere, a difference that is especially striking between latitudes 70° and 85°. Brightness temperatures as low as 125 K were observed at L_S = 295° and 320°; in the south a low of 128 K was recorded several times during the winter seasons. The low-emission events virtually disappeared immediately following the second dust storm of 1975, whereupon a period of minimum brightness temperature ensued. A detailed appraisal of the data revealed that a high frequency of extremely low brightness temperatures was reached around L_S = 300°, just after the peak in dust optical depth recorded by the Viking lander probes.

Thus, there appears to be a correlation between airborne dust and low emission zones that Forget et al. (1996) explained by the infrared emissivity of dust particles, which strongly increase the cooling rate of the polar night atmosphere and thus favour the condensation of CO_2 ice in the atmosphere rather than on the surface. This relationship may also explain the strong hemispheric asymmetry with respect to the low-emission zones.

Other low-emission events have also been recorded near the winter solstices at lower latitudes, these being associated with such features as the permanent cap boundary, Chasma Boreale, and several impact craters. This appears to imply an orographic influence on brightness temperature too. For instance, the relief might set up adiabatic cooling that would enable CO_2 to condense locally.

Extrapolation of the IRTM data enabled Forget et al. to conclude that the low-emission zone – believed to be caused by the radiative properties of fresh CO_2

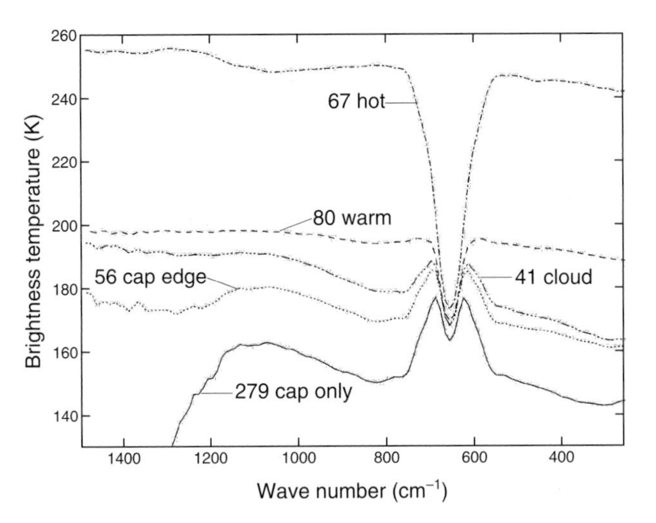

Figure 12.9 Averaged TES spectra for each of several conditions in the south polar region (after Christensen et al. 1998).

snow and precipitating CO_2 clouds – strongly alters the polar radiative budget, introducing major hemispheric inequalities that lead to major changes in the amount of CO_2 that condenses during winter.

Mariner 7 spectra were studied by Calvin & Martin (1994), who found significant spectral variation in the cap interior. They were able to identify distinctive regions at the cap edge, where there was enhanced abundance of water frost, and in the cap interior, which were characterized by spectral features of CO_2 at grain sizes of several millimetres to centimetres. Additionally, they defined an unusual region probably indicative of layered deposits, and a fourth one of thinning frost cover and transparent ice, well into the interior of the seasonal cap.

Mars Global Surveyor TES data have been used to define three spectral end members for Martian materials: solid surface, atmospheric dust and atmospheric water-ice clouds. In low latitudes, TES studies are relatively straightforward, but under the very low temperatures typical of the Martian winter poles, thermal spectroscopy becomes hazardous. However, by synthesizing a seven-band thermal emission mapper, Christensen et al. (1998) were able to identify particulates in the atmosphere in their TES spectra for the southern polar region (Fig. 12.9). The observed consistency in brightness temperature at $670\,cm^{-1}$ indicates that the upper atmospheric temperature drops only about 10 K over the polar ice cap. Atmospheric dust appears in the emission spectrum, as do water-ice clouds. The cap-edge curve represents an average of 56 spectra taken near the southern cap edge, and their warm curve is the average of 80 spectra of frost-free ground surrounding the cap. The cloud curve is the average of 41 spectra, with strong contrast between 11 and 20 μm. The hot curve represents an average of 61 spectra well off the cap area, where temperature is >210 K.

The Christensen et al. analysis indicates that from L_S = 185° to 223° the seasonal south polar CO_2 cap retreated continually, and the atmosphere above the cap gradually warmed up. Temporary dusty incursions took place sporadically, with most atmospheric dust occurring close to the cap periphery. Throughout much of the period, the upper atmospheric temperature had a consistent thermal gradient of about 0.2° per degree latitude away from the pole. The minimum temperature at the pole rose gradually from 148 K at L_S = 185° to 180 K at L_S = 225°. By L_S = 222° the dust opacity had reached 0.3 over nearly the entire cap, and atmospheric temperatures showed a strong diurnal dependence. At this time, the minimum temperature was recorded at 75°S.

The dust opacity appears to reach a maximum near the cap edge and over the adjacent frost-free terrain at the time of the spring recession. The highest opacities probably result from the dust lying at the 3 mb level, the associated spectra indicating that at this time the highest atmospheric temperatures are close to the surface; hence, the dust will be there too.

One suggested scenario for polar cap activity is shown in Figure 12.10.

The polar valleys

MOLA traverses clearly reveal the pattern of canyons and spiralling troughs that cut the upper horizon of layered deposits (Figs 12.1, 12.2, 12.3). In places these are 1 km deep and they penetrate down almost to the level of the surrounding plains. The Global Surveyor TES data indicate that the canyon interiors have significantly lower reflectivity than the exterior, implying that the former are sinks for windblown dust.

Cutts (1973) and Sharp (1973b) suggested that wind was responsible for canyon formation, although they noted that in the north the valleys mostly swirl in the anti-Coriolis direction. In the present state, downslope windflow from the cap tends to spiral westwards to about 80°N because of the Coriolis force, but the troughs trend to the east. Therefore, troughs cannot be explained simply by ablation associated with downslope winds. Additional factors, such as the position of the Sun, may be required to explain trough formation.

Cutts et al. (1979) made the alternative suggestion that the valleys and scarps are not attributable to erosion at all, but to zones where there was no deposition because of higher surface temperatures on the darker sloping terrain. Howard (1978), on the other hand,

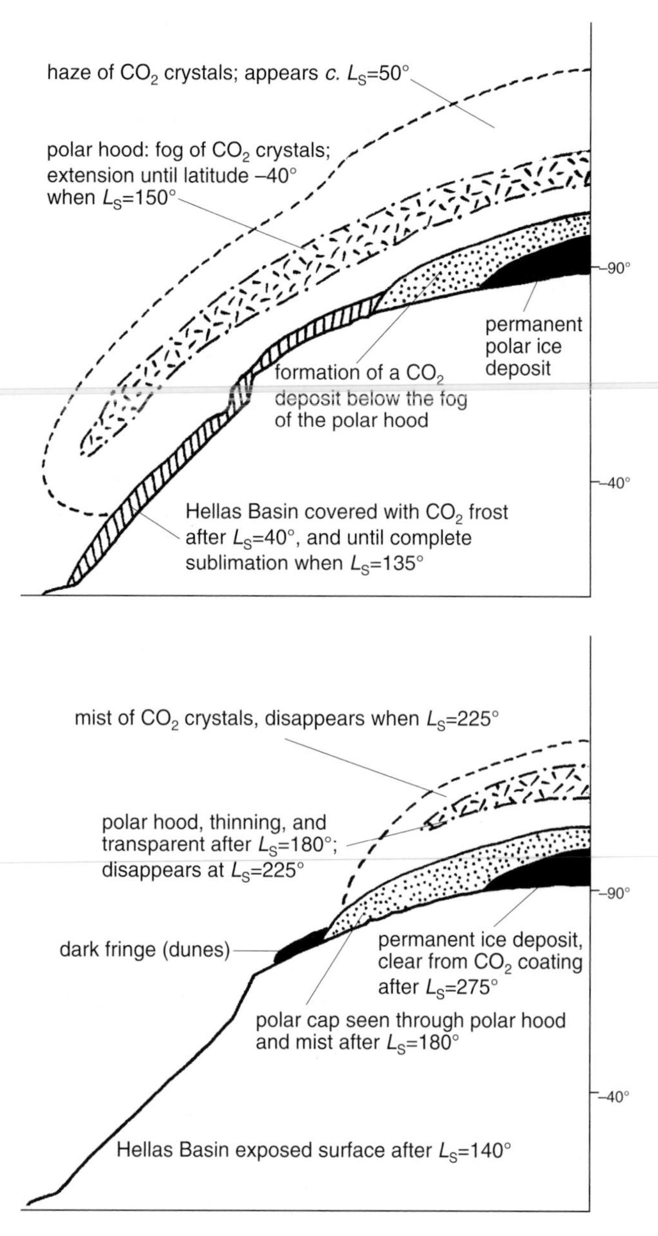

Figure 12.10 Scenario for seasonal activities at the polar caps according to Dollfus et al. (1996).

proposed that the present polar topography is in dynamic equilibrium with the climate, with volatile ablation occurring on the dark scarps and deposition on the icy flats. The ablated dust is believed to be removed by wind. The consequence of this process is a rough linearity and parallelism of escarpments, probably attributable to scarp retreat on a regional slope. The spiralling pattern of valleys is, according to Howard, explained by the more rapid retreat of escarpments facing slightly west of the equatorward meridian, that is, in the direction of greatest solar and atmospheric warming.

The circumpolar plains and dunefields

Surrounding the laminated polar deposits and passing beneath them polewards of latitude 80° are sparsely cratered plains. In the southern hemisphere the plains are etched by pits, which are presumed to have been hollowed out by the wind; they also exhibit aeolian fretting (i.e. etching by abrasion). Large areas are characterized by polygonal fractures, recently imaged at high resolution by Mars Global Surveyor (Fig. 12.11); such patterns are common on Earth in permafrost regions. The polygons, 2–18 km in diameter, are picked out by low-albedo troughs separating rather rounded higher-albedo mounds (Fig. 12.7). Helfenstein & Mouginis-Mark (1980) suggested that these are ice-wedge polygons. On Earth, they form where ice shrinks upon cooling in annual cycles; on Mars the cycles are significantly longer (100 000–1 000 000 years), which may help to explain why they are so much larger than their supposed terrestrial counterparts.

Dunefields and formation of duneforms

Aeolian activity in the northern hemisphere has created vast circumpolar dunefields. Dunes, also, are found close to the southern pole, but are far less extensive. The dunes that do occur tend to be confined to

Figure 12.11 Patterned ground in the Martian south polar region. Frost trapped in cracks, enhances polygonal cracks that typify the plains of the polar regions. MGS image PIA02344.

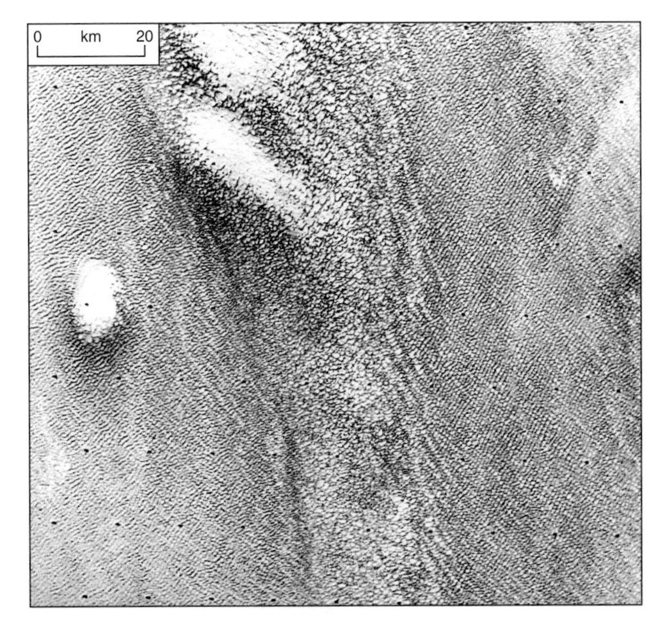

Figure 12.12 Part of the massive northern dunefield. Viking orbiter image 060b02; centred at 80.2°N 176.92°E.

the interiors of impact craters or within the vicinity of escarpments. The northern dune collar is sometimes several hundred kilometres wide and covers an area of approximately a million km^2 (Breed et al. 1979). The predominant landforms within the northern collar are transverse dunes, although crescentic barchans tend to occur towards the edges of major dunefields or near escarpments (Figs 12.12, 12.13). However, longitudinal dunes are relatively rare. On Earth they are typical of depositional sinks, whereas transverse dunes generally occur where sand is being actively moved through a region. Lee & Thomas (1995) argue that, since bimodal and multi-modal transportation regimes are required on Earth to generate longitudinal duneforms, their general absence on Mars is a reflection of the lack of such conditions on Mars.

The great majority of duneforms can be explained quite satisfactorily by current wind regimes, as indicated by such features as wind streaks, which originate down wind of both dunes and crater rims. Thomas & Gierasch (1995) concluded that the approximately concentric arrangement of layered deposits and dunefields surrounding the Martian poles are probably a manifestation of the virtually steady-state dispersal of material released from the polar region during spring and summer (Fig. 12.14). Around the north pole, the effects of winds between 90° and 135° azimuth dominate. Deviations from orientations in accord with this zone are all less than 20°.

Thomas & Gierasch suggested that the largest developments of dunes surrounding the northern polar region appear to be confined by on- and off-pole

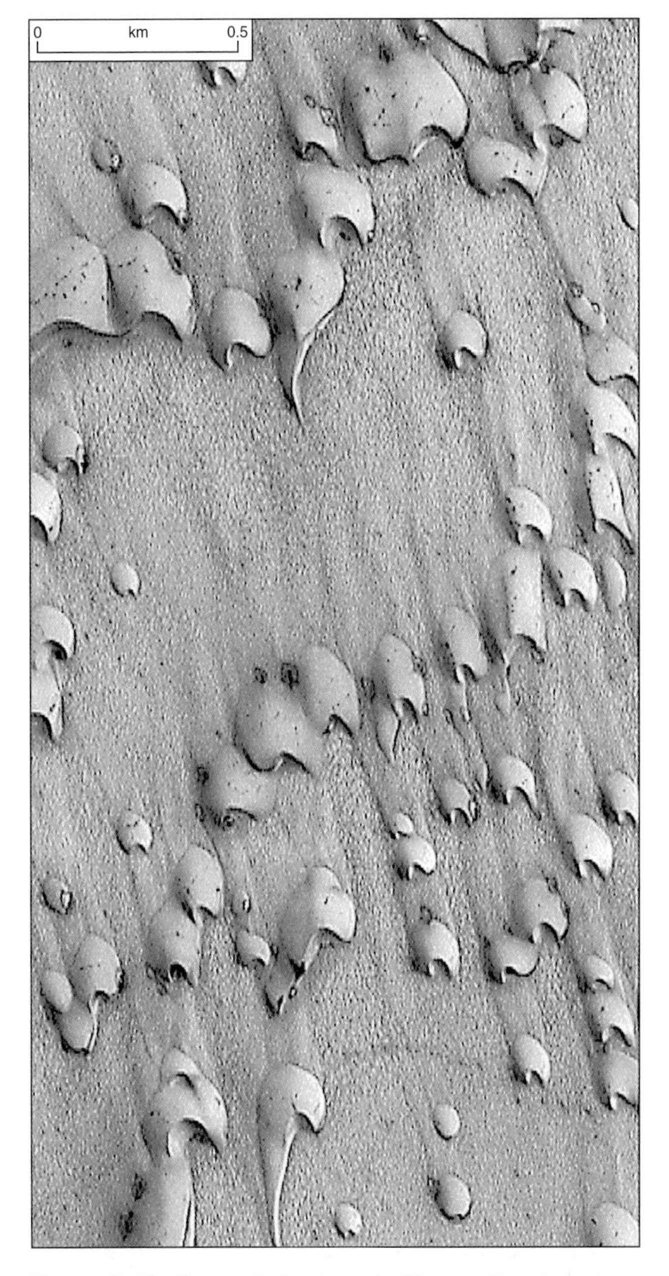

Figure 12.13 Crescentic barchans in Chasma Boreale, a huge trough in the northern cap. It is possible to see dark sand emerging from the bright frost left over from northern winter. MGS image PIA02069; centred at 82°N 52°W.

airflow. They also suggested that the strong albedo contrast between the dunes and adjacent ice-rich areas may generate winds that strengthen as a result of their confinement between duneforms, rather like sea breezes on Earth. This contrasts with projected wind activity at the southern pole, where, although the dunes face more pronounced topographical obstacles to dispersal, the very obstacles prevent confined winds akin to sea breezes being set up, enabling the transported materials to be dispersed more widely.

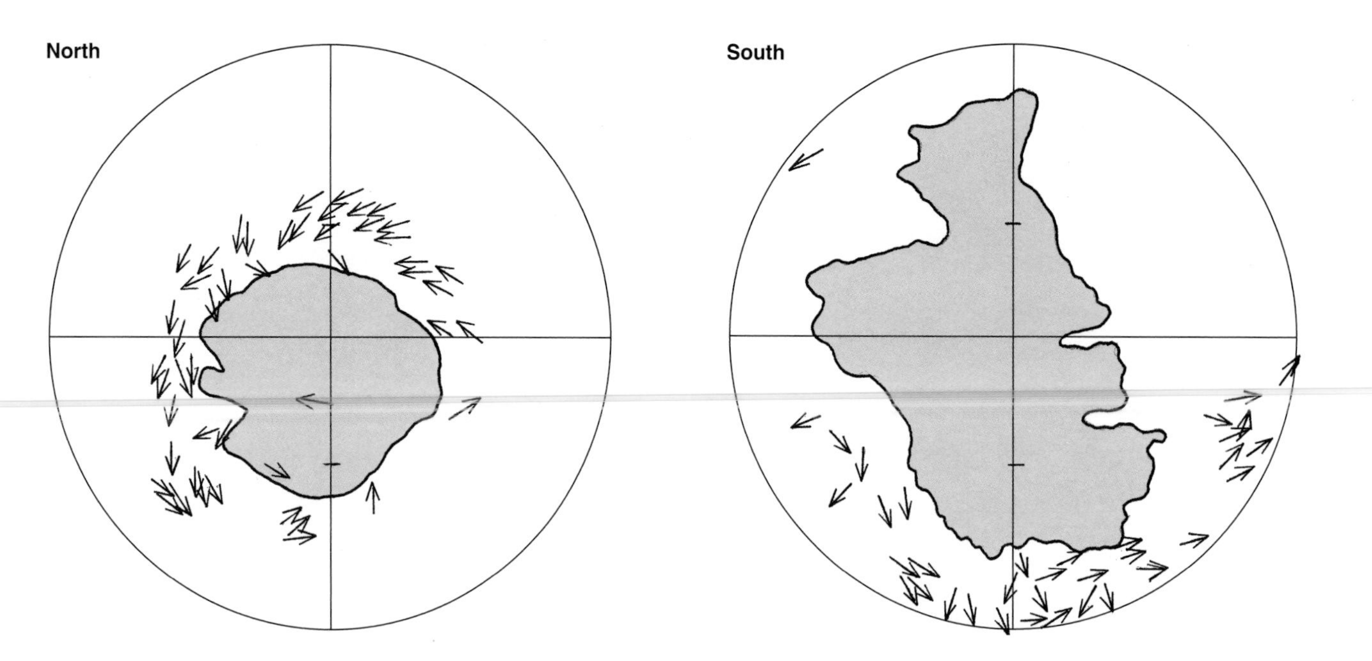

Figure 12.14 Distribution of major deposits in the polar regions and prevailing wind directions as indicated by wind streaks and dune orientations (after Thomas & Gierasch 1995).

Tsoar et al. (1979) found that the majority of the northern dunes showed the effects of multiple wind directions. This has recently been confirmed by Malin et al. (1998), who observed that complex forms imaged by MGS indicate juxtaposition of fresh and degraded dunes. Star dunes and multiple transverse dune sets have been recorded at high resolution for the first time, showing that episodes of multiple wind direction must have operated on the planet, possibly being associated with precessional changes (the 51 000-year perihelion cycle, for instance). MGS has also revealed juxtaposition of dark and bright duneforms, their patterns often crossing or being superposed on one another. Although most dune forms have relatively low albedo, the implication is that a variety of materials is available for aeolian transport in some areas.

Since most terrestrial dunes are composed of quartz or, more rarely, calcite sand, it is interesting to speculate about where the material for the Martian polar dunes came from. Particulate material sampled at both the Viking and Mars Pathfinder lander sites was in the micrometre (clay-size) rather than the millimetre (sand-size) grain size and the particulates did not include quartz. Indeed, it is difficult to envisage much quartz existing on the planet at all, since continental-type silicate rocks appear to be absent. On Earth, clay-size particles do not normally form dunes; thus, there is something of a problem here. There is of course the possibility that the considerable volume of potential fluvial material (from the large flood channels) that drained towards the northern polar regions may have been a major sand source, as are most terrestrial fluvial systems, or the material could be released from the polar deposits.

Physiography of marginal polar regions

We now turn briefly to those areas of the high latitudes that are marginal to the polar caps and adjacent to the circumpolar dunefields. One prominent landform is Chasma Boreale, a large re-entrant in the northern cap at longitude 300°E. It is about 350 km wide at its entrance and 600 km long. MOLA data show that the relief of the chasm is ~1.5 km from the floor at its headwall to the plateau above. It is likely to have been shaped by aeolian processes and currently is a major collecting ground for dark dune material.

Another major feature is Olympia Planitia, a fan-shape dune-covered area detached from the main northern cap at 80–85°N and 140–240°E. MOLA data indicate that this region slopes towards the pole and shows that regional topographic slopes are subtle, but with a much rougher local surface. It is a region characterized by straight-crested transverse dunes with average heights of 24 ± 9 m and crest-to-crest spacing of 2.4 ± 1.3 km. Zuber et al. (1998) estimated the sediment volume in these dunes alone to be of the order of 10 000 ± 300 km^3.

Additionally, the outliers of the residual ice cap that MOLA profiles show exhibit positive topography (10–1000 km) and widths of tens of kilometres. These are clearly too large to be simply seasonal and they must represent remnants of previous polar-ice deposits. The local concentration of irregular depressions within them tends to confirm that this is an area where residual ice melted or sublimed.

MOLA also sampled over 100 impact craters in the north polar region. Virtually all craters within 100 km of the permanent ice cap display a level of infill beyond that which can be explained by isostatic uplift or interior modification by wall slumping. In many cases at least 70 per cent of cavities have been infilled. Generally speaking, those craters nearer to the edge of the permanent ice cap contain greatest fractions of fill. These data support the contention that cavities of these impact features have been filled by non-impact materials, such as ice and dust, or as a result of previous polar-cap advance.

The steepness of the slopes along the periphery of the ice caps and in the spiralling canyons indicates that ablation must be the controlling factor on these aspects of morphology. Thus, at any local point when the temperature is >150 K, CO_2 frost will condense onto the surface. When the temperature rises above that level (and temperatures of 205 K have been recorded in northern summer), the frost rapidly sublimes back into the atmosphere, as may water ice. Warm parts of the northern cap undergo ablation, particularly on south-facing slopes, causing these to be steeper than north-facing slopes. As northern winter approaches, water vapour condenses out onto the ice cap, initially nearest the pole, but eventually over the entire northern polar region throughout the winter. The smooth profile of most of the ice cap may be the result of viscoplastic flowage.

On the basis of topography in Olympia Planitia and the presence of residual ice patches in the mantled plains surrounding it, it is likely that the material underlying the longitudinal dunes is an extension of the polar cap deposits and that patches of residual ice encircling this region represent remnants of the former extent of the polar cap. The continuation of several of the spiral troughs from the residual cap into Olympia Planitia tends to support this conclusion. This implies that the modification and erosion of previous polar deposits make an important contribution to the sedimentary material currently found in the circumpolar mantling deposits.

Finally, in these high latitudes the activity of Martian winds and the availability of copious dune-forming material are nicely illustrated in Figure 12.15, a region close to the southern pole. It shows a 40 km impact crater almost completely filled with a tail of aeolian material, evidently formed under the influence of strong prevailing winds. Towards the bottom left, it also shows, transverse, dark, streaks that must be related to a completely different wind regime. Such is the complexity of modern Martian geology.

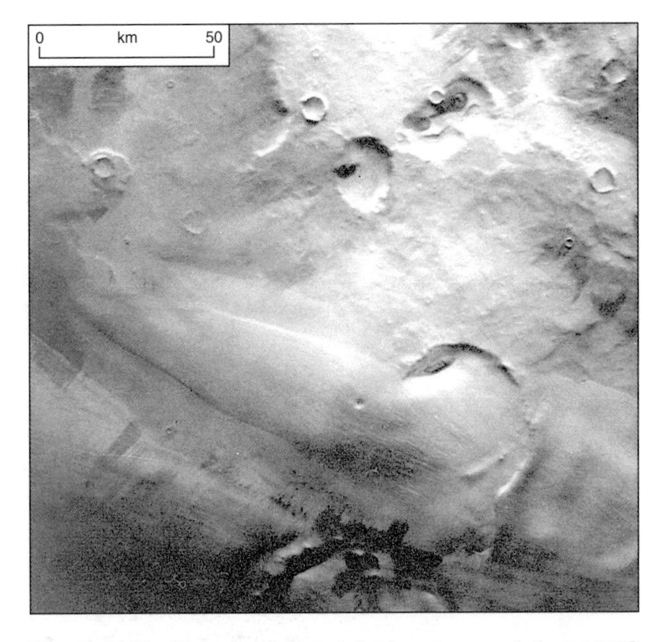

Figure 12.15 Strongly orientated dust/sand accumulation at 70°S almost burying small impact crater. Note also the dark wind streaks crossing another dust accumulation towards the bottom of the image. Viking orbiter image 479b57; centred on 72.03°S 193.84°W.

13 The dichotomy

One of the most striking features of Mars is the difference between the northern third and southern two thirds of the planet (see Plate 7 and Fig. 3.3). The dichotomy boundary is the complex escarpment that defines the line of separation between the cratered uplands to the south and the lowlands to the north. One way to explain the north/south differences is to assume that the whole of the northern hemisphere has been lowered by erosion of the original cratered crust; however, this raises the seemingly insoluble problem of where the eroded material has gone. Alternatively, the low elevation could be the result of lithospheric thinning by vigorous mantle activity early in Mars' history. Indeed, Wise et al. (1979), suggested that convective overturn of the mantle may have led to a lowering of the entire northern hemisphere. However, there is no hard evidence that such a process ever occurred. Another proposal cited several huge impacts into the northern hemisphere as the lowering agency (Wilhelms & Squyres 1984).

More recently, Sleep (1994) invoked a form of early plate tectonics in seeking to explain the boundary, whereas Watters & Robinson (1999) have described lobate scarps that they consider attributable to thrust faulting associated with structural formation of the dichotomy boundary. That there has been some structural control on the form of the boundary seems very likely, but whether or not the boundary itself is a structural feature remains unclear.

It has also been suggested that the northern hemisphere was once covered by an ocean and that the dichotomy represents some kind of huge palaeoshoreline. It is certainly as smooth as Earth's abyssal plains, and the volume of water required to fill such a depression is less than the upper limit of the volume of water calculated to have been available on Mars early in its history. Clearly, this is an exciting hypothesis that merits further consideration here. In this chapter I explore some of the possible explanations.

Physiography of the dichotomy boundary

The altitude difference on either side of the dichotomy is marked by a contrast in the surface roughness. Thus, most of the northern lowlands are composed of Late Hesperian plains of the Vastitas Borealis formation, which is both flat and smooth. The overlying Arcadia formation, of Amazonian age, is also smooth at all scales, and is believed to have either a volcanic or a sedimentary origin. In contrast, the southern hemisphere is composed of Noachian ridged plains, which form locally flattish intercrater areas and more widespread Hesperian ridged plains. Both have relatively rough surfaces.

One of the more distinctive geomorphological features of lengthy parts of the boundary is the regions of fretted channel development (Baker 1982, Carr 1995). The flat-floored and somewhat sinuous fretted channels often penetrate deeply into the upland hemisphere and are characteristic of both Nilosyrtis and Protonilus Mensae (Fig. 13.1). The channels, anything up to 40 km across, are often seen to merge with large impact craters, which are usually highly modified and degraded. Almost everywhere, talus aprons may be seen at the base of escarpments or isolated mesas. In the latter case these may extend for up to 20 km across the surrounding plains. The surfaces of such aprons are finely striated, the striations being arranged normal to the scarp walls. Where the debris is confined, as it is between channel walls, the striae are longitudinal; they diverge and converge around obstacles

Figure 13.1 Lineated debris flooring a fretted channel in northern Arabia Terra. Such a pattern is suggestive of flow in a manner similar to that of glacier ice. Image PIA02075

and, where two valleys meet, the striations merge, rather like those in modern terrestrial glaciers. The very few impact craters observed indicate that they are very young features indeed.

The observed relationships seem to imply that the cratered plateau has been, and probably still is being, eroded by a process of scarp retreat, the eroded material being slowly moved towards lower ground by mass wasting. Squyres (1978) showed that the most prominent debris flows outcrop in two latitude bands 25° wide and centred on 40°N and 40°S. In these zones, ice precipitates out from the atmosphere during mid-winter and must be accumulating on the ground. He postulated that this ice mixes with debris being eroded from the scarp faces, and forms an ice–rock mixture that shares many of the features of terrestrial rock glaciers. Downslope flow is accomplished largely by slow movement of the interstitial ice, the debris glacier slowly creeping towards the northern lowlands. The continual removal of the debris from the scarp faces allows erosion to continue, slowly and inexorably adding to the degradation of the cratered plateau, remnants of which are left as mesa of variable shapes and sizes.

It seems clear, therefore, that today the dichotomy zone is a region of active erosion, where scarp retreat is eating into the older cratered terrain. Such activity must have been going on for a considerable time, as is shown by the extensive occurrence on the plains side of the dichotomy of knobby terrain, which consists of isolated knobs and small mesas that appear to be relicts of the old plateau. This knobby terrain outcrops in two main settings: along one half of the boundary between upland and lowland, and in the lowlands north of it. Since much of this terrain has the knobs arranged in roughly circular patterns, it has long been assumed that it developed from the upland cratered plateau. By studying impact-crater densities within various regions of such modified terrain, Wilhelms & Baldwin (1989) showed that much of the knobby terrain has an Early Hesperian age: north of the line of dichotomy, there are surfaces as old as the upland cratered terrain.

Earlier explanations for the dichotomy boundary

In an attempt to explain the situation, Wilhelms & Squyres (1984) proposed that one of the very earliest events to have affected Mars was a giant impact that gave rise to a vast 7700 km-diameter basin. This supposed Borealis basin was centred at 50°N 190°W. On this basis, the northern lowlands occupy what was the interior of the multi-ring structure. It follows logically that, because this major cosmic event pre-dated all others, Noachian heavily cratered units were emplaced on its floor and therefore must exist north of the lowland/upland boundary.

The notion of a single huge impact early in Martian history has not received universal support. To remove the need for one vast impact, Frey & Schultz (1989) proposed that perhaps several large impacts were responsible for lowering the northern third of the planet and, by invoking several overlapping large basins, they consider that the pattern of knobs within the northern lowlands can more reasonably be explained. The principal problem with this idea is that it is difficult to envisage why concentration of such large basins should be so effectively polarized into one third of the surface area. The latest gravity data weaken the theory and reveal a complete lack of mass concentrations that might be attached to such structures.

McGill (1989) argued that relatively little thinning of the ancient crust by removal of surface materials can have occurred since Middle Noachian times. Because the oldest impact craters now buried beneath the northern plains had became substantially fractured and eroded prior to burial, they survive now as isolated knobs and knob rings. By applying established dimensional equations for craters, McGill showed that a maximum of around 200 m of lowlands-wide erosion of the preplains surface could have occurred. If this is so, then clearly it invalidates some suggestions that 2–3 km of ancient crustal material must have been removed north of the line of dichotomy to produce the northern lowlands. Furthermore, it implies that the present position of the boundary cannot be far removed from its pre-Noachian position. To him it implies that, far from being erosional, the dichotomy scarp so characteristic of the upland/lowland boundary in the eastern hemisphere must be structural in origin, albeit modified since its time of formation.

Of greater popularity has been the hypothesis that the dichotomy is a shoreline related to an ancient Martian ocean or palaeolakes (Parker et al. 1989, Parker et al. 1993, Scott et al. 1995, Clifford & Parker 1999). Parker et al. (1989, 1993) mapped two possible shoreline levels, and Squyres et al. (1992) and Thomas et al. (1992) both described distinctive geomorphological features indicative of potential shoreline levels associated with either palaeolakes or oceans. This notion is worthy of further discussion here, along with a more detailed description of the topography and elevation of the boundary region.

MOLA data and interpretation

MOLA data has confirmed that the northern hemisphere is smooth in the mid- to high latitudes and that the mean elevation north of 50°N is about −4 km. Within this region there is an elevation range of only 3 km. There is also an overall slope of 0.056° from the north pole towards the equator, which is the result of the separation of the centre of mass from the centre of figure[1] of Mars along its polar axis, plus the 1 km increase in the equatorial diameter introduced by the presence of the Tharsis Rise (Smith et al. 1998). In contrast, the southern hemisphere is heavily cratered and lies, on average, about 5 km higher than the northern plains (Fig. 13.2).

The dichotomy boundary zone has the highest regional slopes on the planet, which, on a scale of tens of kilometres, are generally between 1° and 3°. Locally, however, the slopes reach over 20°. Recent reanalysis of occultation data from Mariner 9 and Viking orbiter shows that the hemispheric dichotomy boundary is ~3 km offset from the centre of figure and centre of mass of Mars (Smith & Zuber 1996).

In the quest to establish the true nature of the dichotomy, it is vital to establish how much of the planet's hemispheric elevation difference is attributable to long-wavelength planetary shape as opposed to the boundary scarp. Figure 13.3 presents histograms of planetary topography, elevations after removing the offset between the centre of mass and centre of figure along the polar axis, and the distribution of 100 km-scale slopes calculated from the global topographic grid. The last is seen to peak at about 0.3° and has a long tail because of topography associated with volcanoes, impact basins and tectonic features. The centre histogram clearly shows that, by removing the offset between centre of mass and centre of figure, slope distribution becomes unimodal. Even a large part of the southern hemisphere, away from Tharsis and Hellas, is seen to have derived elevations close to those of the present northern hemisphere. The implication is that the mass/figure offset of Mars accounts for most of the hemispheric elevation difference and that it is a long-wavelength effect.

Another revelation of the new MOLA data is the distinctive circular signature of the buried Utopia basin, unobserved in earlier studies. Even allowing for the effect of Tharsis, which occurred presumably after the dichotomy was formed, the hemispheric boundary is not circular, as would be expected if the lowering of the hemisphere was achieved by a massive impact.

MGS gravity data shows that major impacts such as those that formed Utopia and Hellas are marked by distinctive positive anomalies; however, no other anomalies of comparable scale occur in association with regional topographic lows in the northern hemisphere, except for Isidis, which is located right on the dichotomy boundary. This appears to point to the long-wavelength topographic expression of the northern hemisphere having been moulded by an internal process or processes, thus supporting the contention of Watters & Robinson (1999).

One further factor is Hellas, which MOLA data show to contribute to the boundary topography in its vicinity. Associated with Hellas are high relief, a distinct gravity signature, plus large crustal magnetic anomalies. The present relief is 9 km, which must mean that the lithosphere was of considerable strength when impact occurred. If Hellas is ~90% isostatically compensated, as suggested by gravity data, then a basin relaxation model indicates that the elastic lithosphere at the time of its formation was at least 30 km.

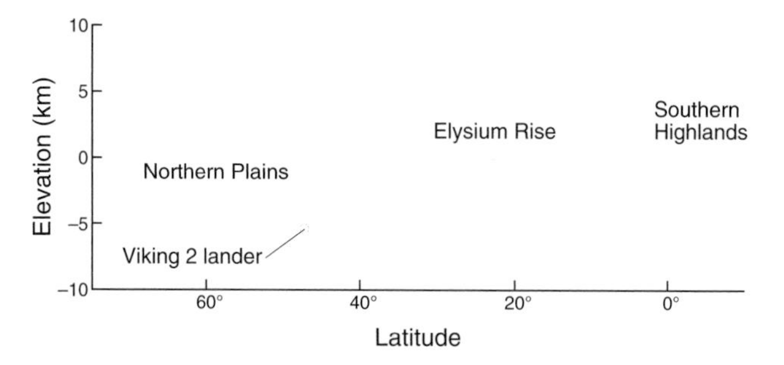

Figure 13.2 MOLA laser altimeter profile across the northern hemisphere of Mars; vertical exaggeration ×100.

1. The centre in terms of the spherical shape of the planet, as distinct from the centre of mass, which is unequally distributed and therefore offset.

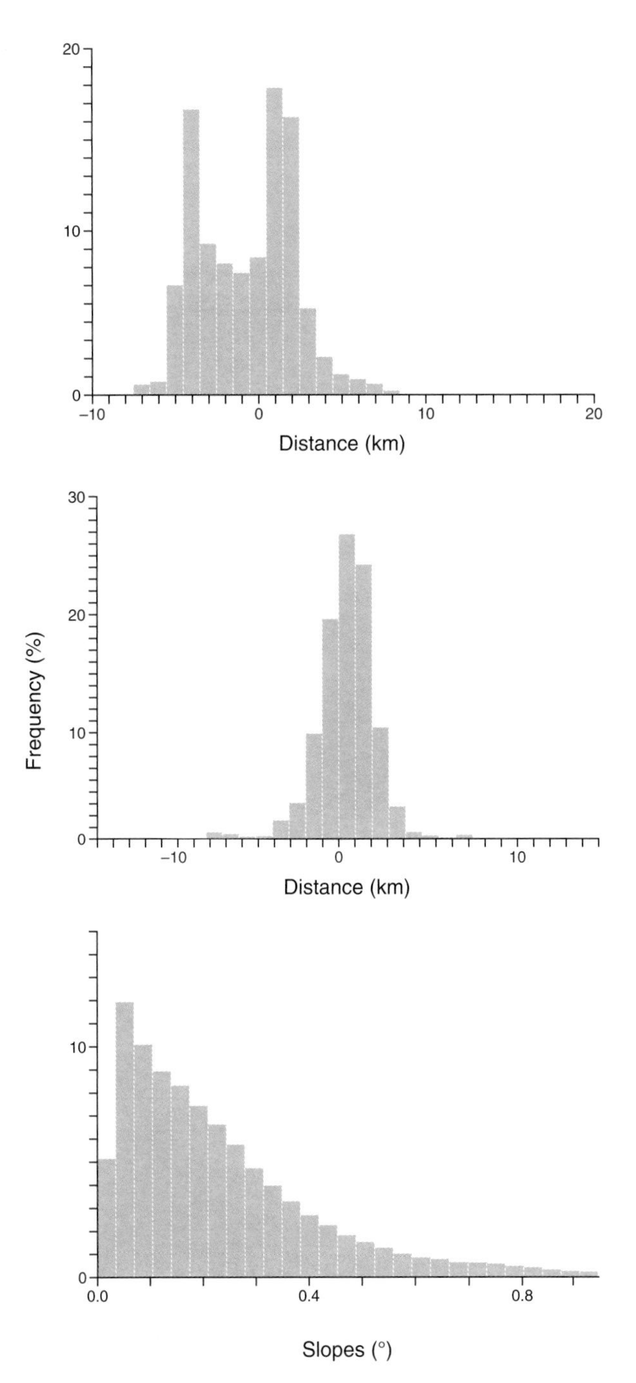

Figure 13.3 Histograms based on MOLA data for **(a)** topography, **(b)** heights with respect to an ellipsoid shifted by ~2.986 along the z axis, and **(c)** 100 km-scale slopes. (After Smith et al. 1999a.)

When all of the data is analyzed, the geological features of the dichotomy boundary and regional topography appear largely to be a manifestation of:

- volcanic construction largely associated with Tharsis
- major excavated deposits approximately circumferential to Hellas, with additional contributions from Isidis and possibly Utopia
- modification of the intervening region by fluvial processes associated with the outflow channels that empty Chryse Planitia.

In other words, the surface is the result of a complex interplay of various mechanisms.

Structural mapping along the boundary between Nilosyrtis Mensae and Aeolis (Maxwell 1989) indicates little correlation between the trends of ridges and scarps and the dichotomy itself. The exception to this is a wide trough extending from Isidis to the large crater Herschel, where scarp orientations are axial-symmetric, suggestive of continued deformation by downfaulting through the period of plains formation. Maxwell concluded that this implies the downdropping of the terrain north of the present dichotomy as a discrete block 300 km wide and he postulated that this could also be the case for the partially buried ground – now represented as the knobby terrain – which outcrops elsewhere to the north of the dichotomy. Although the faulting along this section of the dichotomy can be dated as Late Noachian to Early Hesperian, elsewhere along the boundary fracturing occurred at different times.

A recent analysis of lobate scarps in the region of Terra Cimmeria–Amenthes indicated them to be a manifestation of significant compressional deformation of the heavily cratered highlands close to the dichotomy boundary (Watters & Robinson 1999). They estimated that crustal shortening of the order of 2–3 km occurred locally, and concluded that a tectonic origin is indicated and that any satisfactory explanation must take into account both the lobate scarps and extensional fractures that are a feature of the boundary.

Ancient oceans and lakes on Mars?

The northern lowlands account for about a third of the surface area of Mars and must have played a major role in the planet's hydrological history (Carr 1996). Because most outflow channels flowed towards this region, it has been suggested that it was at least partially covered by an ocean or lakes.

MOLA data reveal that currently there are presently only three major closed basins that can act as sinks for volatiles. This is likely to have been the case over much of Martian history. Of the three, the low-lying northern plains area is by far the largest of the three. Its watershed constitutes about three quarters of the planet's surface area. The other basins are Hellas and that now represented by Argyre and Solis Planum, both in the southern hemisphere.

Although smaller than the northern plains region, Hellas is much deeper and would have a comparable volume. The highest contour is at 1250 m, at which level it breaches into the Isidis basin. Two outflow channels empty into the basin from the east, having probably formed because of interaction between one or more of the nearby large volcanic paterae and ground ice. Thus, lacustrine deposits could be an important contributor to the flat-lying deposits seen partially to fill the cavity.

Argyre has a small volume compared with Hellas, but has a watershed just as large. It breaches Chryse Planitia through a major system of outflow channels. The Solis Planum part of this watershed is structural and thus unusual (the others being impact related) and its drainage area shallow, the mean depth being only 500 m.

The global south–north gradient and transport of volatiles (as indicated by direction of the major outflow channels) towards the northern plains is, as we have seen, explained by the offset between Mars' centre of mass and centre of figure. Thus, if water on the surface of the planet was at any time ubiquitous, most of it would have flowed towards the northern plains, even from high southern latitudes. Furthermore, if the extended period of formation of Tharsis did not significantly modify regional slopes in the southern hemisphere, then Hellas and Argyre would each have collected water from about an eighth of the planet.

Some of the larger impact rings might also have had their own hydrological systems. Grin & Cabrol (1997), for instance, published evidence to suggest that Gusev crater and Ma'adim Vallis formed such a system, and that Gusev was the site of an ice-covered lake. Their analysis suggests that Ma'adim Vallis flooding may have entered the lake as late as Late Hesperian to Early Amazonian time and that the lake may have occupied the cavity for as long as 2 billion years ago. A similar palaeolake appears to have filled the crater Gale during Noachian times (Cabrol et al. 1999). This has left young floor deposits, streamlined terraces, stratified deposits and channelling.

Shoreline tests along the boundary

As we have seen, various workers have reported distinctive geomorphological features associated with the boundary. It remains to be seen whether or not the recent spate of new data shed new light on the past oceans hypothesis. There are two lines of enquiry: analysis of MOLA slopes data in the vicinity of the hemispheric boundary, and searching MGS high-resolution imagery that covers sections of the dichotomy boundary.

Parker et al. (1989, 1993) reported two contacts adjacent to the boundary, which they interpreted as potential shorelines (Fig. 13.4a). Head et al. (1999) tested both contacts, which, to be serious contenders as highstands, should represent equipotential surfaces. Using MOLA elevation data, they found that contact 1 (Fig. 13.4b) was not a good equipotential fit, being high in the Tharsis region, low in Arabia and with a range in elevation of at least 11 km. On the other hand, contact 2 provided a much closer fit, having a mean elevation range of only 3.7 km. The most substantial deviation occurs in Elysium and Arabia, where activity after contact 2 has probably taken place, and in Tharsis. These three areas accounted for 75 per cent of the departure from equipotentiality.

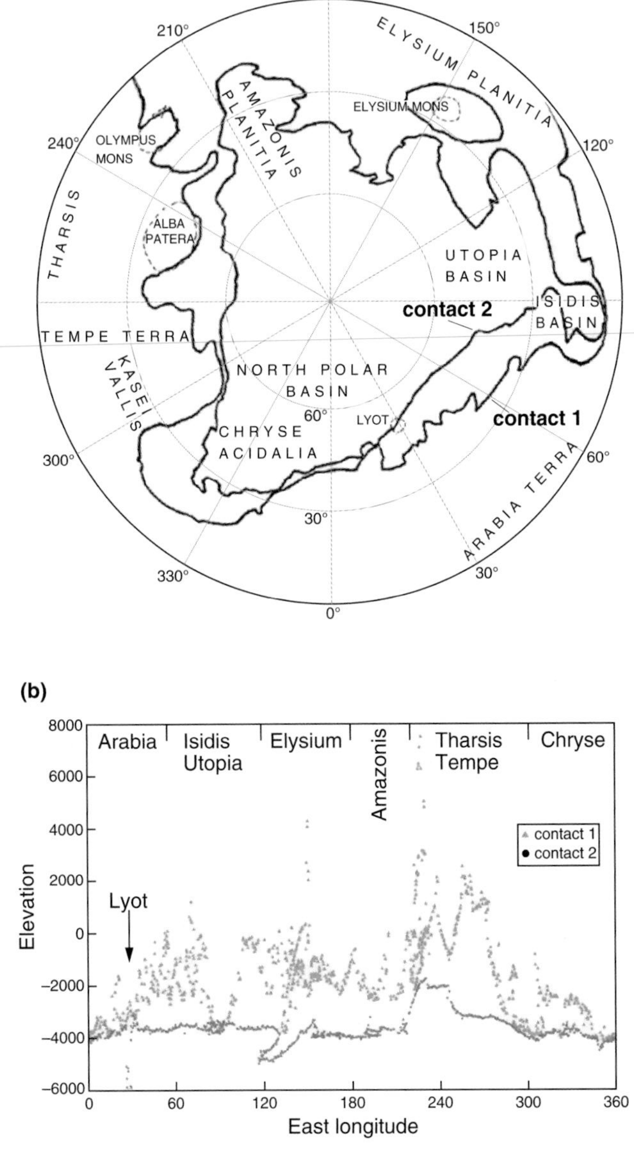

Figure 13.4 **(a)** Position of mapped contacts and major features around the hemispheric boundary; **(b)** elevation of contacts 1 and 2 as a function of longitude. (After Head et al. 1999.)

Parker et al. (1993) argued that there would have been sedimentation in any palaeo-ocean, such that one might therefore be able to identify terrain smoothing representative of originally submarine regions. MOLA, of course, has established the flatness of the northern lowlands, but additionally shows that below the level of contact 2 the surface is smoother at all scales than that between contacts 1 and 2. Perhaps there is some evidence here for such an ocean with contact 1 being at least one of its highstands.

Sedimentation would be a part of ocean activity and various workers undertook mapping of geomorphological features that characterized the dichotomy boundary. For instance, Lucchitta (1986) mapped the location and characteristics of possible sedimentary deposits in the northern lowlands. Head et al. (1999) found that there was a good correlation between polygonal fracturing mapped by Lucchitta and the deeper regions of both Utopia and the North Polar basin; this is consistent with their formation in bodies of standing water.

Another distinctive geomorphological feature is the nature of ejecta blankets surrounding impact craters with diameters greater than 2 km; they are typically lobed or multi-lobed, indicating the entraining of volatiles during emplacement. Kuzmin et al. (1988) assessed the onset diameter of lobate ejecta globally, which Head et al. (1999) re-analyzed. They found that the smallest diameter (< 2 km) craters correlated with the position of the two large basins within the northern hemisphere and seemed to provide support for the notion that they were formed in a region rich in groundwater. On the basis of the area defined by contact 2, they calculated a volume of water below contact 2 equivalent to 14 million km^3, which falls between the minimum and maximum values estimated to have been provided by the Chryse outflow channels. It would correspond to a global layer about 100 m deep.

By experimenting with various scenarios for channel emptying and ocean advance and recession, Head et al. (1999) observed that water would tend to pond. The lowest basin in the region is the North Polar basin, which could accommodate as much as 54 000 km^3 of water before infilling of the adjacent Utopia basin would begin. They found that, at a depth of 500 m, both basins would begin to fill and become interconnected through a narrow trough, whereupon each would hold over a million km^3 of water. Continued filling to the level of contact 2 yields 14 million km^3 and an average depth of about 560 m.

Outflow channels flowing into the low-lying region from both Chryse and Amazonis would flow into the North Polar basin, whereas those originating in western Elysium would have entered Utopia. Although the volumes that are attached to individual outflow flooding events are not well constrained, enough is known to suggest at least eight or nine such events would be needed to fill the North Polar basin to an average depth of 120 m. At least 40 such events would be required completely to fill the northern region up to the level of contact 2. This must have involved lengthy periods of time, and therefore highstands at various levels might be anticipated as water levels alternately receded and advanced. This being so, changes in slope might be expected at a range of levels, for example wave-cut terraces or, if the water margin was frozen, ice-cut features.

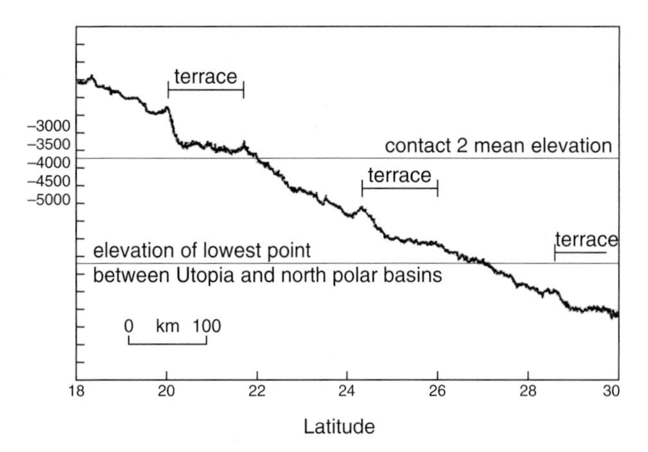

Figure 13.5 MOLA slopes map of the lower northern flanks of Alba Patera.

Head et al. (1999) located several linear slope changes that ran in tandem with one another and the topographic contours, and lay at a level close to contact 2 above. These are most prominent along the northern lower flanks of Alba Patera (Fig. 13.5), but also occur along the southern margin of Utopia. Their detailed assessment of the profiles crossing these linear slope changes indicates that several of the most prominent ones in Utopia are near the position of contact 2 and the approximate level where the two basins might first become interconnected. They could find no processes other than lake or ocean filling that would generate such features.

There has also been a fairly rigorous search of MGS high-resolution imagery for potential shoreline features and topographic benches along the dichotomy boundary. Viking orbiter imagery had led some workers to interpret the boundary northwest of Olympus Mons as two units separated by a cliff-like feature, the cliff facing in the direction of the smooth Amazonis plains. High-resolution images of the contact between Lycus Sulci highlands (hilly ridges located northwest of Olympus Mons) and the Amazonis Planitia lowlands reveal obvious differences in texture between the two units, but no nickpoint; simply a subtle rise across the area (PIA02338). Three more high-resolution images have been obtained, none showing a potential shoreline feature.

MGS also took images of the region of Cydonia, where mesa-like remnants of the upland plateau surface lie along the dichotomy boundary. This region was chosen as it had been suggested that, if the boundary did represent an ancient shoreline, benches might be found around the mesa remnants. None were found.

At first sight, this might seem to cast some doubt on the viability of the ocean hypothesis; however, bearing in mind that such an ocean would have been a fairly old feature of Mars, would we expect to see evidence of it at the present time? We know that Mars is and has been an active planet, with aeolian, fluvial and volcanic activity having spanned its history. I would find it extremely surprising if any high-resolution image reveals such a feature. My own guess is that it would have long since either been covered, intensely modified or removed altogether. The best one could hope to find would be a regional feature, perhaps the one identified by Head and his colleagues. My own view is that continued slope analysis might be the most profitable avenue of research.

The possibility of glaciation

There has been widespread support for the notion that certain kinds of landform have a glacial or fluvioglacial origin. An analysis of cold-climate features was undertaken by Lucchitta (1981), who also discussed the possibility of ice in the northern lowlands (1993) and in certain outflow channels (1982). Carr & Schaber (1977) described permafrost landforms. The distribution of ground ice received attention from Rossbacher & Judson (1981), Squyres & Carr (1986), Squyres et al. (1987) and Mellon & Jakosky (1993). Glaciation and past oceans were discussed by Baker et al. (1991) and glaciation was also discussed by Kargel & Strom (1992), and Mellon & Jakosky (1993).

Kargel et al. (1995) catalogued the many whorled ridges that give rise to thumbprint terrain and which resemble the eskers formed during some glaciations on Earth. Such terrain is quite widespread within the northern plains and was recognized earlier by Scott & Underwood (1991) and Chapman (1994). Similar landforms are also associated with Hellas in the southern hemisphere. Major areas of this type of landform occur within Isidis, along the dichotomy boundary between here, Protonilus Mensae and Deuteronolis Mensae, Arcadia, along the southwestern side of Utopia Planitia, north of Elysium Planitia, and on the east side of Chryse, again near the dichotomy boundary.

The thumbprint landforms mimic the form and scale of recessional moraines deposited by Pleistocene ice sheets on Earth, the major difference being the apparent absence of accompanying drumlins in the Martian development. However, it may be that insufficient till-like debris was available for floodwaters to build such landforms on Mars. Kargel et al. (1995) identify the landforms as ice-margin features and estimate that the largest individual area of such terrain approaches 420 000 km^2. Such landforms would form on Earth by processes set in motion by the action of water in a proglacial environment or in sub-glacial tillite (or both). They infer that thumbprint terrain on Mars was generated with the involvement of liquid water as well as ice. Their analysis concludes that the northern plains terrain were formed by active wet-based continental-style glaciers that appear to have been active in early Amazonian times.

On the assumption that such an interpretation is along the right lines, it is relevant next to note that the glacial terrains are concentrated at or near an elevation of −1 km, thus seeming to imply a connection with bodies of standing water (Luchitta et al. 1986, Chapman 1994). Where the same terrain is found at higher altitudes, for instance near Hellas and Argyre, it is presumably related to ponding within impact basins. One possible terrestrial analog would, in their view, be the Laurentide ice sheet in the Great Lakes region. In this scenario, a mobile ice sheet would entrain rock debris (later to be deposited as eskers and similar landforms). This would be followed by subglacial (thermal) or surface (seasonal insolation) melting, to produce meltwater that was released onto the bed of the glacier, thereby causing the cutting of tunnel channels, probably during stagnation of the ice sheet. Eskers were later built up along the tunnels, followed by deposition of smooth plains deposits in a proglacial lake. Finally, the lake and remains of the stagnant glacier evaporated, leaving glacial landscapes behind, in a kind of "freeze-dried" state.

An alternative scenario takes into account the absence of drumlin-like features from the regions of thumbprint terrain and the apparent absence of post-glacial fluvial modification from the Martian environment. In this version, the −1 km level at which the thumbprint terrain outcrops is connected to the existence of putative palaeolakes or oceans, as described above. Naturally, an ice-covered sea is not a glacier; however, they note that it could develop into one and preserve the geographical relationships characteristic of a former ocean. Assuming outflow channels discharged into an early ocean or lakes, the ocean may have begun freezing almost as soon as it entered the standing water; in this case the sea or lake would freeze quite quickly. On the other hand, if the discharged water originated in a deep aquifer, it would be relatively warm and the freezing time must have been longer. For two widely differing cases:

- initial surface temperature 200 K, initial water temperature 274 K and water depth 100 m
- initial surface temperature 250 K, initial water temperature 293 K and initial water depth 100 m

they calculate that freezing would commence after 19 days in the former case and after about 18 years in the latter.

If this interpretation is correct, it has far-reaching consequences for palaeoclimate. Since there are no ice sheets present today, there must have been a fairly dramatic climatic swing to bring about melting. If the glaciers formed in a way similar to Earth's ice sheets, and within a similar temperature regime, it seems to imply that, during the Martian summer season, liquid water must have been stable at the surface. On the other hand, the putative ice may have formed only on the floor of palaeolakes or a palaeo-ocean, in which case temperatures would presumably be too low to permit surface water. Certainly, the simplest explanation for the style and distribution of the thumbprint landforms is to regard them as having been formed by a single widespread episode of glaciation, peaking during Amazonian times.

Whether or not the glaciation theory proves to be true, is unlikely to be established until such time as a lander probe is able to sample the stuff of potential glacial landforms or perhaps retrieve core from a potential glacial deposit. At present, this can be viewed only as an untested hypothesis. If supporting evidence is forthcoming, the interesting implication is that large volumes of volatiles must have been available, possibly in the form of large bodies of standing water.

14 The interior of Mars

A considerable amount is understood about the surface of Mars, but until very recently relatively little was known about the interior, at least with any degree of certainty. The mean density is known to be 3930 kg m^{-3}, the mass only about a tenth, and the mean surface gravity 2.63 times less than that of Earth. When the density value is corrected for the effect of self-compression[1] in Mars' gravity field, the figure becomes smaller (3730 kg m^{-3}) than that for either Venus or Earth. This density deficit has traditionally been explained in one of two ways: either Mars always contained less iron than Earth (Urey 1952) or the total amount of iron was roughly the same for all the terrestrial planets, except for Mercury and the carbonaceous chondrites,[2] but on Mars it was in a more oxidized state (Ringwood 1966).

Much information relating to Earth's interior is derived from geophysical data, in particular records of seismic activity. We have also learned something of the chemical and physical make up of the upper mantle from geochemical analyses of exotic blocks brought to the surface in vents and diatremes, and from experimental work on igneous rocks. Thus far, only a modest amount of petrological information exists for Mars, and seismic data is minimal, so the best that can be said is that Mars appears to be much less seismically active than Earth. An estimate of the thickness of the Martian crust, derived from two possible seismic events recorded at the Viking 2 lander site in Utopia, gives a figure of 16 km. Current knowledge of the internal structure of Mars largely stems from what has been learned of its figure (body shape), gravity field, moment of inertia, and from modelling exercises, most of which assume Mars to have been formed from primordial material of approximately type 1 carbonaceous chondrite composition.

If the moment of inertia[3] can be calculated, then it is possible to place constraints on the structure of the interior of a planet. In Earth's case, because both the Sun and Moon exert torques on the equatorial rise, its moment of inertia is known with a high degree of accuracy. This does not apply to Mars, and moment-of-inertia calculations are based largely on the effects of the planet on spacecraft trajectories and on its larger moon, Phobos; the most recent figure for this has been derived from the Mars Global Surveyor spacecraft.

The geoid, figure and gravity of Mars

The geoid and gravity field of Mars have already been described in Chapter 9. Mars' mean equatorial radius is 3396.6 km and the polar radius 3376.7 km; this gives an ellipticity of 0.005 (Earth is 0.003). Precise determinations of the orbital parameters and rotational period of Mars, and of its two small moons, allow for calculation of the mass (6.418×10^{23} kg) and the hydrostatic approximation of dynamical flattening (5240). The latter has consistently been reported as being smaller than the optical flattening (the difference between equatorial and polar radii, divided by the polar radius (5860), which implies that the planet cannot be in isostatic equilibrium (they should be

1. That is, the effect of the planet's mass on the density value.
2. A group of primitive meteorite containing significant quantities of carbon. Chondrites all have ages around 4.6 billion years and are believed to represent the most primitive Solar System materials left for study.

3. A property that allows for modelling of the internal distribution of density within a planet.

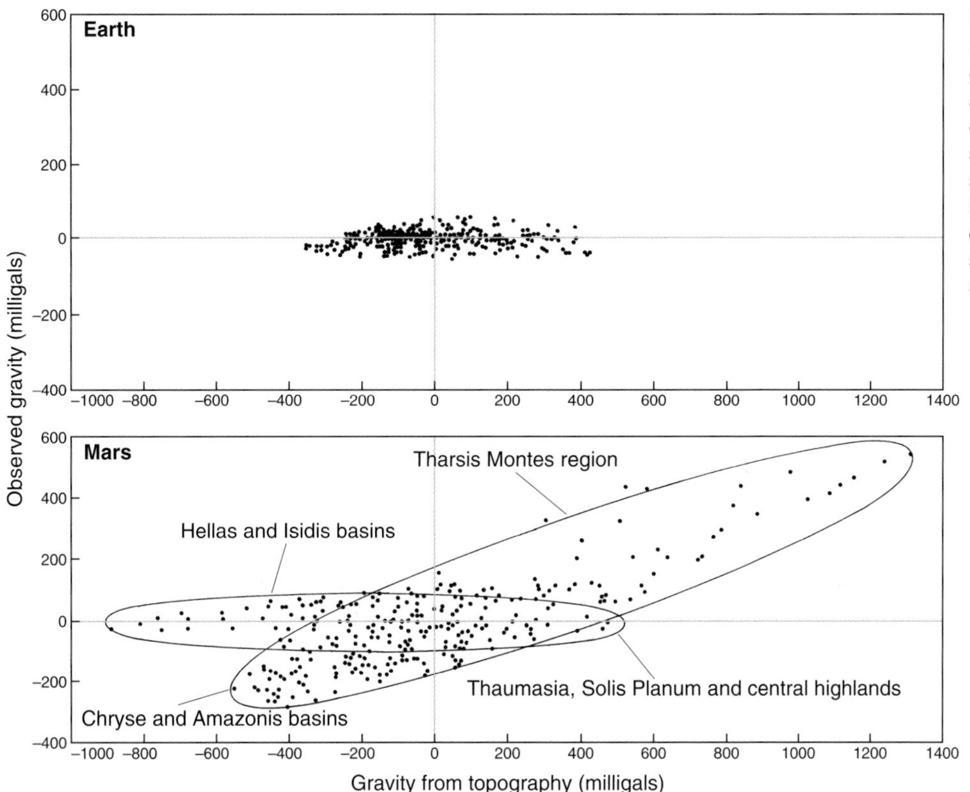

Figure 14.1 Scatter diagrams for Earth and Mars, showing observed gravity and the gravitational equivalent of topography. On Earth the observed and computed gravities show no correlation because of isostatic compensation, whereas on Mars the strong correlation indicates a lack of compensation at shallow depths. (After Phillips & Saunders 1975.)

equal). The difference between the two is about 25 km; from this figure Urey (1952) suggested that Mars might have an equatorial belt of isostatically compensated low mountains, producing the observed discrepancy without requiring large non-hydrostatic stresses.

Most geophysical modelling exercises pre-suppose that Mars, like Earth, has a crust, mantle and core; furthermore, that the crust is a shell of approximately uniform density and thickness, not in hydrostatic equilibrium, and making a small contribution to the total mass and moment of inertia. The non-equilibrium state of the Martian crust was confirmed by the Mariner 9 spacecraft, which experienced marked variations in acceleration as it orbited the planet. The gravity measurements made by the Mariner, Viking and Mars Global Surveyor spacecraft indicate that the elevated region of Tharsis and the adjacent basins of Chryse and Amazonis are associated with large free-air gravity anomalies The long-wavelength free-air positive gravity anomaly over the Tharsis region is 3000 mgal, and there are large anomalies associated with Elysium and Isidis. Large negative anomalies exist over Chryse and Amazonis (Lorrel et al. 1972, Sjogren et al. 1975, Smith et al. 1999b). Because of the distinct regional correlation between gravity and topography (Fig. 14.1) – something absent on Earth – it is generally accepted either that there is incomplete isostatic compensation over these extensive regions, or that compensation is achieved at depths greater than 1000 km (Phillips & Saunders 1975). Elsewhere, gravity and topography do not correlate and it has to be assumed that compensation is achieved at relatively shallow depths.

By calculating the gravitational effects of the observed topography and subtracting these from the gravity as measured by spacecraft, it is possible to arrive at Bouguer anomalies. In undertaking such calculations, Phillips et al. (1973) and Phillips & Saunders (1975), recognized a large negative Bouguer anomaly beneath Tharsis and corresponding positive anomalies beneath the adjacent lowlands. Evidently, at least partial compensation has been achieved, and Phillips et al. suggested that the anomaly associated with Tharsis can be accounted for by variations in crustal thickness, from 20 km beneath basins to 130 km beneath Tharsis. Subsequent geophysical modelling by Comer et al. (1985), who determined the thickness of the elastic lithosphere of Mars at different sites by analyzing the tectonic effects of major volcanic loads, yielded a range of 20–150 km over the Tharsis region, and a figure in excess of 120 km beneath Isidis.

The more accurate data collected by Mars Global Surveyor established that the largest free-air anomalies (3000 mgal) are associated with Olympus Mons and the Tharsis Montes, with lesser ones attached to the volcanic rise of both Elysium and Isidis (Smith et al. 1999b). The new data also resolve significant individual anomalies associated with the Tharsis Montes, Olympus Mons, and with Isidis and Valles Marineris (Plates 8 and 9).

Isidis has the largest impact-related positive anomaly (600 mgal), with Utopia coming next (350 mgal). The latter, from MOLA topographic data, is now identified as a 1500 km-diameter depression, whereas the gravity signature is somewhat diffuse, with no clear centre. Of course, Utopia lies beneath the resurfacing of the northern hemisphere and evidently has been infilled – presumably to give an excess of mass; but the size of the anomaly suggests it may have been of roughly the same size as Hellas. Hellas, characterized by a gravity anomaly of –150 mgal, has not been completely filled and, as has been pointed out by Smith et al. (1999b), the gravity data may shed light on the density of the material that infilled Utopia (i.e. the material of the northern plains).

Gravity data for the polar regions are interesting too. Several positive anomalies are revealed over the northern pole, but they have no obvious connection with topography. In contrast, a positive anomaly lies immediately over the south pole, presumably representing the load attached to the permanent ice cap. The absence of a similar signature above the northern pole may imply that the south cap is more youthful and has had insufficient time to react isostatically, or that the layered deposits around the southern pole contain a relatively high proportion of dust.

Internal density profile

The mean density of Mars is 3930 kg m^{-3} (about 75 per cent that of Earth), implying that, compared to Earth, Mars is apparently deficient in heavy elements such as metallic iron. Nevertheless, the density at the centre of Mars is likely to be close to 8000 kg m^{-3}. Early work by Sir Harold Jeffreys (1970) assumed that the interior of the planet was composed mainly of the mineral olivine, with only a small iron core, and that density variations with depth are largely the result of olivine–spinel phase changes. The pressure at the centre of Mars must approach 1011 Pa, which is well above the pressure at which terrestrial-type materials would show phase changes of this type.

Since the magnetometer on board Mariner 4 was unable to detect a magnetic field, it was assumed either that Mars lacks a metallic Ni–Fe core or that one is present but is convecting too slowly to act as a dynamo and thus generate a field. Various geochemical, geophysical and geological constraints suggest the latter to be the more likely. Calculations by Cole

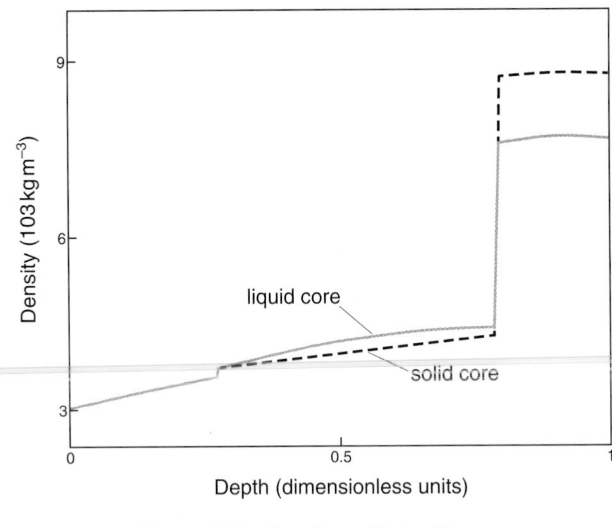

Figure 14.2 Density profile for Mars.

(1978), based on the moment of inertia, suggested that, if an iron core does exist, its radius must approach 0.33 that of the whole planet; this would give it about 6 per cent of the planetary mass, compared with 32 per cent for Earth's iron core. Johnston & Toksoz (1977) and Solomon (1979) calculated the core to have a radius of 1400 km and 2000 km respectively, accounting for 7–21 per cent of the total volume. More recent analysis of the tracking of the Pathfinder lander further constrained the central metallic core of Mars to a radius of between 1300 and about 2000 km (Golombek et al. 1999). Geophysical and geochemical constraints also mean that the mantle of Mars must be enriched in FeO compared with Earth, perhaps by as much as three times (McGetchin et al. 1981), with a density of 3500–3600 kg m^3. A density profile for Mars, based on the Cole (1978), is presented as Figure 14.2.

Doppler and range measurements to the Mars Pathfinder lander have been combined with measurements from the Viking landers to estimate improved values of the precession of Mars' pole of rotation and the variation in its rotation rate (Folkner et al. 1997). The observed precession of –7576 ± 35 milli-arc seconds of angle per year implies a dense core and constrains possible models of internal composition. The estimated annual variation in rotation is in good agreement with a model of seasonal change in mass exchange of CO_2 between the atmosphere and polar caps. On this basis, the non-hydrostatic component of the degree 2 gravity field must be axisymmetric about an equatorial axis through Tharsis. Indeed, it might deviate in the unexpected direction of an extra contribution to the equatorial moment of inertia, which would imply that the planet is either in a rotationally unstable state or close to it.

However, as explained by Bills & James (1999),

Table 14.1 Representative interior model for Martian structure (after Bills & James 1999).

Layer	Name	R (km)	ρ (kg m^{-3})	μ (GPa)
1	Crust	3390.1	2800	45
2	Upper mantle	3292.5	3550	80
3	Middle mantle	2360.0	4050	120
4	Lower mantle	2030.0	4200	155
5	Core	1469.9	7200	0

knowing the inertia tensor for Mars does not provide any one unique solution to the internal density distribution; however, one potential solution is shown in Table 14.1.

Bills & James (1999), in analyzing the data via a variety of methods, investigate what type of solutions are arrived at by partitioning the observed gravitational oblateness into hydrostatic and non-hydrostatic components, as was done by Reasenberg (1977), Kaula (1979), Kaula et al. (1989) and Zuber & Smith (1997). Identifying certain weaknesses in this argument, they suggest that, in non-hydrostatic bodies, such as Mars, the moment of inertia and gravitational flattening would effectively be decoupled. They therefore see the elastic lithosphere as an important player in modifying the rotational flattening and rescuing the planet from apparent rotational instability. Indeed, the addition of a mass anomaly to the planet's surface would generally cause it to reorientate itself.

Although Mars appears to be rotationally stable in the long-term, periodic and variable mass accumulations in the polar regions during the precessional cycle would tend to move Mars temporarily into an unstable condition. To see whether or not such a situation could arise, Bills & James estimated how much mass would need to be moved into and out of the polar regions. It transpires that the non-hydrostatic excess is equivalent to 4.8 million kg; this is equivalent to a global ice layer 30 m thick, something that is both plausible and available. It is possible, therefore, that rotational stability could be closely tied in with cyclical processes operating at the Martian poles.

Composition of the interior

Many models have been proposed for the Martian interior, most having been developed since the 1960s, at which time the relationships between high-pressure mineral phases became better understood (Kovach & Anderson 1965, Binder 1969, Johnston & Toksoz 1977). Ringwood & Clark (1971) investigated the properties of various bulk compositions similar to dehydrated type 1 carbonaceous chondrites, with iron in the oxidized state. The most plausible of their models contained 60 per cent olivine +19 per cent pyroxene +21 per cent magnetite, a composition which, at P–T conditions appropriate to the Martian interior, showed two major series of phase changes, at depths of 1200 and 2000 km. As the calculated density of a homogeneous Mars of this composition was 3970 kg m^{-3} (in good agreement with observation), at first this seemed an appropriate model; however, the moment-of-inertia coefficient for such a composition turned out to be 0.391, much greater than that observed. Consequently, Ringwood & Clark modified the model to assume that there had been extensive melting and differentiation, with the result that the high-pressure form of magnetite could segregate into a core of radius 1640 km (Fig. 14.3).

Thus, the calculated density of 3940 kg m^{-3} and inertia coefficient of 0.373 fell into excellent agreement with observation. Although this may not be the only plausible or even the actual model for the interior of Mars, it can account for the density and moment of inertia and it has the essential characteristic of a Martian mantle that contains substantially more FeO than Earth's. Thus, it has become widely accepted that Mars is much more highly oxidized than Earth.

More recently, studies of SNC meteorites (now

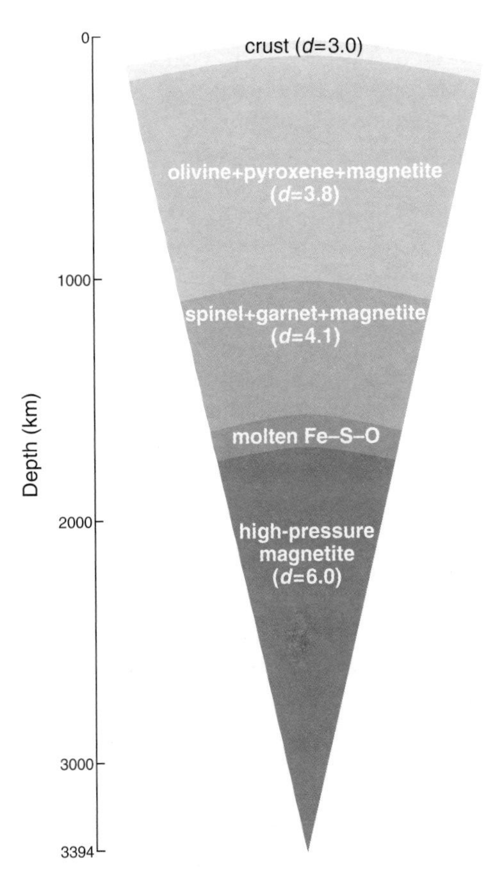

Figure 14.3 Model for Martian interior after core segregation (after Ringwood & Clark 1971).

Table 14.2 Bulk composition of Mars derived from SNC meteorites (after Chicarro et al. 1989)

Mantle+crust (%)		Core (%)	
SiO_2	44.4	Fe	77.8
Al_2O_3	3.02	Ni	7.6
FeO	17.9	CO	0.36
MgO	30.2	S	14.24
CaO	2.45	Core mass	
TiO_2	0.14	= 21.7%	
Na_2O	0.50		
P_2O_5	0.16		
Cr_2O_3	0.76		
K (ppm)	305		
Ni (ppm)	400		

widely attributed to Mars) have led to newer experimental and modelling exercises, some of which have been reviewed by Kerridge & Matthews (1988) and also by Holloway (1990). Such work indicates that the parent magmas of SNC meteorites must have been low in Al_2O_3, high in CaO or CO_2, and depleted in light rare-earth elements; therefore, multiple melting events have to be invoked to account for the phase relationships seen. The bulk compositions of the mantle, crust and core of Mars, derived from SNC meteorite studies, are shown in Table 14.2.

One of the more surprising things about Mars is the very long-term stability of volcanism, which appears to have spanned at least 3 billion years. Breuer et al. (1996) suggest that it is primarily related to the past and present structure of the planet's mantle convection and that there must have been relatively few strongly upwelling mantle plumes. To explain how these might come about, they invoke that both olivine →ß-spinel and ß-spinel→γ-spinel phase transitions, occurring above the core/mantle boundary, could readily generate vigorous superplumes.

Both geophysical and geochemical constraints indicate that the mantle of Mars must be relatively enriched in iron when compared with Earth, perhaps by as much as a factor of three (McGetchin et al. 1981). This means that the first 1–2 per cent of partial melt to form, from a composition consistent with all available data, will be relatively enriched in iron and depleted in magnesium compared to a similar melt on Earth (Holloway & Bertka 1989). The result is that Martian primitive magmas are likely to be very low in silica and alumina, and very rich in iron. This makes for lavas with very low viscosity and high density.

Several workers have suggested that primitive Martian magmas (i.e. those that were generated during terrestrial Archaean times) were like terrestrial komatiites, which have similar properties to the above and, in addition, are rich in sulphur, like the fines at the Viking lander sites. It may well be, therefore, that komatiite-type magmatism was prevalent on Mars during its early history and, indeed, since the SNC meteorites have a crystallization age of 1.3 billion years, may have persisted into more recent times (Burns & Fisher 1989). Such magmas require (on Earth, at least) very high pressures in the melting zone (40–140 kbar). However, as has been mentioned by Breuer et al. (1996), the high iron content of the Martian mantle (Dreibus & Wänke 1985) would be expected to reduce the pressure at which komatiitic magma can form.

Global Surveyor magnetic-field measurements

If Mars had a liquid iron-rich core like that of Earth, it would be expected to show evidence of a global magnetic field. This had not been detected unequivocally prior to 1997. Then, vector magnetic field observations of the crust were acquired by Mars Global Surveyor's magnetometer experiment during aerobraking and also when the spacecraft was 100–200 km above the planet's surface. Initial excitement was generated during aerobraking, in mid-September 1997, when for the very first time the craft recorded a weak dipole field, with an estimated upper limit of $\sim 2\times10^{18}$ A-m^2, or roughly 1/800th that of Earth. At the time it was not clear whether this meant that Mars still has a very weak central dynamo or just a remnant of an originally more vigorous field, now completely decayed. Subsequent measurements then identified much stronger localized magnetic fields, largely within the ancient cratered terrain. This appeared to imply that the planet did once have an active internal dynamo, now extinct.

Dense spatial coverage above the north pole during the early phase of exploration recorded the presence of several rather weak but quite discrete magnetic source regions centred at about 330°W and extending from 60°N to 90°N latitude (Acuña et al. 1999). The average field strength of these dipole sources is ~100 000 A-m^2. It is unclear with what these fields are associated, since there are no recognizable geological, topographical or gravity features with which they can readily be correlated. However, there are some significant gravity anomalies present in the region and these may be connected with some deep-seated iron-rich subcrustal concentrations. No magnetic sources were found elsewhere in the northern lowlands.

As the spacecraft traversed more of the planet, and as the periapsis progressed farther south, it recorded more frequent and stronger crustal magnetization sources. In particular these were recorded in the region between 120°W and 210°W and 30°S to 85°S, where total fields as strong as ~160 000 A-m^2 were observed (Plate 10). One particular source was also reported near 32.8°N, 23.6°W, on the southeast side of Acidalia, and apparently associated with an extended linear feature striking west–east.

A synthesis of the earlier data revealed quite clearly that the majority of crustal magnetic sources lie south of the line of dichotomy, in regions of high-impact crater density in the ancient cratered terrain. No magnetic sources were found beneath Tharsis, Elysium or Valles Marineris, and there are no sources associated with large basins such as Hellas and Argyre. The most intense magnetic sources were detected in the Terra Sireum region, where Mars Global Surveyor measured total field intensities of over 1500 nT. The estimated total net field in this region was calculated to be $\sim 1.3 \times 10^{17}$ A-m^2 (Acuña et al. 1999). This would be a strong enough moment to deflect the solar wind locally as it reached the ionospheric level.

Subsequent measurements (Connerney et al. 1999), made at altitudes varying between 200 km and 103 km, revealed that the most important crustal sources are arranged in extensive west–east striking linear features or stripes (Plate 5). The map is seen to consist of a series of roughly N/S strips that represent those parts of each orbit where the spacecraft dropped below the 200 km level. The colour of each segment represents the radial component of the local field, ranging from 1389 nT (blue) to 1476 nT (red). One very prominent radial feature, centred at 53°S, 180°W, can be traced for over 2000 km, and subparallel features nearby are traceable for over 1000 km in some cases. The intense positive magnetic records are separated by equally intense negative ones, reminiscent of terrestrial magnetic striping.

As Connerney et al. (1999) argue, measurements of the Martian field differ from those of the terrestrial one, since on Earth, because of the strong dynamo action of the core, it is difficult to measure the relatively small signal related to discrete crustal sources. On Mars, in contrast, where there is no induced field generated by a currently active core, the vector field of discrete crustal sources can be measured directly. This means that the Martian sources can unequivocally be tied to remanent magnetization.

On the basis that the magnetized zones are represented by about 20 uniformly magnetized strips of crust, each about 200 km in width and 30 km in depth, Connerney et al. were able to deduce that the Martian highland crust is, on average, one order of magnitude more intensely magnetized than Earth's continental crust. This appears to imply that, whatever the crustal mineralogical conditions, their thermo-remanent magnetization must be high. Obvious mineral candidates must include haematite and pyrrhotite, the former of which has been identified at the Pathfinder landing site. It is also consistent with the high iron contents of soils analyzed at the Pathfinder site (around 17 wt % Fe) and in meteorites of supposed Martian provenance.

Terrestrial magnetic striping is associated with the creation of new oceanic crust by sea-floor spreading and is characterized by mirror symmetry about a spreading axis. Thus far, such a pattern has not been recognized in the Mars stripes, which are also much broader and less regular. Although the most obvious interpretation of the Martian features may seem to be that they formed during the earlier part of Noachian times by some form of plate tectonics, there are still several questions to be answered (Kerr 1999). One problem is that, if the magnetized slabs are 30 km thick, it would takes tens of millions of years for them to cool enough to lock in a remanent magnetization.

Following the argument through, it would seem logical to conclude that the early crust may have been equivalent to Earth's oceanic crust, therefore preserving a record of the early thermal history of Mars. What at first might seem is odd is that there is a lack of evidence of any large-scale tectonic features in the region where the most intense magnetization is found. However, any original structures would almost certainly have been obliterated by impact bombardment and might be detected deep within the subcrust only by, say, high-resolution gravity surveys. This remains an intriguing area of future research.

Another facet of the magnetometer data is that the southerly extent of the magnetization appears to correspond with the maximum extent of the region of destruction or modification of magnetized rocks by the massive impacts that produced the Hellas and Argyre basins. There is a clear, if discontinuous, outer ring of magnetization to the northwest and east of the Hellas basin, indicating the maximum extent of metamorphism of the ancient crust by the impact event; there is also a clear imprint in the shape of the crustal dichotomy boundary associated with the Isidus structure. This adds weight to the hypothesis that the dichotomy was formed after the cessation of dynamo activity (i.e. in the Lower Noachian).

15 Geochemistry

One vital aspect of the geology of any solid body is the chemistry of its surface and, should there be an opportunity to find such, samples that have been brought to the surface from deeper down. Geochemistry has become increasingly important in the quest to understand Mars and during the past five years much new data has arrived. Prior to the landing of Mars Pathfinder in 1997, there were only two major sources of data: XRF analyses of rocks at the Viking lander sites and analyses of certain meteorites that are widely considered to have a Martian provenance. This archive has now been expanded by geochemical analyses of rocks and regolith ("soils") obtained in the vicinity of the Sagan Memorial Station at the mouth of Ares Vallis. These were collected via an APXS mounted on the Sojourner roving vehicle.

Although the two Viking landing sites (Utopia and Chryse) were 6500 km apart, the composition of the regolith analyzed in each place was very similar and was interpreted to have formed by the breakdown of mafic igneous rocks. The so-called Martian meteorites, members of the SNC family, are also mafic to ultramafic, but more variable in composition than the rocks at the Viking site. Mars Pathfinder's APXS instrument measured a wider range of rock compositions than previously encountered, of particular interest being rocks with a broadly andesitic chemistry.

Spectral characteristics and chemistry

Viking soils and fines

The general aspect at the Viking 1 lander site in Chryse Planitia was of a rock-strewn desert with a gently rolling topography. Images revealed fine-grain drifts that had accumulated, and sometimes become encrusted, in the lee of large, generally angular boulders. At the Viking 2 site in Utopia, the ground was somewhat flatter, lacked drifts, and was covered in angular pitted boulders. The fine-grain regolith, despite lacking any organic content (the prerequisite for a terrestrial soil), has been widely described as "soil" and, because this term features in the literature it is used here also; "regolith" is more correct.

The chemical composition of fine-grain drift materials was measured by XRF spectrometry, which produced surprisingly similar results from the widely separated sites (Table 15.1). Martian soils are characterized by low SiO_2 (43–45 wt%) and high iron (18–20%), suggestive of derivation from mafic igneous rocks such as basalts. However, the low CaO (5.0–5.6%) and very low Al_2O_3 (5–6%) are difficult to reconcile with typical terrestrial mafic volcanic rocks. Amounts of SO_3 (6.5–9.5%) and Cl (0.6–0.9%) present indicate a significant volume of volatiles. Toulmin et al. (1977) suggested that the soils may be a mixture of

Table 15.1 Chemical composition of soils at Viking lander sites.

Major elements	Viking 1	Viking 2
SiO_2	44.7	42.8
Al_2O_3	5.7	20.3
Fe_2O_3	18.2	1.0
TiO_2	0.9	5.0
MgO	8.3	<0.3
CaO	5.6	6.5
K_2O	<0.3	0.6
SO_3	7.7	
Cl	0.7	
Trace elements (ppm)		
Rb	<30	<30
Sr	60 ± 30	100 ± 40
Y	70 ± 30	50 ± 30
Zr	<30	30 ± 20

Fe-rich smectite clay, with ferric oxides, carbonates and sulphates. Fe-rich clay minerals are typical as weathering products of basaltic volcanic glasses.

It was also suggested that S and Cl were introduced into the soils by the interaction between the mafic species and volcanic gases, producing sulphates and chlorides (Clark & Baird 1979). Whether or not salts such as carbonates and nitrates are present remains unknown.

Earth-based spectroscopic observations have also contributed to our understanding of Martian materials. A strong absorption band below 0.5 µm, and a weaker ones between 0.7 and 0.95 µm, are consistent with the presence of ferric oxides (Singer et al. 1979). The latter group suggests that a good fit with the Mars data can be achieved with a thin ferric oxide coating on mafic igneous rocks. Rather weak absorptions around 1 µm suggest Fe^{+2} absorptions in silicates such as pyroxene and olivine. On the basis of spectroscopic results and surface chemistry, many potential iron-bearing species have been suggested, including goethite, limonite, smectite clay, weathered palagonite and maghemite. More precise spectral measurements indicate several absorption bands around 0.68 µm and 0.85 µm, which are attributed to crystalline haematite (Bell et al. 1990).

Pathfinder soils and fines

Mars Pathfinder landed at the mouth of Ares Vallis on 4 July 1997, at latitude 19.28°N, 33.22°W. The view from the lander probe is depicted in Plate 6. The APXS on board the Sojourner rover gathered analyses of six soils and five rocks. Soil analyses are similar to those at the Viking sites, but the rocks are close in composition to terrestrial andesites. On this basis, addition of a mafic component and reaction products of volcanic gases would be necessary before local rocks could source local soils (Rieder et al. 1997b). Soils and rocks were also analyzed photometrically, spectrally and physically, providing the most detailed analysis of Martian surface materials to date. A full preliminary report of the mission may be found in Golombek et al. (1999).

The site is partly covered by thin drifts of fine material that overlie soil-like particulates admixed with rocks. The soils range from dark grey to bright red, and are similar in many ways to moderately dense soils on Earth, such as clayey-silt with embedded sand grains and pebbles. They have an average density of 1520 kg m^{-3}. Most soils have the reddish colour typical of those at the Viking landing sites; the grey drifts may derive their colouration from being armoured with dark basaltic dust. One of the sampling sites, Scooby-Doo, may represent an indurated soil akin to a hardpan. Most of the soils are also rather cloddy (again like those at the Viking sites), and the analysis of wheel-track experiments indicates that they contain much very fine dusty material (Rover Team 1997).

Multi-spectral imaging of soils and rocks at the landing site was achieved by the IMP, using all 15 geological filters at 12 wavelengths (Smith et al. 1997). It resolved four spectral classes (Fig. 15.1, Plate 8):

- *Dark soil* This typically occurs in dune forms, such as those near Mermaid rock. It is the lowest reflectivity material at the MPF site. It is less oxidized than the bright red soil.
- *Bright soil* This is typically found in shallow aeolian deposits near and around rocks (e.g. Cradle

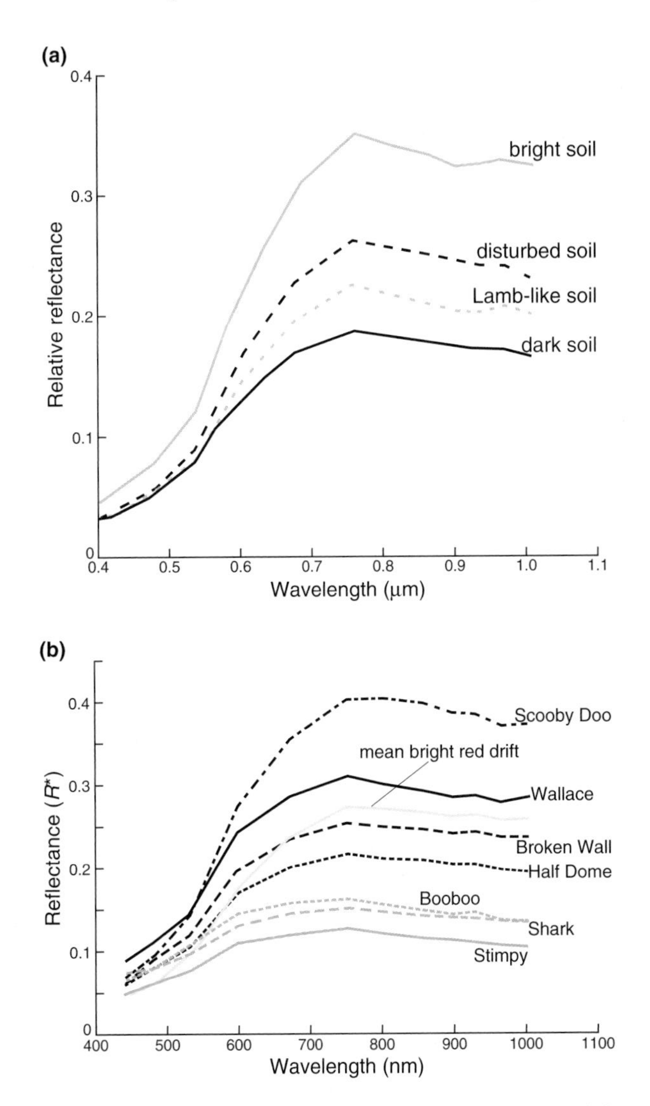

Figure 15.1 **(a)** Relative reflectance spectra of Pathfinder soils (after Smith et al. 1997). **(b)** Representative spectra of grey, red and pink spectral classes at the Pathfinder site, showing primary spectral trend (after McSween et al. 1999).

Table 15.2 Chemical composition of six soils at Mars Pathfinder landing site, in weight % normalized to a sum of 98%. "OS" is the original sum of the oxides prior to normalization. (After Rieder et al. 1997a.)

Name	Na_2O	MgO	Al_2O_3	SiO_2	SO_3	Cl	K_2O	CaO	TiO_2	FeO	OS
A-2	2.3 ± 0.9	7.9 ± 1.2	7.4 ± 0.7	51.0 ± 2.5	4.0 ± 0.8	0.5 ± 0.1	0.2 ± 0.1	6.9 ± 1.0	1.2 ± 0.21	6.6 ± 1.7	68.6
A-4	3.8 ± 1.5	8.3 ± 1.2	9.1 ± 0.9	48.0 ± 2.4	6.5 ± 1.3	0.6 ± 0.2	0.2 ± 0.1	5.6 ± 0.8	1.4 ± 0.21	4.4 ± 1.4	78.2
A-5	2.8 ± 1.1	7.5 ± 1.1	8.7 ± 0.9	47.9 ± 2.4	5.6 ± 1.1	0.6 ± 0.2	0.3 ± 0.1	6.5 ± 1.0	0.9 ± 0.11	7.3 ± 1.7	89.1
A-8	2.0 ± 0.8	7.1 ± 1.1	9.1 ± 0.9	51.6 ± 2.6	5.3 ± 1.1	0.7 ± 0.2	0.5 ± 0.1	7.3 ± 1.1	1.1 ± 0.21	3.4 ± 1.3	99.2
A-10	1.5 ± 0.6	7.9 ± 1.2	8.3 ± 0.8	48.2 ± 2.4	6.2 ± 1.2	0.7 ± 0.2	0.2 ± 0.1	6.4 ± 1.0	1.1 ± 0.21	7.4 ± 1.7	92.9
A-15	1.3 ± 0.7	7.3 ± 1.1	8.4 ± 0.8	50.2 ± 2.5	5.2 ± 1.0	0.6 ± 0.2	0.5 ± 0.1	6.0 ± 0.9	1.3 ± 0.21	7.1 ± 1.7	98.9

and Yogi) and as thin ground coverings around the site. The spectra peak at 750 nm, with a high red: blue ratio. It has the characteristics of an oxidized ferric-rich material.

- *Lamb-like soil* This occurs primarily near Lamb rock, has a moderate red:blue ratio and reflectance.
- *Disturbed soil* Created largely by rover wheel tracks and has red:blue ratios intermediate between the dark and Lamb-type soils. This is typical of disturbed soils, which seem to be darker, although their spectral type remains unaltered. This may be a compaction effect.

The Pathfinder APXS instrument for deriving rock chemistry used three different kinds of interactions: Rutherford backscattering (alpha mode), reactions between alpha particles with some light elements (proton mode), and production of X-rays through ionization by alpha-particles (X-ray mode). As a result of these, three different energy spectra were obtained, each being recorded in 256 channels. By using the different modes, the instrument was able to measure a range of elements with a high degree of accuracy, indeed virtually all elements except H and He. Each analysis was performed on a sample 50 mm in diameter, penetrating to depths of between several to tens of centimetres. All relevant elements were measured and then their concentrations normalized to 100 per cent (McSween et al. 1999)

APXS analyses of six soils are presented in Table 15.2. Sample A-2 was the first to be analyzed after deployment of the Sojourner rover, A-4 was a light soil adjacent to the rock Yogi, A-5 is a dark sample next to Yogi, A-8 is the sample Scooby Doo, A-10 was located next to the rock Lamb, and A-15 is a sample from Mermaid dune. The aluminium contents are almost identical to those of the nearby rocks (see Table 15.3).

Pathfinder rocks

The rocks present at the Pathfinder site are somewhat different from those observed by Viking. Although the rock concentrations at the three sites were rather similar, the shapes proved to be somewhat different. At the Sagan Memorial Station, in addition to angular blocks and fragments, several rounded ones were observed. Do they represent impact melt droplets, lava spatter, igneous nodules, or pebbles and cobbles weathered out of sedimentary deposits? Golombek et al. (1997) favoured the last explanation.

Two of the larger rocks imaged, Shark and Half Dome, appear to be conglomeratic. They show rounded knobs 3–4 cm across that could be pebbles. One particular rock, Squash, has knobby protrusions fully 10 cm across that are darker in albedo than the rest of the material. This could be a conglomerate or autobrecciated lava. Other rocks, such as Stimpy, have a columnar structure, like terrestrial basalt or andesite lavas. Many rocks are strongly pitted or fluted and it has been argued that these resemble ventifacts (Bridges et al. 1999). The many elongated cavities seen on several rocks may indeed be wind-abraded features, but could be elongated vesicles that have undergone surface weathering and abrasion. However, it would be surprising if the activity of Martian winds had not left its mark somewhere.

One or two rocks, for instance Chimp and Zebra, have a banded appearance that may represent sedimentary layering, igneous layering, aeolian etch marks or faint joint planes. The camera was unable to resolve the precise nature of the layer boundaries in most cases.

IMP spectra were obtained for several rocks, and four spectral classes have been recognized:

- *Grey* Detected in rocks of all shapes and sizes. The type specimen is the rock Shark. The reflectance peaks at 750 nm and the red:blue ratio is low. The spectra are consistent with weakly weathered rocks containing Fe^{2+}.
- *Red* This class pertains to rocks of all sizes, often on rocks that have one grey face. Compared to drift, red rocks are of higher reflectance at short wavelengths but comparable at longer wavelengths. The reflectance peak is at 750 nm. These properties suggest the presence of more ferric minerals than in grey rocks.
- *Pink* These are found on several tabular rocks that are partially buried by drift and also as soil crusts

near Yogi and on some pebbles. Scooby Doo is the type specimen. Relative to drift, this class has higher reflectance at all wavelengths, especially short ones. The reflectance peak is at 750–800 nm. They are identified as encrusted drifts, i.e. duricrust-type deposits.

- *Maroon* This class is typical of large, rather rounded boulders (including Yogi) that lay distant from the lander. Compared to drift, maroon rocks are darker in all wavelengths. The reflectance peak is at 800 nm and there is a weak ferric absorption band. These characteristics are taken to originate in ferric-rich coatings.

The primary spectral classes appear to be attributable to thin ferric coatings of windblown dusts on darker rocks. These have been investigated in the laboratory by Fischer & Pieters (1993) and Shelfer & Morris (1998). A secondary spectral trend is recognized, and may be relate to coating of rocks by a different material, possibly maghemite. A chronology based on spectral characteristics suggests that rounded maroon boulders constitute the oldest petrological unit, succeeded by smaller cobbles, and followed by aeolian erosion and deposition. Nearly linear chemical trends in the APXS analyses are interpreted as mixing lines between rock and the adhering dust, a conclusion supported by a correlation between sulphur abundance and red:blue spectral ratio.

The preliminary APXS analyses of five rocks from the Pathfinder site, and the calculated composition and norm of a sulphur-free rock, are presented in Table 15.3. As can be seen from the normative values, the rocks are saturated with respect to silica.

Since all of the Pathfinder soils are enriched in SO_3, the rock compositions can be obtained by plotting the oxides against sulphur, assuming they are igneous in origin. When the rock compositions are plotted on two-component diagrams of this kind, they plot as roughly linear arrays, with soil analyses falling consistently at one end (Fig. 15.2). because the solubility of S in magmas at reasonable oxidation states is small, and accepting that the S contents of typical igneous rocks are around 0.2 wt per cent at maximum, it is possible to extrapolate linear regression lines through the data to zero sulphur and thus estimate the chemistry of the rock end member. The sulphur-free rock arrived at in this way is also shown in Figure 15.2. As has been observed by McSween et al. (1999), the high Fe/Mg and low Al/Si ratios of this rock is a characteristic it shares with the shergottites, nakhlites, chassigny and ALH84001 (Rieder et al. 1997b). However, it has significantly higher SiO_2 and K_2O than SNC meteorites, indicating that it is considerably more highly differentiated than, say, basaltic shergottites.

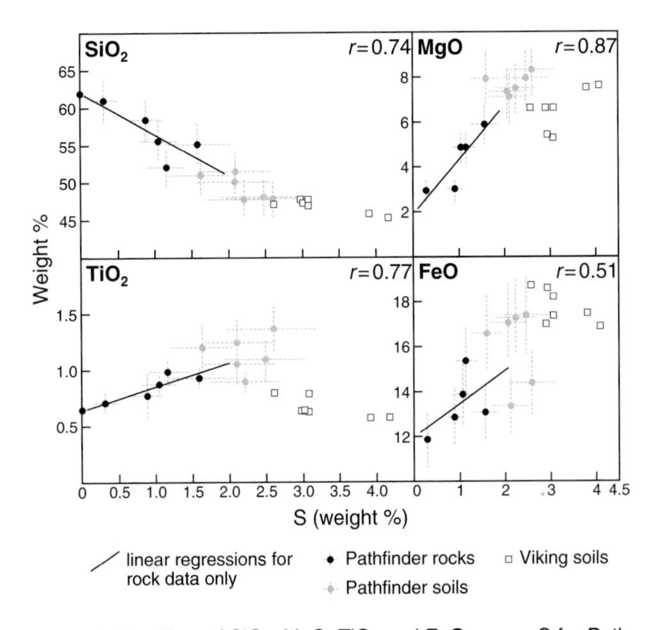

Figure 15.2 Plots of SiO_2, MgO, TiO_2 and FeO versus S for Pathfinder rocks and soils. All data are normalized to 98% total oxides, whereas Viking XRF soil analysis are normalized to 95.4% oxides. (After McSween et al. 1999)

On the basis of its chemistry and norm, the sulphur-free rock plots within the field of andesite on the traditional alkali versus silica diagram (Fig. 15.3).

Table 15.3 Preliminary APXS analyses of five rocks (wt %) from the Pathfinder landing site, plus calculated composition and normative mineralogy of a sulphur-free rock (after Rieder et al. 1997a).

	Rock	Na_2O	MgO	Al_2O_3	SiO_2	SO_3	Cl	K_2O	CaO	TiO_2	FeO	OS
A-3	Barnacle Bill	3.2 ± 1.3	3.0 ± 0.5	10.8 ± 1.1	58.6 ± 2.9	2.2 ± 0.4	0.5 ± 0.1	0.7 ± 0.1	5.3 ± 0.8	0.8 ± 0.2	12.9 ± 1.3	92.7
A-7	Yogi	1.7 ± 0.7	5.9 ± 0.9	9.1 ± 0.9	55.5 ± 2.8	3.9 ± 0.8	0.6 ± 0.2	0.5 ± 0.1	6.6 ± 1.0	0.9 ± 0.1	13.1 ± 1.3	85.9
A-16	Wedge	3.1 ± 1.2	4.9 ± 0.7	10.0 ± 1.0	52.2 ± 2.6	2.8 ± 0.6	0.5 ± 0.2	0.7 ± 0.1	7.4 ± 1.1	1.0 ± 0.1	15.4 ± 1.5	97.1
A-17	Shark	2.0 ± 0.8	3.0 ± 0.5	9.9 ± 1.0	61.2 ± 3.1	0.7 ± 0.3	0.3 ± 0.2	0.5 ± 0.1	7.8 ± 1.2	0.7 ± 0.1	11.9 ± 1.2	88.3
A-18	Half Dome	2.4 ± 1.0	4.9 ± 0.7	10.6 ± 1.1	55.3 ± 2.8	2.6 ± 0.5	0.6 ± 0.2	0.8 ± 0.1	6.0 ± 0.9	0.9 ± 0.1	13.9 ± 1.4	92.6
Calculated sulphur-free		2.6 ± 1.5	2.0 ± 0.7	10.6 ± 0.7	62.0 ± 2.7	0	0.2 ± 0.2	0.7 ± 0.2	7.3 ± 1.1	0.7 ± 0.1	12.0 ± 1.3	
Provisional CIPW norm for sulphur-free rock: Q 21.0, Or 4.1, Ab 22.0, An 15.2, Di 18.2, Hy 15.6, Il 1.3, Mt 0.5.												

Key: OS = Original sum of oxides prior to normalization. All iron is reported as FeO. Calculated wt % norm assumes a molar Fe_2O_3/FeO ratio of 0.026 (average Shergottite value of McSween & Jarosewich 1983).

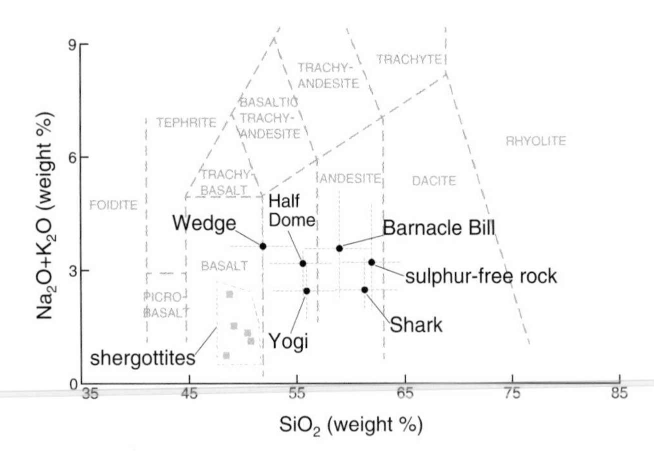

Figure 15.3 Chemical classification diagram for volcanic rocks, based on alkali versus silica contents (after Le Bas et al. 1986).

Barnacle Bill and the dust analyzed at Yogi also plot in this field; the rock Yogi appears to be basaltic. Half Dome and other dust-free rocks lie in the field of basaltic andesite. All of the shergottite meteorites fall well within the basalt field. Of course, there is no proof that any of the rocks are igneous, although the presence of vesicle-like pits in several implies that at least some are. As we have seen, several rocks are composite and may be of sedimentary or volcaniclastic origin. Even if this is so, the chemical data suggest that the clasts they contain are likely to be volcanic in type.

Because multi-spectral and chemical (APXS) measurements were made at identical locations (McSween et al. 1999), it was possible to show the relationship between red:blue spectral ratio and chemistry. Figure 15.4 shows the relationship between SO_3 content and the red:blue ratios for both rocks and soils at the Pathfinder site. The data are plotted in this way

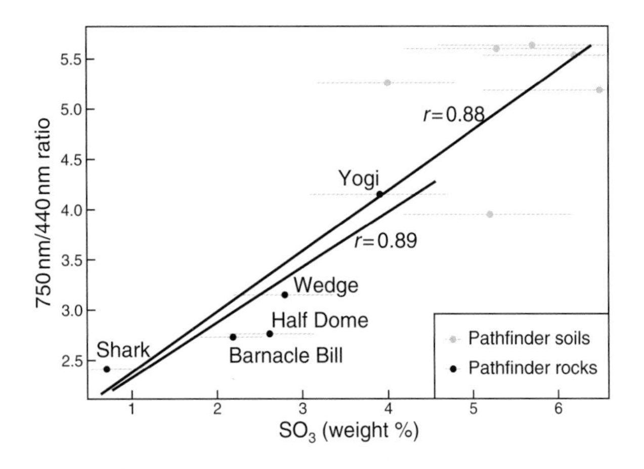

Figure 15.4 Plot of SO_3 contents of Pathfinder rock analyses with red:blue ratios of their spectra. Rocks with higher sulphur content have high ratios. The long regression line was calculated for both rocks and soils, the shorter one for rocks alone. (After McSween et al. 1999.)

because the soils are very rich in sulphur and red relative to the rocks. It can be seen from the diagram that APXS soil sites are generally redder than those of rocks, the exception being Mermaid Dune, which has a lower red:blue ratio than the rock Yogi. It will also be noted that the red:blue ratios for all samples correlate well with SO_3. This is interpreted as meaning that the Pathfinder rock compositions reflect an admixture of rock plus adhering soil or dust.

SNC meteorites

SNC has been adopted as an unofficial term to group nine meteorites that fell in Shergotty (India), Nakhla (Egypt) and Chassigny (France). Subsequently, this group has been extended to various other falls in Africa, Antarctica, the USA and Brazil. They are closely related to the achondrite group, bearing minerals such as olivine, augite, pigeonite and feldspar. There is a widespread view that meteorites of this group have a Martian origin, although there is no direct proof that this is so. The facts are as follows.

Whereas the majority of meteorites yield very ancient radiometric ages (>4 billion years ago), the SNCs characteristically have young ages, ranging from 1.3 billion to 200 million years. This led to the interpretation that they must have formed on a planetary body that was of sufficient mass to sustain volcanic activity until geologically recent times. Further investigation revealed that their oxygen isotope ratios were similar to one another, but very different from all other meteorites. In particular, the SNC group contains a small proportion of their indigenous iron in the ferric form (Fe^{3+}), which implies that they must have been formed in an oxidizing environment. Therefore, they could not have come from the Moon, where the iron is in the ferrous state. Stolper et al. (1979) pointed out that several other of their features indicate that they crystallized from parent magmas more alkaline and more oxidized than lunar magmas.

The final piece of evidence that swayed many workers into concluding that the SNC bodies originated on Mars came from analysis of the gas content of dark glass inclusions that are thought to have formed through impact melting. The trapped gases in shergottite EETA79001 (from Antarctica) proved to be a nearly perfect match for those in the Martian near-surface atmosphere; further analysis of these gases and nitrogen indicated that the match was unmistakable (Bogard et al. 1984). It was almost as if someone had just plucked out the current Martian atmospheric composition. The impact glass is believed to have been produced during a very energetic impact into the

Martian surface. Some workers have undertaken a search for the parent craters; Mouginis-Mark et al. (1997) identified 25 candidate craters in the Tharsis region alone.

Cosmic ray exposure ages (which indicate how long the objects have been in space) show that the SNC objects travelled through space in much the same form as now, for between 10 and 3 million years; this gives a possible range of times for the ejection events. Those of the group that fell in Antarctica apparently arrived there about 100 000 years ago.

Most of the debate concerning a Martian origin for these meteorites centres on the problem of ejecting them from Mars. The dynamics of such a process once seemed impossible. However, in recent years it has been suggested that a very low angle impact, or a large 100 km-size crater-forming event, would provide the required energy and conditions (Nyquist 1983, Vickery & Melosh 1987). On the assumption that a Martian origin is acceptable, what do SNCs tell us about Mars?

The SNCs are all mafic igneous rocks generated by melting and crystallization. Most are cumulate rocks. Calculated parent magma compositions show a similarity with basaltic komatiite, but with rather different CaO and trace-element compositions (Longhi & Pan 1989). The 1300-million year ages of the chassignites and nakhlites imply that Martian volcanism was taking place as recently as that. If the shergottite ages are also igneous ages (they may be metamorphic shock ages), then this date can be brought forward to as recently as 200 million years ago. The presence within them of mineral species such as amphibole and mica indicates that the magmas they represent were also hydrated. The oxygen isotope analysis of water residing in clays and salts within the Martian surface rocks, and extracted from the SNCs by thermal methods, clearly points to an extraterrestrial origin. On the assumption that they are of Martian provenance, they have been widely compared with the surface rocks and soils on Mars, and, as will be seen in the next section, have been used in arguments concerning the planet's differentiation.

New data and Martian differentiation

Ares Vallis was chosen as the Pathfinder landing site because geological observations indicated that it ought to provide samples of a variety of rocks brought down into the northern hemisphere from the southern highlands by massive floods. In broad terms, Barnacle Bill, Shark and the calculated "soil-free rock" are likely to be more representative samples of that ancient crust than, for instance, any SNC meteorite. With their broadly andesitic composition, they are not that far removed chemically from the average composition of Earth's crust. However, the iron content of the Martian rocks is significantly higher, which is presumably a reflection of the higher Fe content of the Martian mantle source regions.

On this basis, if Barnacle Bill's composition is taken to represent the Al-rich end-point of a hypothetical Martian mantle–crust fractionation trend, with Chassigny and ALHA77005 meteorites on the Al-poor side of the line (Fig. 15.5), then the basaltic shergottites (QUE94201, Shergotty and Zagami) are seen as forming a second fractionation line and could be rocks derived from younger intrusions into the ancient Martian crust. Since all of these meteorites are assumed to come from the same source at the same time, the most obvious explanation for their differences is that they were derived from a body of magma that became stratified by crystal fractionation and accumulation.

It is also possible to deduce that Martian soil cannot be directly produced from rocks of the Barnacle Bill type of chemistry, even if weathering and gases such as SO_2 and HCl were added from a volcanic gas fraction. One way to generate the observed soil composition would be to be to add Mg- and Fe-rich material, such as that observed in Martian meteorites (Fig. 15.6). Another would be to add ferromagnesian minerals, preferentially weathered out of the surface rocks and concentrated in the soil. However, since the aluminium contents of soils are almost identical with the adjacent rocks, it appears more likely that the soils are an admixture of locally derived fines, with components brought in from the weathering of mafic rocks elsewhere on Mars, presumably by the wind.

On the assumption that most of the rocks analyzed

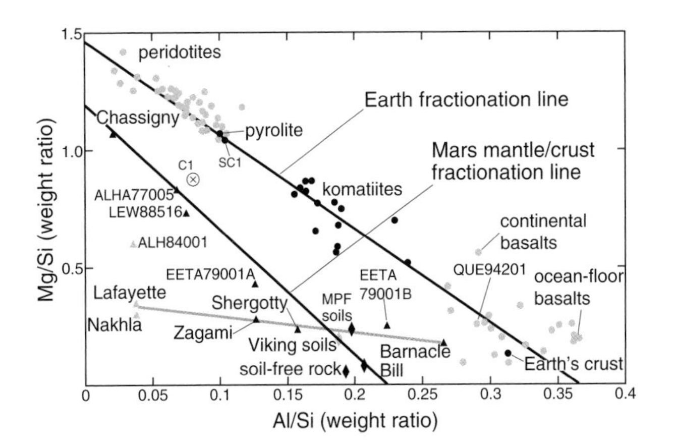

Figure 15.5 Mg/Si versus Al/Si plot of SNC meteorites (black triangles), mean values of Martian soils (grey diamonds) and Pathfinder soils (labelled MPF soils), plus Barnacle Bill and calculated "soil-free rock", together with plots for terrestrial samples (after Rieder et al. 1997a).

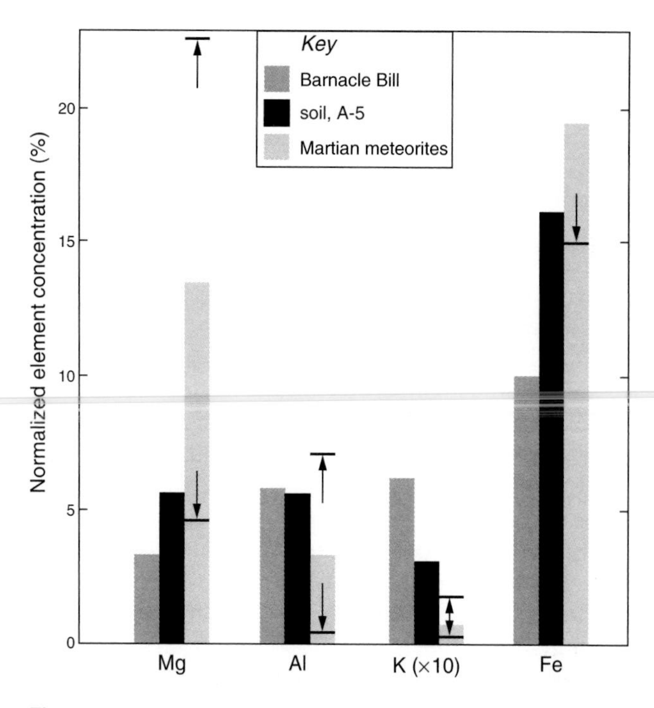

Figure 15.6 Histogram of selected element concentrations in Barnacle Bill, Pathfinder soil A-5 (filled bar) and Martian meteorites (open bar). All data are normalized to the SiO_2 content of Barnacle Bill. (After Rieder et al. 1997a.)

are either volcanic or composed predominantly of volcanic fragments, the question of rock petrogenesis is clearly pertinent. On Earth, andesite is the second most abundant volcanic rock type and the commonest lava type found around the Pacific, which is circumscribed by convergent plate margins. In such locations the rocks are normally termed orogenic andesites. The modern view is that most terrestrial andesites are formed by fractional crystallization of basalt, generally following one of two paths: tholeiitic, characterized by fractionation at low pressure under anhydrous conditions; or calc-alkaline, typified by higher pressures and a hydrous environment (Grove & Kinzler 1986). Traditionally such trends are distinguished on standard AFM diagrams (Fig. 15.7).

On such a diagram, all of the rocks, together with the SNCs melts, plot well within the tholeiite field. Although the plot for the shergottite melt appears to fall at a point suitable for a Mars andesite parent magma (close to the Mg/Fe baseline), it is unlikely that this is the case. For a start, the rocks at the Pathfinder site are almost certainly derived from the southern highlands, which are thought to be significantly older than the SNC meteorites, and, secondly, the Al_2O_3 content of the Martian rocks indicates that their parent magma must have been significantly more aluminous than any shergottite melt. In this sense, the source rock probably lies closer to the end of the anorogenic icelandite trend – terrestrial rocks formed by fractional crystallization of basalt under anorogenic conditions. McSween et al. (1999) hypothesized that the implicated source magma composition may indicate derivation from a primitive mantle source that had not yet been depleted in Al_2O_3 by partial melting. This appears to imply that these rocks may be relatively ancient, perhaps having their origin in the ridged plains of the cratered highlands to the south.

The chemistry of the rocks at the Pathfinder site does not accord with their having been produced under orogenic conditions; and therefore, at the present time, there is no strong petrological evidence in favour of terrestrial-style plate tectonics. It will be interesting to see what analyses from other landing sites add to this picture.

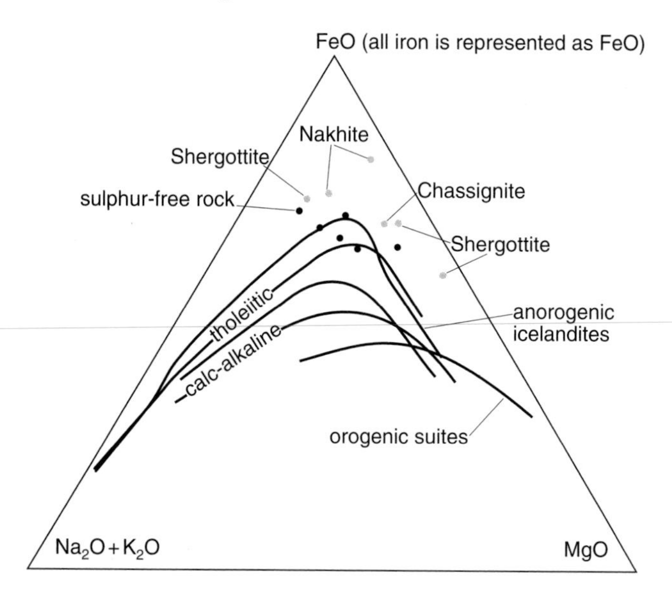

Figure 15.7 AFM diagram showing tholeiitic and calc-alkaline terrestrial differentiation trends, together with Mars sulphur-free rock, APXS analyses of rocks and SNC melts. Orogenic suites represent trends of samples from the Lesser Antilles and California; anorogenic icelandites represent trends of samples from Iceland and the Galapagos. The • symbol denotes Pathfinder rocks. (After McSween et al. 1999.)

16 Phobos and Deimos

Mars has two tiny moons, both of which are considered to be captured asteroids. They are composed of material stronger than their own gravitational force and therefore they have irregular shapes, rather like cosmic potatoes. Of the two, Phobos is the larger, has a longest dimension of 27 km and a mean density of 2200 kg m^{-3}. Deimos, the smaller moon, measures 11–15 km across and has a mean density of only 1700 kg m^{-3}. Both moons were discovered by Asaph Hall in 1877.

Phobos

Phobos was first imaged by Mariner 6 in 1969. It orbits Mars at a distance of 9378.5 km and has a mean diameter of 22.2 km, measuring 13.3×27.1×9.3 km. Its orbit appears unstable and it has been calculated that it will impact Mars within 100 million years.

The first detailed images were returned by Mariner 9 during late 1971. Most of what we knew of the appearance prior to Mars Global Surveyor stemmed from Viking orbiter imagery. It was the intention of the Soviet Phobos mission to study the larger moon in detail and to bounce a "hopper" probe on its surface to conduct detailed chemical analysis (Sagdeef & Zakharov 1989). Unfortunately, the mission was prematurely terminated, and detailed chemical results were not forthcoming. However, the magnetometers aboard both spacecraft recorded the flux, elemental percentages and spectra of electron and ion radiation from the Sun in the environment of Mars. Unexpectedly large electron densities were measured. Phobos 2 obtained just 37 images of the moon, but was able to record spatial variation in colour, and an improved dataset for the shape and volume.

Phobos has a low albedo (6%) and low density (1.9 g cm^{-3}), the latter being somewhat unusual for a body of believed meteoritic parentage. Its spectrum is similar to that of certain kinds of carbonaceous chondrites. The mass of the moon was established at 1.08×10^{16} kg, which suggests that it is either somewhat porous or holds ice beneath the surface. It would require porosities of about 20–25% if it proved to be of carbonaceous chondrite composition. Ice may be present but would be difficult to prove or disprove.

The surface of Phobos is brecciated and heavily pockmarked with impact craters, the largest being Stickney, 10 km in diameter (Fig. 16.1). Like most asteroids, Phobos has a regolith that may be as much as 100 m thick. The bolide that formed Stickney must have been close to the critical size beyond which Phobos would have been entirely disrupted. This weird structure covers a fifth of the satellite and has associated with it a series of 100–200 m-wide grooves, thought to have formed in response to the Stickney-forming event. Some of these have a beaded structure.

Mars Global Surveyor captured some fine images of the moon, showing the crater Stickney with individual boulders on the rim of the crater. These are presumed to be ejectamenta from the Stickney event. Some are enormous, measuring more than 50 m across. Also visible are many shallow elongated grooves, which may also be associated with Stickney's formation. The MGS thermal emission spectrometer (TES) also measured the intensity of thermal radiation at the same time as the camera captured the images. The temperatures are superimposed on the image. Temperatures are highest (–4°C) on the most illuminated slopes, and lowest (–112°C) where the surface is in deep shadow. The juxtaposition of such highs and lows indicates that the surface is covered by very fine-grain material.

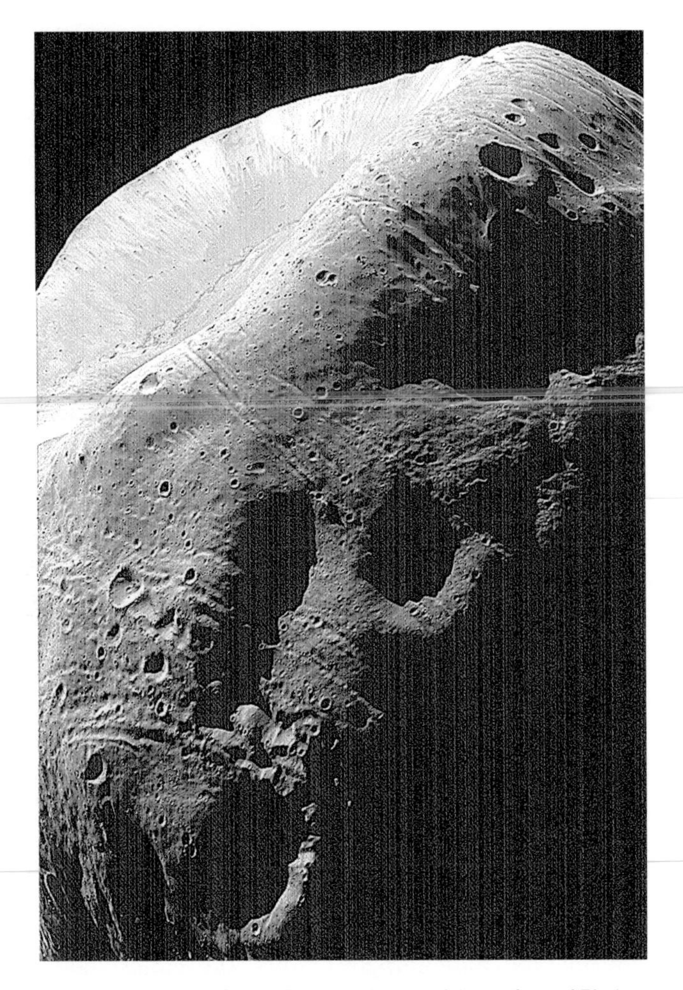

Figure 16.1 Mars Global Surveyor image of the surface of Phobos with TES temperature overlay (PIA01333).

A particularly good image was obtained of the far wall of Stickney, showing light and dark streaks on the inner slopes. The presence of material of different albedo implies that the moon is heterogeneous. Another image showed details of the outer rim, including huge boulders, small craters and grooves (Fig. 16.2).

Deimos

Deimos is rather different from its companion. It has a period of 30.3 days and it orbits Mars at a mean distance of 23 459 km. It has a mean diameter of 12.4 km, measuring 7.5×6.1×5.2 km. It too is densely cratered, but it is generally far smoother than Phobos (Fig. 16.3). It has a lower albedo and seems to be mantled in a thick layer of dust that has buried many of the original craters. There are distinctive albedo markings, brightish streamers seeming to emanate from shallow impact craters.

Deimos has a visual albedo of 6 per cent and its

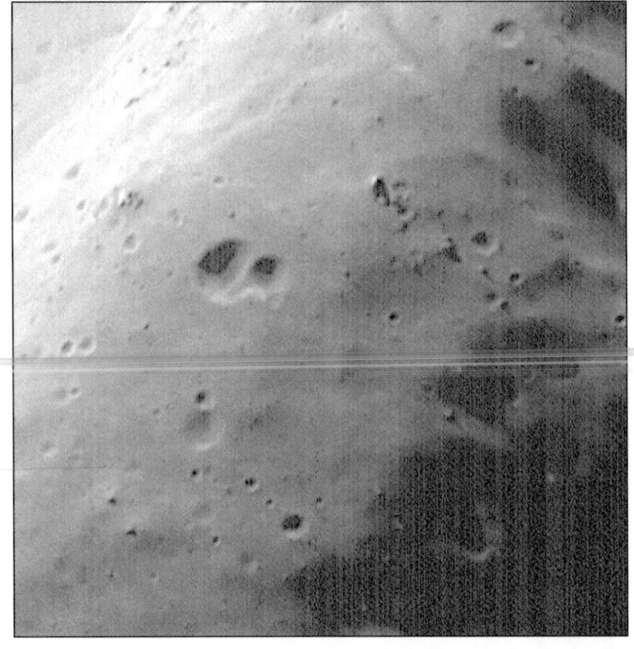

Figure 16.2 High resolution (4 m per pixel) image of the outer rim of Stickney, showing large boulders and trails (PIA01336).

spectrum is flat at wavelengths of greater than 0.35 μm, but falls off sharply at shorter wavelengths. Like Phobos, it is assumed to be of carbonaceous chondrite composition, but this could be misleading since its density (1.8 g cm^{-3}) is rather low for any known meteorite. Perhaps, like Phobos, it too is porous. If so, porosities of between 30 and 50 per cent are required to match carbonaceous chondrite composition.

Figure 16.3 Montage of Phobos (right), Deimos (left) and Gaspra, shown to same scale and nearly the same lighting. Gaspra (top) is 17 km long. All have suffered catastrophic conditions (PIA00078).

17 Life on Mars?

It is generally agreed that during the initial few hundred million years of Solar System evolution the compositions and pressures of the terrestrial planets' atmospheres were predominantly conditioned by volcanic outgassing of CO_2, N_2 and H_2O. It is possible, therefore, that Earth, Venus and Mars may at this time have experienced very similar surface conditions. Mars may have had a relatively dense CO_2 atmosphere in the distant past, with liquid water at the surface, rather like the primeval Earth.

We know that, on the early Earth, atmospheric CO_2 became locked into carbonate rocks that eventually were recycled into the atmosphere. During Mars' earlier history, the carbonate rocks could have been similarly recycled by volcanism, but as this became less intense and more localized the CO_2 recycling process probably decayed, slowly depleting the Martian atmosphere.

Early observations

Ever since telescopic observers in the eighteenth and nineteenth centuries recorded the seasonal waxing and waning of Mars' dark polar collar (known as the "wave of darkening"), they began to speculate on the possibility of life there. The seasonal wave of darkening was thought by many to represent growth of vegetation after release of water from the ice caps during spring. This debate peaked at the close of the nineteenth century when, in 1877, the astronomer Schiaparelli reported the presence of linear canali[1] crossing the red surface. Initially, he believed these to be entirely natural, but because the Italian word translated into "canals" in English, the American Percival Lowell hurried the completion of his (now famous) observatory at Flagstaff, Arizona, and began mapping "canals".

Lowell firmly believed that these features had been dug by intelligent Martians and he wrote several books about his theory of Martian life. Briefly, he did not assert that the artificial canals themselves could be seen from Earth, but rather the vegetation that sprung up adjacent to the water courses as they crossed the Martian desert. Once the canals had been recognized and afforded a certain degree of notoriety, the notion that life once existed on Mars has been almost permanently in the literature.

Not all observers could see the so-called canals, but the debate raged on well after Lowell's death in 1916. The matter was finally resolved when the first spacecraft imaged the surface in detail and no such features were to be seen. The canals turned out to be optical effects, probably because of the way the human eye, stretched to the limit of its resolution, tends to join up individual dark markings. The whole matter was beautifully reviewed by Sheehan (1988).

Viking observations

Although the idea that the canals had been dug by intelligent creatures has long been discounted, the possibility of finding some form of life, however primitive, has been kept alive. With this in mind, various experiments were put on board the Viking landers, mainly to analyze the Martian regolith, but one aspect of the mission was to seek evidence for microbial life forms within the surface layer. The

1. An Italian word whose primary meaning is "channels" but which also means "canals".

pyrolitic release experiment measured carbon assimilation through the regolith by using an isotope of radioactive carbon. The labelled release experiment moistened a sample of the soil with radioactively labelled organic matter, so that if life were present it would consume the organic matter and eventually release gases plus radioactive carbon. The gas exchange experiment moistened the soil with amino acids, salts and vitamins, and tested for the gases evolved.

All of the experiments worked well, but the consensus view was that all of the observed reactions could be explained by reactions in the highly oxidizing conditions that prevailed at the Martian surface, without the need to imply living organisms. It was concluded that no organic molecules were found at either of the sites. However, this does not rule out the possibility that life developed on Mars and that evidence may be found elsewhere.

On Earth, primitive algae proliferated about 3.8 billion years ago. Could such organisms have developed on Mars, and under what conditions? This is a potent question and a fascinating one. Certainly, if oceans and lakes did once exist on Mars, life may have developed there, possibly protected from the harsh exterior environment – that of the Viking landers – by a protective carapace of ice. Thus, if it proved possible to identify either marine or lacustrine sediments on the surface, these would become prime targets in the search for a Martian biota, probably via a lander probe of some kind. Although the search for a biota has generally focused on the surface of Mars, the first really exciting scrap of potential Martian life came from a completely unexpected source: a meteorite found in Antarctica.

Meteorites and possible Martian life

In 1969, Japanese scientists discovered concentrations of meteorites in Antarctica. Most of these had fallen onto the ice during the past million years or so and, because of the nature of the environment, were well preserved. Subsequently, the collection and curation of these precious samples from space became the collective responsibility of NASA, the US National Science Foundation and the Smithsonian Institution. In recent years this collection has proved invaluable in learning more about early planetary history

Meteorite ALH84001

In late 1993 a paper was published that described one particular meteorite, ALH84001. The ALH prefix indicates that it had been collected from the Allan Hills Field in the far western icefield of Antarctica. The sample was an achondrite and most of it was covered in a dull black fusion crust. What was unusual about this specimen was that inside it were abundant small grains of chromite and larger crystals of orthopyroxene and high-temperature feldspar. The sample also appeared to hold some carbonates. It was classified immediately as a new kind of SNC meteorite; further analysis established it to have a Martian provenance.

Later, in the summer of 1996, the scientific community was rocked by the publication of a research paper in *Science* magazine describing evidence for bacteria-like structures in ALH84001 (McKay et al. 1996). It was reported that polycyclic aromatic hydrocarbons (PAHs) had been discovered on fractured surfaces within it. PAHs are large complex molecules that are widely distributed in cosmic dust particles and many organic-rich chondrite meteorites. They are believed to be the residue from non-organic reactions between simpler carbon compounds.

There was also a development of carbonate globules along the fractures and, by studying these with very high-resolution scanning and transmission electron microscopy, the team discovered that inside several of the globules were fine-grain growths of magnetite and iron sulphides, very similar to associations found in some terrestrial bacterially induced carbonate precipitates. Furthermore, within some patches of carbonate were tiny worm-like structures about 50 μm across, which resembled micro-organisms. The possibility that Martian life had been discovered had been reborn.

Sophisticated laser techniques were used to lift and separate organic molecules from the sample by Clemett & Zare of Stanford University in 1995. These workers and others subsequently claimed an extraterrestrial origin for the polycyclic aromatic hydrocarbons they identified (Clemett et al. 1998). The key issue was to establish whether or not the PAHs were indigenous to the meteorite or whether they could have originated by terrestrial contamination.

It is probably true to say that most researchers believe that the carbonate globules were deposited by liquid water (Romanek et al. 1994, Treiman 1995, Wentworth & Gooding 1995). However, Harvey & McSween (1996) reported evidence suggesting that the carbonate globules must have formed at very high temperatures (>650°C). Such conditions would mean it was extremely unlikely that any form of life could have survived inside the meteorite.

Clemett et al. appear to have convinced most scientists that the PAHs are indeed indigenous, primarily because there are no PAHs within the fusion crust, and a lack of such molecules in other Antarctic meteorites subsequently studied. John Kerridge of UCLA, a seasoned meteorite expert, agreed that Clemett's team have done a good job in convincing the community of the PAH's Mars meteorite origin, but goes on to point out that such molecules could just as easily have been formed by non-organic processes. This alternative interpretation has been at the hub of all subsequent, often heated, debate.

The team who reported the material in the first place was careful to state that everything they had seen could also have been produced by inorganic processes. They preferred the organic origin because of the combined evidence of the carbonate growths, the organic residue and the metallic grains. It is necessary, therefore, to delve a bit deeper into what was found.

As far as McKay and his team were concerned, the critical evidence was to be found within the carbonate globules. Tiny particles (< 100 nm across) of magnetite and pyrrhotite occurred there. The former, they insist, bear a strong resemblance to what are called magnetofossils. These are left in terrestrial sediments by the action of bacteria, which use the magnetic particles to guide them along magnetic field lines. Furthermore, the iron particles are located where acids appear to have pitted the surface of the globules. This combination of facts indicated to the team that only a biological origin seemed tenable.

However, there is one more claim, even more astounding. This involves the detection, using very high-resolution electron microscopy, of tiny worm-like growths, a mere 20–100 nm long, These are claimed to be the remains (not the actual organisms) of ancient Martian bacteria-like organisms. Not everyone agrees with this interpretation. For instance, Schopf (1999) points out that these objects are a hundred times smaller than any ancient bacteria found on Earth. Indeed, a variety of scientists claim that blob-like structures have been shown to form in rocks by a wide range of different inorganic means. There are many sceptics, to this day. Currently, there is an intensive search for cell structure within the putative fossils, and for the presence of amino acids in the sample. This question is not going to be resolved quickly.

The object crystallized from magma 4.5 billion years ago, presumably on the surface of Mars. Radiometric dating shows. This must make it one of the oldest rocks on the planet. Around 3.6 billion years ago, a major impact shattered the rock, leaving fractures, along which grew minerals and, if the theory proves true, also traces of putative life. Much later, another impact hurled it away from Mars, whereupon it sped through space for a further 16 million years or so before impacting Earth. ALH84001 lay buried in the Antarctic ice for a further 13 000 years.

The question of life on Mars is not dead, however, neither has it been established beyond reasonable doubt that the objects and minerals within ALH84001 are bacterial in origin. Yet again, Mars remains an object of fascination and the quest for proof of past life persists. Debate about the interpretation of ALH84001 continues, as does the search for evidence of past life on Mars. The planet remains fascinating, as it has since the days of Schiaparelli and Lowell.

For an excellent summary of recent papers and commentary by Allan Treiman of the Lunar and Planetary Institute, see *Recent scientific papers on ALH84001 explained, with insightful and totally objective commentaries*, which can be found at website: http://cass.jsc.nasa.gov/lpi/meteorites/.

18 Today on Mars

In the preceding chapters I have presented an interpretation of what we have learned in recent years by studying the surface and atmosphere of Mars. I hope this narrative will have provided insight into its geological and atmospheric evolution, and unfolded a story that is much closer to being a true one than it was a decade ago. This leads me on to discussing what is happening on Mars today. In essence this means looking at aeolian activity and what is happening in the polar regions.

Wind activity on Mars

The action of global and local winds can be observed on Mars at the present time. Indeed, the HST has been monitoring global weather for several years and we have recent data about day-to-day conditions at the Pathfinder landing site, and observations made at the Viking sites during the late 1970s. Wind erosion and transport are two processes that currently can be observed on Mars.

Viking landing-site observations

Meteorological instruments aboard the Viking landers on the plains of Chryse and Utopia measured windspeed and direction over more than one Martian year. Measurements were made at a height of 1.3 m above the ground surface and showed that nocturnal wind velocities were around 2 m s^{-1}, but that windspeed increased near sunrise, to reach around 7 m s^{-1}; occasional gusts registered at 25 m s^{-1}. At sunset each day, wind direction changed from east to southwest; then, at sunrise, it swung towards the south-southwest.

Wind is very effective at moving fine-grain sedimentary material across the Martian surface, but there is scant evidence to suggest that it is an effective eroder of consolidated rocks. For instance, images of the ancient Hesperian rocks of the Chryse basin reveal perfectly preserved lava flows and crater details. On this basis, Arvidson (1979) estimated that erosion rates cannot be more than around 1000 μm per year.

Mars Pathfinder landing-site measurements

Windstreaks imaged prior to landing in Ares Vallis show them about 80 km northeast of the landing site; many of these are associated with impact craters. These range in length between 1 km and 10 km, and strike between 202° to 225°.

The general circulation model (Pollack et al. 1981) enables near-surface wind patterns to be predicted as a function of season and location. For the cell within the GCM that covers the MPF site, the strongest winds blow from the north-northeast at L_S = 285°, with an azimuth of 198°. Thus, the measurements made at the landing site accord well with model predictions and show that they are the result of aeolian activity at the present time.

The aeolian landforms and other features observed at the site include windtails, duneforms, ripple-like patterns and ventifact flutes. Measurements of their orientations were made by Greeley et al. (1999). For windtails, the azimuths ranged from 171° to 260°, with an average of 217°. The axes of duneforms and ripple-like patterns strike northwest to southeast, that is, approximately orthogonal to the wind directions. The orientation of barchan dunes is also consistent with winds blowing from the northeast.

Bridges et al. (1999) measured flutelike features on rocks around the site; these were orientated at ~280°, indicative of winds blowing from the east at the time of their formation. However, the meaning of these data is unclear, since the rock may not always have been in its current location.

Also observed at the MPF site were flutes and grooves, believed to represent ventifacts. These indicate that deflation has occurred at the site in relatively recent times. The change in albedo observed near the base of several large rocks suggested to Greeley et al. (1999) that the regolith level was formerly higher, once standing 5–7 cm above the modern level. This provides further evidence for recent deflation.

Martian winds and windstreaks

Particulate material is moved by being suspended in the atmosphere, by saltation (bouncing across the surface), or by creep (gradual slow movement induced by saltation). On Earth, the optimum particle size for movement by wind is 0.08 mm. Wind-tunnel experiments conducted by Greeley et al. (1980), have shown that this figure is nearer 0.1 mm on Mars. Threshold wind velocities required to move particles are, however, somewhat higher than on Earth, where the figure is around $0.2\,\mathrm{m\,s^{-1}}$ mainly because of the thin-ness of the Martian air. Thus, for an atmospheric pressure of 1 mbar the speed required to lift a 0.1 mm clast is $2.4\,\mathrm{m\,s^{-1}}$; for a 5 mbar pressure, it is $4\,\mathrm{m\,s^{-1}}$.

These figures are those that apply right on the surface. At the 1.6 m height of the Viking instrument boom, calculations suggest they should be approximately one half of this. It is quite simple, therefore, to see why little particulate material was seen in the air around the lander sites, even during the stronger gusts. However, threshold velocities would be reached in the source areas of the major dust storms; indeed, it appears that only during such activity can large volumes of particulate matter be transported by the atmosphere. Dust storms appear to be the major redistributors of fine-grain sediment over Mars' surface.

Returning to the matter of aeolian landforms, there are widespread temporary albedo features that cover quite extensive parts of Mars. Particularly good examples of these outcrop in Syria Planum, where sharply defined albedo differences (which show little correlation with topography) can be traced over large distances. The most common of the albedo markings are splotches and windstreaks, and these do tend to have a relationship with local relief. Streaks are also widespread on the surface of Syrtis Major. Three types have been noted (Sagan et al. 1973, Greeley et al. 1978, Thomas & Veverka 1979, Veverka et al. 1981) and each is believed to be the result of the movement of fine-grain particulate material across the surface. Bright streaks are depositional and show a preferential distribution in the lee of obstacles such as craters and knobs; dark streaks are erosional in origin and seem to represent areas where the thin veneer of light dust has been removed; dark splotches generally occur within impact craters and probably represent deposits that have been deflated from interior dunes or sheet deposits. The dark streaks tend to be less stable than the light ones.

By mapping the planet-wide orientation and distribution of such features both during and after the 1971 and 1977 dust storms, it has been possible to gain some insights into the windflow over the planet (Fig. 18.1). Thus, between the equator and 30°N, the windflow is from the northeast, whereas south of the equator winds tend to be more northerly and swing around between latitudes 20°S and 30°S to a more northwesterly direction. At higher latitudes, the winds tend mostly to blow from east to west.

Aeolian erosional landforms

Some of the best-developed wind-erosional features are in southern Amazonis, among the putative pyroclastic plains deposits described in Chapter 7. Here, in a series of what appear to be readily erodible strata, the wind has etched families of yardangs – elongate, streamlined ridges – which lie parallel to the dominant wind direction. Similar landforms are widespread on Earth and they develop where abrasion and deflation combine together in major deserts (Ward 1979). In this area, too, are many deflation hollows, fluted scarps and pedestal craters. Pedestal craters are common in latitudes higher than about 40° and represent a crater modification produced by deflation. At these higher latitudes, impact craters are often surrounded by a relatively broad (2–6 crater radii wide) pedestal, this seeming to be a rather broad ejecta blanket. Since there is very little reason to expect ejecta blanket diameter to be latitude dependent, another explanation for the pedestals is required.

Arvidson et al. (1976) have suggested that, in high latitudes, aeolian dust once blanketed the surface over very large regions. Subsequently, much of this has been removed by deflation, but around impact craters the surface was protected by blocky ejecta, which retarded the rate of erosion locally. The remnant pedestals are viewed, therefore, as residual landforms that may or may not coincide with the extent of the original ejectamenta.

Deflation also appears to have affected crater deposits in higher latitudes. Many of the larger craters exhibit a concentric crater fill pattern and it has been suggested that this results from centripetal movement of sedimentary debris from the crater rim towards the interior. However, high-resolution imagery shows that

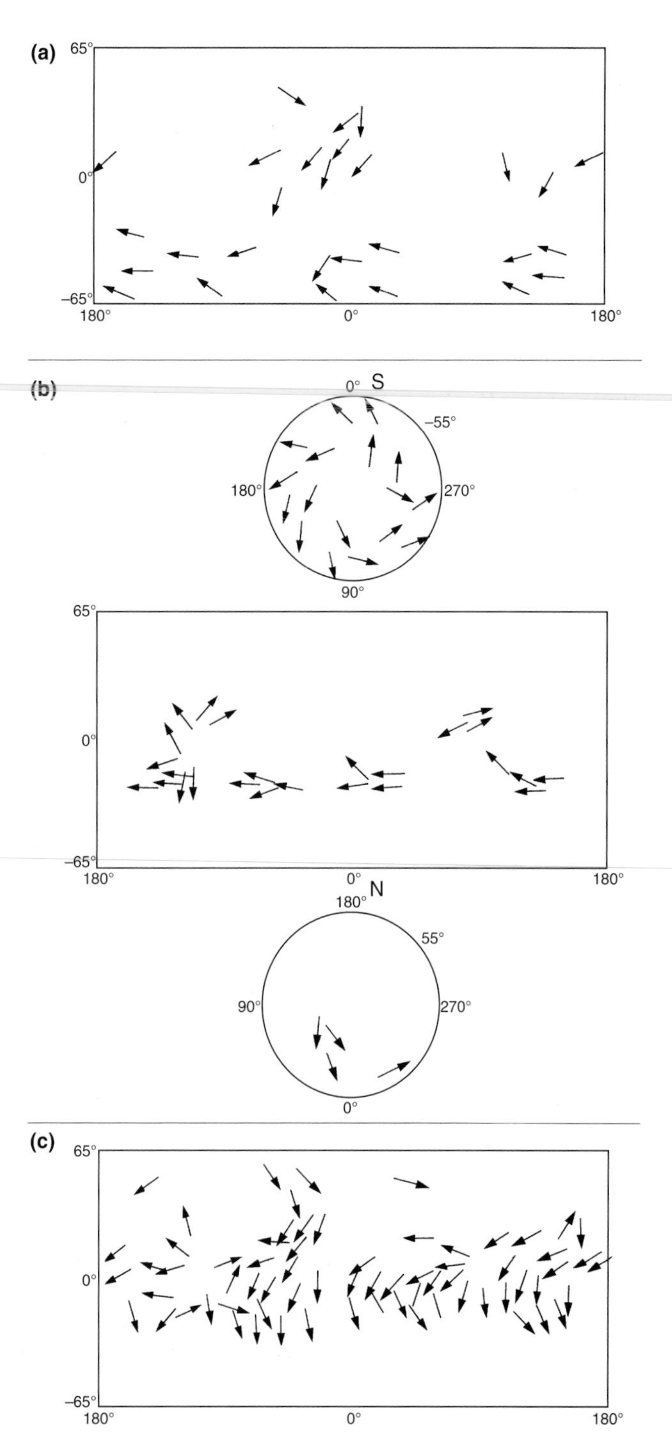

Figure 18.1 Global distribution of windstreak directions from Viking post-duststorm data (after Thomas & Veverka 1979): **(a)** dark splotch associated streaks, **(b)** dark erosional streaks, **(c)** bright streaks.

there are sharp changes in slope, both inside and outside such craters, that are inconsistent with slow downslope creep (Zimbelman et al. 1988). It is more likely that what is being observed in these locations is the slow stripping of stratified aeolian sedimentary rocks by current wind activity.

Viking IRTM data has shown that a quarter of all craters larger than 25 km diameter, located between latitudes 30° north and south, have high-inertia dark deposits on their floors. Inertias are significantly higher than their surroundings and are darker than the majority of the Martian surface. The data imply grain sizes in the range 0.1 mm to 1 cm (which puts the deposits well into sand and pebble grade) with a <16% block cover. Christensen (1983) suggests that this material is probably being actively reworked, either being sediment that has become trapped inside craters or via deflation of *in situ* materials. Because of its relatively coarse grain size, it would seem logical to conclude that it is being transported only by the strongest winds currently blowing across the surface.

In summary, Martian winds seem to be much better transporters of fine sediment than efficient eroders of coherent rocks. Few prominent erosional aeolian landforms are seen in low- or medium-resolution images. However, it should be noted that, where higher-resolution imagery is available, many surfaces that appear unaffected by wind in lower-resolution frames are seen to exhibit etched and pitted surfaces in the 10–20 m scale range. This is born out by the observation of ventifacts at the Pathfinder landing site. In consequence, our current inventory of aeolian landforms may be incomplete because of a lack of global high-resolution imagery. Mars Global Surveyor is currently doing its best to change this situation.

Depositional landforms

The most obvious evidence for movement and deposition of aeolian debris is to be found in the high-latitude dunefields (Ch. 12) and in the major dust storms that have been observed from Earth, the HST and from approaching spacecraft.

Duneforms are also found within many large impact craters and valleys. Recently Mars Global Surveyor obtained a magnificent image of linear dunes on the floor of Echus Chasma (Fig. 18.2). Many of the new images show development of dunes in low-lying areas, such as channels, craters and canyons.

In the context of the palaeolake debate, Greeley & Williams (1994) have discussed what, on Earth, are termed parna deposits. These occur both as dunes and in sheet form, and they mantle older terrain. They consist of a mix of sand, silt and clay, and generally have their origin in lacustrine environments, such as playa lakes. After heavy rainfall and disintegration, pellets of this material may survive for quite lengthy periods. Since there is growing evidence for past bodies of standing water on Mars, Greeley & Williams suggested

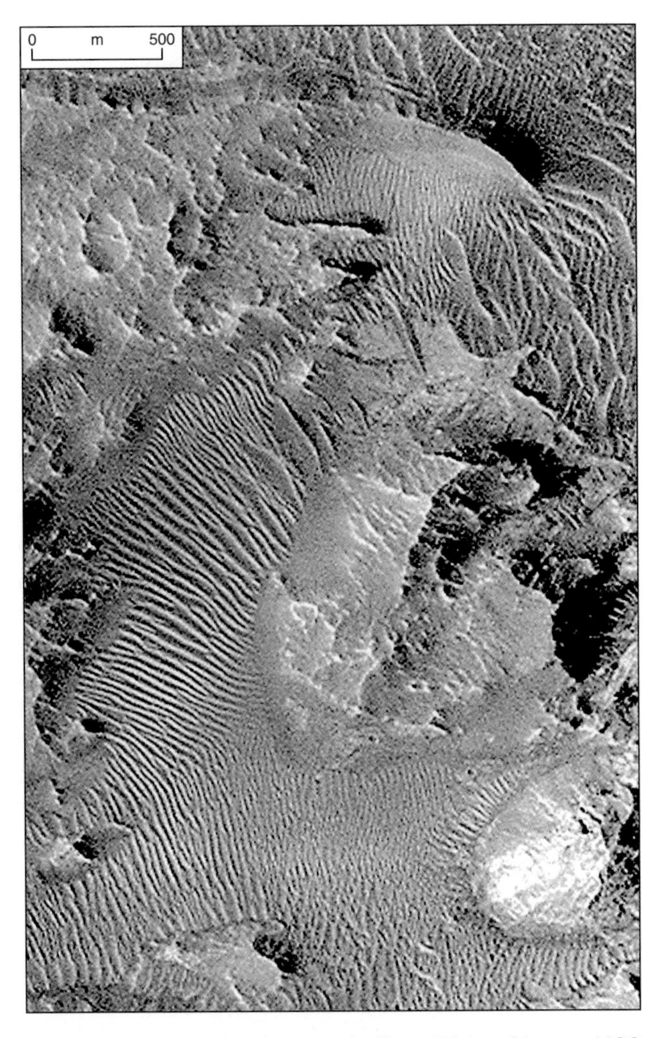

Figure 18.2 Linear duneforms on the floor of Echus Chasma. MGS image PIA00803; centred at 0.8°S 76.3°W.

that there is a potential for lacustrine deposits of this type on Mars and that, although no dune-like deposits of this type have so far been recognized, certain units identified near White Rock (8°S, 335°W) and Mamers Valles (34°N, 343°W) may represent sheet parna. High-resolution imagery may be able to resolve this in due course.

An observation from the same Viking IRTM data described on p. 24 is that the high-inertia material diagnosed by the data is not found in the interior of large low-inertia regions such as Arabia, but is recognized around their perimeters. In this case, the darker coarser deposits may have been mantled by finer-grain windblown dust, implying that currently these regions are expanding. This is further evidence that fine-grain materials are being moved by the wind.

Hubble space telescope observations of dust storms

The HST has proved invaluable in updating our synoptic view of Martian weather and climate. Thus, as Mars Pathfinder approached Mars, HST found cold cloudy conditions prevailing. This provided mission controllers with advance warning of what to expect at the landing site.

Among the most interesting observations made by the HST have been those of dust storms. For instance, pictures taken on 27 June 1997 show a dust storm churning through the deep canyons of Valles Marineris, something never witnessed before. At the same time, it showed cirrus clouds forming over Ares Vallis, where Mars Pathfinder was shortly to land. Between 18 September and 15 October 1998, HST revealed a Texas-size dust storm churning near edge of north polar cap. Since then, many such observations have been made and collated. Indeed, during 1996–7, observations covered the whole of one Martian spring, when conditions were ripe for viewing the north polar region. HST observed that local dust activity occurred near the cap edge as it sublimed during the early spring season, with local dust clouds rising above the surface. It also imaged the development of a hitherto unseen arc-shape dust feature whose origin remains obscure.

The many observations of the cap regions unequivocally confirm the hypothesis that the dust storms are the result of winds associated with fronts moving across the seasonal cap. These lift dust exposed on the cap during CO_2 sublimation (James et al. 1999).

HST also observed the region of Cerberus, which is one of the classical dark albedo features. Once a large dark area of fairly constant dimensions, this now consists of only three dark splotches, indicating a major change in recent years. This has presumably been achieved by recent aeolian activity, which has shifted large quantities of bright dust across the surface.

The telescope also collects data in the near-infrared. A series of such measurements made between February 1995 and June 1997 have given scientists the highest spatial-resolution global near-infrared data to date (Bell et al. 1999). They show that substantial changes have occurred in the surface albedo over the past two decades. Thus, several dark classical (one being Cerberus) have become lighter and a few formerly lighter regions have darkened. This has to be a manifestation of current and recent aeolian activity on the planet.

Finally, Martin (1995), using Viking IRTM data collected during the 1977 dust storm, found a mass total of 43 000 g was suspended in the air, this being equivalent to 43 000 g cm^2, or a layer 1.4 µm thick. Also, during a local storm near Solis Planum, approximately

1.3 million kg of dust were often equivalent to a layer 6 μm thick in the vicinity.

Pathfinder landing site weather conditions

Mars Pathfinder made a series of daily observations of conditions in Ares Vallis, giving us an up-to-date record of local conditions at this location. The ASI/MET experiment measured surface meteorology for 83 sols and found it to be similar to that observed at the Viking 1 site, 21 years before. However, there were differences in diurnal patterns. Thus, it observed a cold night-time temperature minimum, light slope-controlled winds, and, for the first time, dust devils in Ares Vallis.

Pressure ranged during the first 30 sols from between 0.2 and 0.3 mbar, while temperature ranged from 263 K at just after midday to 197 K just before sunrise. Diurnal temperatures near the bottom mast were observed to be more extreme than those at the top (1.1 m above the ground). In the same period, wind direction rotated clockwise through all points of compass during the course of each sol. Most scatter was seen during the daytime, with recurrent southerly winds being recorded from late evening through morning. This is consistent with a drainage flow through Ares Vallis, which slopes up hill south of the lander, and the northerly wind seen in the afternoon is indicative of flow up the valley.

Compared with measurements made at the Viking 1 site, recorded windspeeds of 5–10 m s^{-1} were similar at both sites (Golombek et al. 1997); however, temperatures were slightly higher at the Pathfinder location (Schofield et al. 1997). The higher temperatures are consistent with a lower surface albedo at the MPF site compared with that of Viking 1.

Annual data show that the pressure cycle reached a minimum at about sol 20, corresponding to the greatest mass of the southern polar cap. This is precisely what would be expected.

Plans for the future

The Hubble space telescope continues to make observations of global changes at a variety of wavelengths and will do so into the future, barring unforeseen problems. It should, therefore, continue to monitor the weather, the movement of dust and record the seasonal changes in the polar regions.

It had been hoped that both Mars Polar orbiter and Mars Polar lander would currently be contributing to our understanding of the poles; regrettably, this has not happened for reasons now well known. Future plans are for the continuation of the Global Surveyor programme with a view to sample collection and return by the year 2005. Whether or not the failure of the two polar missions will affect this schedule is, at present, unclear. Both Japan and ESA have scheduled missions to Mars. Only time will tell if these meet with more success than the most recent two US spacecraft.

Fortunately, Mars Global Surveyor continues to send back wonderful images and, day by day, it reveals more about the details of Martian geomorphology. This archive will take many years to collate and study. As usual, although it has answered many questions about the geology of Mars, it has posed as many new ones. Certainly, it has not resolved the question of ancient lakes and oceans on the planet; high-resolution images along the boundary between the northern plains and southern uplands do not reveal ancient shorelines. It will take much more painstaking study of the imagery before we know with any degree of certainty that such bodies of standing water once existed.

In some ways, the future of Mars exploration hangs in the balance; recent failures may affect budgets for those future missions planned. If it is secure, then we should see landers on the surface, collection of rock samples and return of this material for analysis on Earth during the next five years. There will doubtless be further analysis of meteorites believed to have their origin on Mars. Many scientists will continue to search for signs of past Martian life

Mars remains the one rocky planet upon whose surface we may yet find signs of past life. It also remains the one inner planet that may have had large areas of standing water, albeit in the distant past. Certainly, it has had a very dynamic geological and climatic history and I for one wish all future missions and mission scientists all the luck in the world, in their efforts to explore our neighbour world and find answers to the major questions that remain: warmer wetter climate, lakes, oceans, glaciers, ice – who knows? The mystery is still unfolding . . .

Appendix

Model chronology

No radiometric ages exist for any Martian rocks. Thus, we can estimate the ages of surface units only by predicting the relationship between crater densities and absolute age, using data derived in lunar studies. This necessitates certain assumptions, in particular that of the ratio between the cratering fluxes of Mars and the Moon. The flux for Mars is not well constrained, thus the results from the two most widely used chronologies, those of Neukum & Wise (1976) and Hartmann et al. (1981), differ considerably. In Table A.1 my own preferred chronology is presented.

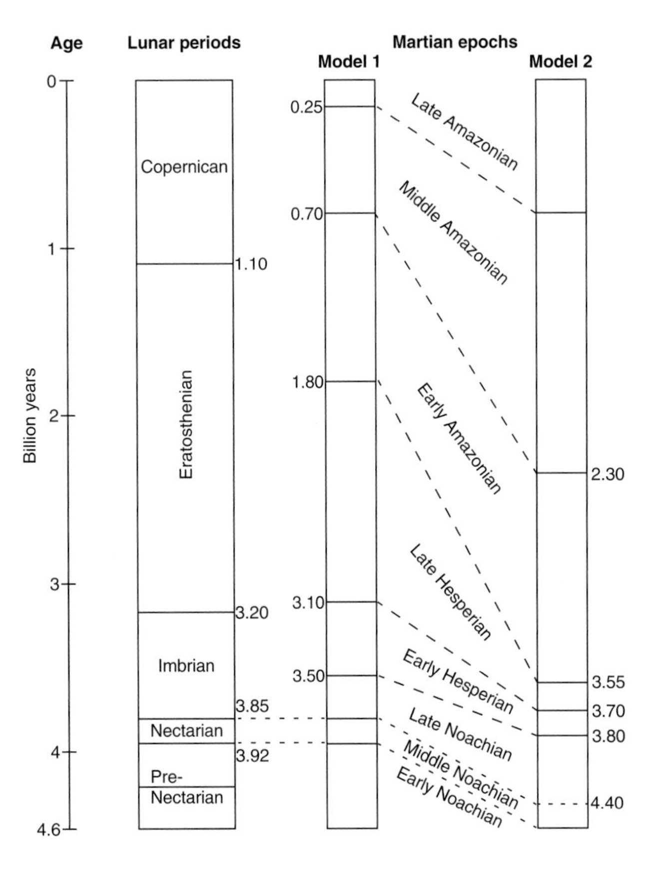

Astronomical data

Diameter	6787 km
Mass	6.4185×10^{23} kg
Volume	162.6×10^{12} km^{-3}
Mean density	3933 kg m^{-3}
Centre of mass/figure offset	2.50 ± 0.07 km
Surface gravity	3.71 m s^{-2}
Escape velocity	5.02 km s^{-1}
Axial inclination	23.98°
Visual geometric albedo	0.16
Rotation period	24.623 h
Orbital period	686.98 d
Mean orbital velocity	24.13 km s^{-1}
Maximum solar distance	249 million km
Minimum solar distance	206 million km
Length of sidereal day	24 h 27 m 22 s
Length of mean solar day	88 775.2 s
Elipticity	0.0059
Eccentricity	0.0934
Surface temperature range	148–310 K
Mean surface pressure	6.1 mbar
Magnetic field strength	1/800th that of Earth

Table 1.1 Chronology for Martian units (after Hartmann et al. 1987).

Geological province	Crater density relative to average lunar mare	Estimated crater minimum likely	Retention age best estimate	Maximum likely (billions of years)
Central Tharsis volcanic plains	0.1	0.06	0.3	1.0
Olympus Mons volcano	0.15	0.1	0.4	1.1
Extended Tharsis volcanic plains	0.49	0.5	1.6	3.3
Elysium volcanic rocks	0.68	0.7	2.6	3.5
Isidis Planitia	0.76	0.8	2.8	3.6
Solis Planum volcanics	0.90	0.9	3.0	3.7
Chryse Planitia volcanic plains	1.1	1.2	3.2	3.8
Lunae Planum	1.2	1.3	3.2	3.8
Noachis ridged plains	1.3	1.7	3.3	3.8
Tyrrhena Patera	1.4	1.8	3.4	3.8
Tempe Fossae faulted plains	1.6	2.3	3.4	3.8
Plains south of Hellas rim	1.7	2.6	3.5	3.8
Alba Patera volcano	1.8	2.6	3.5	3.8
Hellas floor	1.8	2.6	3.5	3.8
Syrtis Major Planitia	2.0	2.0	3.6	3.9

Websites about Mars

Here is a brief listing of some websites that I have found useful. Several of these give extensive links to other Mars and planetary sites.

Table 1.2 Websites about Mars.

Ares Valles area	http://space.magnificent.com/sol/Pathfinder/LandingSite/
HST images of Mars	http://www.seds.org/~spider/spider/Mars/mars_hst.html
HST images of Mars	http://photojournal.jpl.nasa.gov/cgi-bin/uncgi/
Index of Mars – special images and maps	http://www.atmos.washington.edu/mars/special/
Mars fact sheet	http://nssdc.gsfc.nasa.gov/planetary/planets/marspage.html
Mars Global Surveyor – Welcome to Mars	http://www.bchip.com/mars/mgs/index.html
Mars Global Surveyor MOC images	http://ida.wr.usgs.gov/
Mars Pathfinder image thumbnails	http://mars.jpl.nasa.gov/mgs/pdf/pdf-mgs.html
Mars Pathfinder site	http://mars.jpl.nasa.gov/MPF/sitemap/
Malin Space Science Systems image release directory	http://www.msss.com/mars_images/
Malin Space Science Systems – MGS MOC image release directory	http://www.msss.com/mars/global_surveyor/camera/images/
Mars Exploration education program home page	http://marssnts.jpl.nasa.gov/education/index-education.html
NASA Mars Links	http://www.tui.edu/STO/SolSys/Mars/NASAMarsLinks.html
PDS planetary image front page	http://www-pdsimage.wr.usgs.gov/ATLAS.html
NASA Planetary Photojournal site	http://photojournal.jpl.nasa.gov/
Planetary sciences at NSSDC	http://nssdc.gsfc.nasa.gov/planetary/planetary_home.html
USGS Flagstaff planetary map index	http://wwwflag.wr.usgs.gov/USGSFlag/space/mapbook/mars
Mars image mosaics and meteorology index	http://www.atmos.washington.edu/mars.html
Mars Pathfinder science results	http://mars.jpl.nasa.gov/MPF/science
Mars meteorite homepage	http://www.jpl.nasa.gov/snc
ALH84001 – Recent scientific papers explained	http://www.jpl.nasa.gov/lpi/meteorites/alhnpap.html

Bibliography

Acuña, M. H. and 12 co-authors 1999. Global distribution of crustal magnetization discovered by the Mars Global Surveyor MAG/ER experiment. *Science* **284**, 5415–27.

Anders, E. T. & T. Owen 1977. Mars and Earth: origin and abundance of volatiles. *Science* **198**, 453–65.

Anderson, S. & R. E. Grimm 1998. Rift processes at the Valles Marineris, Mars: constraints from gravity modelling on necking and rate-dependent strength evolution. *Journal of Geophysical Research* **103**, 11113–24.

Antoniadi, E. M. 1975. *The planet Mars* (translated by P. Moore). Newton Abbot, Devon: Reid.

Arvidson, R. E. 1979. A post-Viking view of Martian geologic evolution. See Boyce & Collins (1979: 80–81).

Arvidson, R. E. and 7 co-authors 1976. Latitudinal variation of wind erosion of crater ejecta deposits on Mars. *Icarus* **27**, 503–516.

Baker, V. R. 1982. *The channels of Mars*. Austin, Texas: University of Texas Press.

Baker, V. R. & D. J. Milton 1974. Erosion by catastrophic floods on Mars and Earth. *Icarus* **23**, 27–41.

Baker, V. R. & R. C. Kochel 1978. Morphometry of streamlined forms in terrestrial and Martian channels. *Proceedings of the Ninth Lunar and Planetary Science Conference*, 3193–203. New York: Pergamon.

Baker, V. R. & J. B. Partridge 1986. Small Martian valleys: pristine and degraded morphology. *Journal of Geophysical Research* **91**, 3561–72.

Baker, V. R., R. G. Strom, V. C. Gulick, J. S. Kargel, G. Komatsu, V. S. Kale 1991. Ancient oceans, ice sheets, and the hydrological cycle on Mars. *Nature* **352**, 589–94.

Balmino, G., B. Moynot, N. Vales 1982. Gravity field model of Mars in spherical harmonics up to degree and order eighteen. *Journal of Geophysical Research* **87**, 9735–46.

Baloga, S. M. & D. C. Pieri 1985. E Estimates of lava eruption rates at Alba Patera, Mars. In *Report of the Planetary Geology and Geophysics Program – 1984*, J. Boyce (ed.), 245–7. Technical Memorandum 87563, NASA, Washington DC.

Banerdt, W. B., R. J. Phillips, N. H. Sleep, R. S. Saunders 1982. Thick shell tectonics on one-plate planets: applications to Mars. *Journal of Geophysical Research* **87**, 9723–33.

Barlow, N. G. 1988. The history of Martian volcanism determined from a revised relative chronology [abstract]. See Zimbelman et al. (1988: 20–21).

Barlow, N. G. & T. L. Bradley 1990. Martian impact craters: correlation of ejecta and interior morphologies with diameter, latitude, and terrain. *Icarus* **87**, 156–79.

Barnes, J. R., J. B. Pollack, R. M. Haberle, C. B. Leovy, R. W. Zurek, H. Lee, J. Schaeffer 1993. Mars atmospheric dynamics as simulated in NASA Ames general circulation model, 2: transient baroclinic eddies. *Journal of Geophysical Research* **101**, 3125–48.

Barnes, J. R., T. D. Walsh, J. R. Murphy 1996. Transport timescales in the Martian atmosphere: general circulation model simulations. *Journal of Geophysical Research* **101**, 16881–90.

Bell, J. F., T. B. McCord, P. D. Owensby 1990. Observational evidence of crystalline oxides on Mars. *Journal of Geophysical Research* **95**, 14447–61.

Bell, J. F. and 7 co-authors 1999. Near-infrared imaging of Mars from HST: surface reflectance, photometric properties and implications for MOLA data. *Icarus* **138**, 25–35.

Belton, M. J. S., A. L. Broadfoot, D. M. Hunten 1968. Abundance and temperature of CO_2 on Mars during the 1967 opposition. *Journal of Geophysical Research* **73**, 4795–806.

Betts, B. H., B. C. Murray, T. Svitek 1995. Thermal inertias in the upper millimetres of the Martian surface derived using Phobos' shadow. *Journal of Geophysical Research* **100**, 5285–96.

Bibring, J-P. and 7 co-authors 1990. ISM observations of Mars and Phobos: first results. *Proceedings of the Twentieth Lunar and Planetary Science Conference*, G. Ryder (ed.), 461–71. Cambridge: Cambridge University Press (on behalf of the Lunar and Planetary Institute, Houston, Texas).

Biemann, K. and 8 co-authors 1977. The search for organic substances and inorganic volatile compounds in the surface of Mars. *Journal of Geophysical Research* **82**, 4641–58.

Bills, B. G. & T. S. James 1999. Moments of inertia and rotational stability of Mars: lithospheric support of subhydrostatic rotational flattening. *Journal of Geophysical Research* **104**, 9081–9096.

Bills, B. G. & R. S. Nerem 1995. A harmonic analysis of Martian topography. *Journal of Geophysical Research* **100**, 26317–38.

Binder, A. B. 1969. Internal structure of Mars. *Journal of Geophysical Research* **74**, 3110–18.

Bishop, J. L., H. Froschl, R. L. Mancinelli 1998. Alteration processes in volcanic soils and identification of exobiologically important weathering products on Mars using remote sensing. *Journal of Geophysical Research* **103**, 31457–76.

Blasius, K. R. 1976. *Topical studies of the geology of the Tharsis region of Mars*. PhD thesis, California Institute of Technology.

Blumsack, S. L. 1971. On the effects of topography on planetary circulation. *Journal of Atmospheric Science* **28**, 1134–43.

Bogard, D. D. 1997. A reappraisal of the Martian $^{36}Ar/^{38}Ar$ ratio. *Journal of Geophysical Research* **102**, 1653–63.

Bogard, D. D., L. E. Nyquist, P. Johnson 1984. Noble gas contents of shergottites and implications for the Martian origin of SNC meteorites. *Geochemica et Cosmochimica Acta* **48**, 1723–39.

Boyce, J. M. 1979. A method for measuring heatflow in the Martian crust using impact crater morphology. See Boyce & Collins (1979: 114–18).

Boyce, J. M. & D. J. Roddy 1978. Martian rampart craters: crater processes that may affect diameter-frequency distributions. *Reports of Planetary Geology Program*, 162–5. Technical Memorandum 79729, NASA, Washington DC.

Boyce, J. & P. S. Collins (eds) 1979. *Reports of the planetary geology program*. Technical Memorandum 80339, NASA, Washington DC

Brackenridge, G. R. 1987. Intercrater plains deposits and the origin of Martian valleys. In *MEVTV workshop on Mars: evolution of volcanism, tectonics and volatiles*, Napa, California, 19–21.

Breed, C. S., M. J. Grolier, J. F. McCauley 1979. Morphology and distribution of common "sand" dunes on Mars: comparison with Earth. *Journal of Geophysical Research* **84**, 8183–204.

Breuer, D., H. Zhou, D. A. Yuen, T. Spohn 1996. Phase transitions in the Martian mantle: implications for planet's volcanic history. *Journal of Geophysical Research* **101**, 7531–42.

Bridges, N. T. 1994. Elevation-corrected thermal inertia and derived particle size on Mars and implications for the Tharsis Montes. *Geophysical Research Letters* **21**, 785–8.

Bridges, N. T. and 6 co-authors 1999. Ventifacts at the Pathfinder landing site. *Journal of Geophysical Research* **104**, 8596–616.

Burns, R. G. & D. S. Fisher 1989. Sulfide mineralization related to early crustal evolution of Mars. See Frey (1989: 20–22).

Cabrol, N. A., E. A. Grin, H. E. Newsom, R. Landheim, C. P. McKay 1999. Hydrogeologic evolution of Gale crater and its relevance to the exobiological exploration of Mars. *Icarus* **139**, 235–45.

Calvin, M. H. 1997. Variation of the 3-μm absorption feature on Mars: observations over eastern Valles Marineris by the Mariner 6 infrared spectrometer. *Journal of Geophysical Research* **102**, 9097–9107.

Calvin, W. M. & T. Z. Martin 1994. Spatial variability in the seasonal south polar cap of Mars. *Journal of Geophysical Research* **99**, 21143–52.

Cantor, B. A., M. J. Wolff, P. B. James, E. Higgs 1998. Regression of Martian north polar cap: 1990–1997 Hubble space telescope Observations. *Icarus* **136**, 175–91.

Carr, M. H. 1973, Volcanism on Mars. *Journal of Geophysical Research* **78**, 4049–4062.

— 1974. The role of lava erosion in the formation of lunar rilles and Martian channels. *Icarus* **22**, 1–23.

— 1979. Formation of Martian flood features by release of water from confined aquifers. *Journal of Geophysical Research* **84**, 2995–3007.

— 1981. *The surface of Mars*. New Haven, Connecticut: Yale University Press.

— 1984. Mars. See Carr et al. (1984: 207–263).

— 1987. Water on Mars. *Nature* **326**, 30–35.

— 1995. The Martian drainage system and the origin of networks and fretted channels. *Planetary and Space Science* **44**, 1411–23.

— 1996. *Water on Mars*. New York: Oxford University Press.

Carr, M. H., L. S. Crumpler, J. A. Cutts, R. Greeley, J. E. Guest, H. Masursky 1977. Martian impact craters and emplacement of ejecta by surface flow. *Journal of Geophysical Research* **82**, 4055–65.

Carr, M. H. & J. G. Schaber 1977. Martian permafrost features. *Journal of Geophysical Research* **82**, 4039–54.

Carr, M. H. & G. D. Clow 1981. Martian channels and valleys: their characteristics, distribution and age. *Icarus* **48**, 91–117.

Carr, M. H., R. S. Saunders, R. G. Strom, D. E. Wilhelms 1984. *The geology of the terrestrial planets*. Special Publication 469, NASA, Washington DC.

Cattermole, P. J. 1986. Linear volcanic features at Alba Patera, Mars: probable spatter ridges. *Journal of Geophysical Research* **92**, 159–65.

— 1987. Sequence, rheological properties and effusion rates of volcanic flows at Alba Patera, Mars. *Journal of Geophysical Research* **92**, 553–60.

— 1989a. Volcanic flow development at Alba Patera, Mars. *Icarus* **83**, 453–93.

— 1989b. *Planetary volcanism*. Chichester, England: Ellis Horwood.

— 1996. *Planetary volcanism* (2nd edn). Chichester, England: John Wiley.

Chapman, C. R. & K. L. Jones 1977. Cratering and obliteration history of Mars. *Annual Reviews of Earth and Planetary Science* **5**, 515–40.

Chapman, M. G. 1994. Evidence, age and thickness of frozen palaeolake in Utopia Planitia, Mars. *Icarus* **109**, 393–406.

Chapman, M. G. & J. S. Kargel 1999. Observations at the Mars Pathfinder site: do they provide "unequivocal" evidence of catastrophic flooding? *Journal of Geophysical Research* **104**, 8671–8.

Chicarro, A. F. 1989. Towards a chronology of compressive tectonics on Mars. See Frey (1989: 23–5).

Chicarro, A. F., G. E. N. Scoon, M. Coradini 1989. *Mission to Mars*. Special publication 1117, European Space Agency, Paris.

Christensen, E. H. 1989. Lahars in the Elysium region of Mars. *Geology* **17**, 203–206.

Christensen, E. J. 1975. Martian topography derived from occultation, radar, spectral and optical measurements. *Journal of Geophysical Research* **80**, 2909–913.

Christensen, P. R. 1983. Eolian intracrater deposits on Mars: physical properties and global distribution. *Icarus* **56**, 496–518.

Christensen, P. R. and 11 co-authors 1998. Results from the Mars Global Surveyor thermal emission spectrometer. *Science* **279**, 1686–92.

Clancy, R. T. & H. Nair 1996. Annual (perihelion–aphelion) cycles in the photochemical behaviour of the global Mars atmosphere. *Journal of Geophysical Research* **101**, 12785–9.

Clancy, R. T., M. J. Wolff, P. B. James, E. Smith, Y. N., Billawalla, S. W. Lee 1996. Mars ozone measurements near the 1995 aphelion: HST UV spectroscopy with the faint object spectrograph. *Journal of Geophysical Research* **101**, 12777–84.

Clark, B. C. & A. K. Baird 1979. Volatiles in the Martian regolith. *Geophysical Research Letters* **6**, 811–4.

Clemett, S. J., M. T. Dulay, J. S. Gilette, X. D. F. Chillier, T. B. Mahajan, R. N. Zare 1998. Evidence for the extraterrestrial origin on polycyclic aromatic hydrocarbons in the Martian meteorite ALH84001. *Chemistry and physics of molecules and grains in space*, 417–36. Faraday Discussion 109, Royal Society of Chemistry, London.

Clifford, S. M. 1981. A model for the climatic behaviour of water on Mars [abstract]. Third International Colloquium on Mars. Contribution 441, Lunar and Planetary Institute, Pasadena.

— 1993. A model for the hydrologic and climatic behaviour of water on Mars. *Journal of Geophysical Research* **98**, 10793–11016.

Clifford, S. M., R. Greeley, R. M. Haberle 1988. Mars: evolution of climate and atmosphere. *Eos* **69**, 1596–6.

Clifford, S. M. & J. R. Zimbelman 1988. Softened terrain on Mars: the ground ice interpretation revisited. *Abstracts of the Eighteenth Lunar and Planetary Science Conference*, 199–200. Lunar and Planetary Institute, Houston, Texas.

Clifford, S. M. & T. J. Parker 1999. Hydraulic and thermal arguments regarding the existence and fate of a primordial Martian ocean. *Abstracts of the Thirtieth Lunar and Planetary Science Conference*, 1619–20. Lunar and Planetary Institute, Houston, Texas.

Cole, G. H. A. 1978. *The structure of the planets*. London: Wykeham.

Collins, S. A. 1971. The Mariner 6 and 7 pictures of Mars. Special publication SP-263, NASA, Washington DC.

Comer, R. P., S. C. Solomon, J. W. Head 1985. Mars: thickness of the lithosphere from the tectonic response to volcanic loads. *Reviews of Geophysics and Space Physics* **23**, 61–92.

Connerney, J. E. P. and 9 co-authors 1999. Magnetic lineations in the ancient crust of Mars. *Science* **284**, 794–8.

Craddock, R. A., T. A. Maxwell, A. D. Howard 1997. Crater morphometry and modification in the Sinus Sabaeus & Margaritifer Sinus regions of Mars. *Journal of Geophysical Research* **102**, 13321–40.

Croft, S. K. 1989. Spelunking on Mars: the carbonate-tectonic hypothesis for the origin of Valles Marineris, Mars [abstract]. *MEVTV workshop on tectonic features on Mars*, T. R. Watters & M. P. Golombek (eds), 22–3. Technical Report 89-06, Lunar and Planetary Institute, Houston, Texas.

— 1990. *Geologic map of Hebes Chasma quadrangle, VM500K 00077*. Technical Memorandum 4210 (pp. 539–41), Report of

Planetary Geology and Geophysics Program 1989, NASA, Washington DC.
Crown, D. A. & R. Greeley 1990. Styles of volcanism, tectonic association, and evidence for magma–water interactions in eastern Hellas, Mars. *Lunar and Planetary Science* **XXI**, 250–51.
— 1993. Volcanic geology of Hadriaca Patera and the eastern Hellas region of Mars. *Journal of Geophysical Research* **98**, 3431–51.
Crown, D. A., K. H. Price, R. Greeley 1990. Evolution of the eastern rim of Hellas basin, Mars. *Lunar and Planetary Science* **XXI**, 252–3.
Crumpler, L. S. & J. C. Aubele 1978. Structural evolution of Arsia Mons, Pavonis Mons and Ascaeus Mons, Mars [abstract]. *Lunar and Planetary Science* **21**, 252–3.
Cutts, J. A. 1973. Nature & origin of the layered deposits in the Martian polar regions. *Journal of Geophysical Research* **78**, 4231–49.
Cutts, J. A., K. R. Blasius, W. J. Roberts 1979. Evolution of Martian polar landscape: interplay of long-term variation in perennial ice caps and dust-storm activity. *Journal of Geophysical Research* **84**, 2975–94.

de Hon, R. A. 1988. The Martian sedimentary record. *MEVTV workshop on the nature and composition of surface units on Mars*, J. R. Zimbelman, S. C. Solomon, V. L. Sharpton (eds), 13–15. Technical Report 88-05, Lunar and Planetary Institute, Houston, Texas.
de Hon, R. A. & E. A. Pani 1993. Duration and rates of discharge: Maja Valles, Mars. *Journal of Geophysical Research* **98**, 9129–38.
Dohnanyi, J. S. 1972. Interplanetary objects in review: statistics of their masses and dynamics. *Icarus* **17**, 1–48.
Dollfus, A., S. Ebisawa, D. Crussaire 1996. Hoods, mists and ice caps at the poles of Mars. *Journal of Geophysical Research* **101**, 9207–226.
Dreibus, G. & H. Wänke 1985. Mars: a volatile-rich planet. *Meteoritics* **20**, 367–81.
Dzurisin, D. & K. R. Blasius 1975. Topography of the polar layered deposits of Mars. *Journal of Geophysical Research* **82**, 4225–48.

Edgett, K. S., B. J. Butler, J. R. Zimbelman, V. E. Hamilton 1997. Geologic context of the Mars radar "Stealth" region of southwestern Tharsis. *Journal of Geophysical Research* **102**, 21545–67.
Espenak, F., M. J. Mumma, T. Kstiuk, D. Zipoy 1991. Ground-based infrared measurements of the global distribution of ozone in the atmosphere of Mars. *Icarus* **92**, 252–62.
Evans, J. E. & E. W. Maunder 1903. Experiments as to the actuality of the "canals" of Mars. *Monthly Notes of the Royal Astronomical Society* **63**, 498–507.

Fanale, F. P. 1976. Martian volatiles: their degassing history and geochemical fate. *Icarus* **28**, 179–202.
Fanale, F. P., J. R. Salvail, A. P. Zent, S. E. Postawko 1986. Global distribution and migration of subsurface ice on Mars. *Icarus* **67**, 1–18.
Farquhar, J., M. H. Thiemens, T. Jackson 1998. Atmosphere–surface interactions on Mars; $\Delta^{17}O$ measurements of carbonate from ALH84001. *Science* **280**, 1580.
Finnerty, A. A. & R. J. Phillips 1981. A petrologic model for an isostatically compensated Tharsis region of Mars. Paper presented at the Third Mars Colloquium, Pasadena, 1981.
Finnerty, A. A., R. J. Phillips, W. B. Banerdt 1988. Igneous processes and closed system evolution of the Tharsis region of Mars. *Journal of Geophysical Research* **93**, 10225–35.
Fischer, E. & C. M. Pieters 1993. The continuum slope of Mars: bidirectional reflectance investigations and applications to Olympus Mons. *Icarus* **102**, 185–202.
Folkner, W. M., C. F. Yoder, D. N. Yuan, E. M. Standish, R. A. Preston 1997. Interior structure and seasonal mass redistribution of Mars from radio tracking of Mars Pathfinder. *Science* **278**, 1749–52.
Forget, F., G. B. Hansen, J. B. Pollack 1995. Low brightness temperatures of Martian polar caps: CO_2 clouds or low surface emissivity? *Journal of Geophysical Research* **100**, 21219–35.
Forget, F. & J. B. Pollack 1996. Thermal observations of the condensing Martian polar caps: CO_2 ice temperatures and radiative budget. *Journal of Geophysical Research* **101**, 16865–79.
Forget, F., F. Hourdin, O. Talagrand 1996. Simulation of the Martian atmospheric polar warming with LMD circulation model. *Annals of Geophysics* **14**, C797.
Forget, F. & R. T. Pierrehumbert 1997. Warming early Mars with carbon dioxide clouds that scatter infrared radiation. *Science* **278**, 1273–6.
Forsythe, R. D. & C. R. Blackwelder 1998. Closed drainage basins of the Martian highlands: constraints on the early Martian hydrologic cycle. *Journal of Geophysical Research* **103**, 31421–32.
Frey, H. 1979. Thaumasia: a fossilized early-forming Tharsis uplift. *Journal of Geophysical Research* **84**, 1019–1023.
— (ed.) 1989. *MEVTV workshop on the early tectonic and volcanic evolution of Mars*. Technical Report 89-04, Lunar and Planetary Institute, Houston, Texas.
Frey, H. & R. A. Schultz 1988. Large impact basins and the mega-impact origin for the crustal dichotomy on Mars. *Geophysical Research Letters* **15**, 229–32.
— 1989. Origin of the Martian crustal dichotomy. See Frey (1989: 35–7.

Gault, D. E. & B. S. Baldwin 1970. Impact cratering on Mars – some effects of the atmosphere. *Eos* **51**, 342.
Gault, D. E. & R. Greeley 1978. Exploratory experiments of impact craters formed in viscous-liquid targets: analogs for Martian rampart craters? *Icarus* **34**, 486–95.
Geissler, P. E., R. B. Singer, B. K. Lucchitta 1990. Dark materials in Valles Marineris: indications of the style of volcanism and magmatism on Mars. *Journal of Geophysical Research* **95**, 14399–413.
Golombek, M. P. and 13 co-authors 1997. Overview of the Mars Pathfinder mission and assessment of landing site predictions. *Science* **278**, 1765–8.
Golombek, M. P. and 54 co-authors 1999. Overview of the Mars Pathfinder mission: launch through landing, surface operations, datasets, and science results. *Journal of Geophysical Research* **104**, 8523–55.
Greeley, R., R. Papson, J. Veverka 1978. Crater streaks in the Chryse Planitia region of Mars: early Viking results. *Icarus* **34**, 556–67.
Greeley, R., R. Leach, J. White, J. Iverson, J. Pollack 1980. Threshold wind speeds for sands on Mars: wind tunnel simulations. *Geophysical Research Letters* **7**, 121–4.
Greeley, R. & P. D. Spudis 1981. Volcanism on Mars. *Review of Geophysics and Space Physics* **19**, 13–41.
Greeley, R. & J. E. Guest 1987. *Geologic map of the eastern hemisphere of Mars*. Map I-1802b, USGS, Flagstaff, Arizona.
Greeley, R. & D. A. Crown 1990. Volcanic geology of Tyrrhena Patera, Mars. *Journal of Geophysical Research* **95**, 7133–49.
Greeley, R. & S. W. Williams 1994. Dust deposits on Mars: "parna" analog. *Icarus* **110**, 165–77.
Greeley, R. and 8 co-authors 1999. Aeolian features and processes at the Mars Pathfinder landing site. *Journal of Geophysical Research* **104**, 8573–84.
Grin, E. A. & N. A. Cabrol 1997. Limnologic analysis of Gusev crater palaeolake, Mars. *Icarus* **130**, 461–74.
Grove, T. L. & R. J. Kinzler 1986. Petrogenesis of andesites. *Annual Reviews of Earth and Planetary Science* **14**, 417–54.
Gulick, V. C. & V. R. Baker 1990. Origin and evolution of valleys on Martian volcanoes. *Journal of Geophysical Research* **95**, 14325–44.

Haberle, R. M. 1998. Early Mars climate models. *Journal of Geophysical Research* **103**, 28467–79.
Haberle, R. M. & B. M. Jakosky 1991. Atmospheric effects on the remote determination of thermal inertia on Mars. *Icarus* **90**, 187–204.
Haberle, R. M., H. C. Houben, R. Hertenstein, T. Herdtle 1993. Comparison with Viking lander and entry data. *Journal of Atmospheric Science* **50**, 1544–60.
— 1997. Meteorological predictions for the Mars Pathfinder lander. *Journal of Geophysical Research* **102**, 13301–311.
Harris, S. A. 1977. The aureole of Olympus Mons, Mars. *Journal of Geophysical Research* **83**, 3099–3107.
Hartmann, W. K. 1973a. Martian cratering, 4: Mariner 9 initial

analysis of cratering chronology. *Journal of Geophysical Research* **78**, 4096–4116.

Hartmann, W. K. and 12 co-authors 1981. Chronology of planetary volcanism by comparative studies of planetary cratering. See McGetchin et al. (1981: 1051–76).

Harvey, R. P. & H. Y. McSween 1996. The petrogenesis of the nahklites: evidence for cumulate mineral zoning. *Geochimica et Cosmochimica Acta* **56**, 1655–63.

Hayashi, J. N., B. M. Jakosky, R. M. Haberle 1995. Atmospheric effects on the mapping of Martian thermal inertia and thermally derived albedo. *Journal of Geophysical Research* **100**, 5277–84.

Head, J. W. & L. Wilson 1981. Ascent and eruption of basaltic magma on the Earth and Moon. *Journal of Geophysical Research* **86**, 2971–3001.

Head, J. W., H. Hiesinger, M. A. Ivanov, M. A. Kreslavsky, S. Pratt, B. J. Thomson 1999. Possible ancient oceans on Mars: evidence from Mars orbiter laser altimeter data. *Science* **286**, 2134–40.

Helfenstein, P. & P. J. Mouginis-Mark 1980. Morphology and distribution of fractured terrain on Mars. *Lunar and Planetary Science* **XI**, 429–31.

Hess, S. L., R. M. Henry, C. B. Leovy, J. A. Ryan, J. Tillman 1977. Meteorological results from the surface of Mars: Viking 1 and 2. *Journal of Geophysical Research* **82**, 4559–74.

Hess, S. L., R. M. Henry, J. E. Tillman 1979. The seasonal variation of atmospheric pressure on Mars as affected by the south polar cap. *Journal of Geophysical Research* **84**, 2923–7.

Hess, S. L., J. A. Ryan, J. E. Tillman et al. 1979. The annual cycle of pressure measured on Mars measured by Viking 1 and 2. *Geophysical Research Letters* **7**, 197–200.

Hodges, C. A. & H. J. Moore 1979. The subglacial birth of Olympus Mons and its aureole. *Journal of Geophysical Research* **84**, 8061–8074.

Holloway, J. R. 1990. Martian magmas and mantle source regions: current experimental and petrochemical constraints. Lecture delivered at the Twenty-first Lunar and Planetary Science Conference.

Holloway, J. R. & C. M. Bertka 1989. Chemical and physical properties of primary Martian magmas. See Frey (1989: 43–5).

Houck. J. R. and 7 co-authors 1973. High-altitude infrared spectroscopic evidence for bound water on Mars. *Icarus* **18**, 470–80.

Howard, A. D. 1978. Origin of the stepped topography of the Martian poles. *Icarus* **34**, 581–99.

— 1991. *Role of groundwater in formation of Martian channels.* Technical Memorandum 4300 (pp. 117–9), Reports of Planetary geology and Geophysics Program 1990), NASA, Washington DC.

Hutchins, K. S. & B. M. Jakosky 1996. Evolution of Martian atmospheric argon: implications for source of volatiles. *Journal of Geophysical Research* **101**, 14933–50.

Hviid, S. F. and 11 co-authors 1997. Magnetic properties experiments on the Mars Pathfinder lander: preliminary results. *Science* **278**, 1768–80.

Ivanov, A. B., D. O. Muhleman, A. R. Vasavada 1998. Microwave mapping of the Stealth region of Mars. *Icarus* **133**, 163–73.

Jakosky, B. M. & M. H. Carr 1985. Possible precipitation of ice at low altitudes on Mars during periods of high obliquity. *Nature* **315**, 559–61.

James, P. B., J. L. Hollingsworth, M. J. Wolff, S. W. Lee 1999. North polar dust storms in early spring on Mars. *Icarus* **138**, 64–73.

Jeffreys, Sir H. 1970. *The Earth* (5th edn). London: Cambridge University Press.

Johnston, D. H. & M. N. Toksoz 1977. Internal structure and properties of Mars. *Icarus* **32**, 73–84.

Jones, K. L. 1974. Evidence for an episode of Martian crater obliteration intermediate in Martian history. *Journal of Geophysical Research* **79**, 3917–32.

Kargel, J. S. 1989. Morphologic variations of Martian impact crater ejecta and their dependencies and implications [abstract]. *Abstract and program, Fourth International Conference on Mars*, 132–3, Tucson. Lunar and Planetary Institute, Houston, Texas.

Kargel, J. S. & R. G. Strom 1992. Ancient glaciation on Mars. *Geology* **20**, 3–7.

Kargel, J. S. and 7 co-authors 1995. Evidence of ancient continental glaciation in the Martian northern plains. *Journal of Geophysical Research* **100**, 5351–68.

Kaula, W. M. 1979. The moment of inertia of Mars. *Geophysical Research Letters* **6**, 194–6.

Kaula, W. H., N. H. Sleep, R. J. Phillips 1989. More about the moment of inertia of Mars. *Geophysical Research Letters* **16**, 1333–6.

Keating, G. M. and 27 co-authors 1998. The structure of the upper atmosphere of Mars: *in situ* accelerometer measurements from Mars Global Surveyor. *Science* **279**, 1672–6.

Kerr, R. A. 1998a. Surveyor shows the flat face of Mars. *Science* **279**, 1634.

— 1998b. Signs of plate tectonics on an infant Mars. *Science* **279**, 1605–607.

Kerridge, J. F. & M. S. Matthews 1988. *Meteorites and the early Solar System.* Tucson: University of Arizona Press.

Kiefer, W. S. & B. H. Hager 1989. The role of mantle convection in the origin of the Tharsis and Elysium provinces of Mars. See Frey (1989: 48–50).

Kieffer, H. H., B. M. Jakosky, C. W. Snyder, M. S. Matthews (eds) 1992. *Mars.* Tucson: University of Arizona Press.

Kieffer, H. H., T. Z. Martin, A. R. Peterfreund, B. M. Jakosky 1977. Thermal and albedo mapping of Mars during the primary Viking mission. *Journal of Geophysical Research* **82**, 4249–91.

Kieffer, H. H. & F. D. Palluconi 1979. The climate of the Martian polar caps. Conference Publication 2072 NASA, Washington DC.

King, E. A. 1978. *Geologic map of the Tyrrhena Patera quadrangle of Mars.* Map I-1073, USGS, Flagstaff, Arizona.

King, E. S. & J. R. Riehle 1974. A proposed origin for the origin of the Olympus Mons escarpment. *Icarus* **23**, 300–317.

Kliore, A. J., G. Fjeldbo, B. L. Seidel, I. Rasool 1969. Mariners 6 and 7: occultation measurements of the atmosphere of Mars. *Science* **166**, 1393–7.

Komatsu, G., P. E. Geissler, R. G. Strom, R. B. Singer 1993. Stratigraphy and erosional landforms of layered deposits in Valles Marineris, Mars. *Journal of Geophysical Research* **98**, 11105–121.

Komatsu, G. & V. R. Baker 1997. Palaeohydrology and flood morphology at Ares Vallis. *Journal of Geophysical Research* **102**, 4151–60.

Kovach, R. L. & D. L. Anderson 1965. The interiors of the terrestrial planets. *Journal of Geophysical Research* **70**, 2873–82.

Krasnopolsky, V. A., M. J. Mumma, G. Randall Gladwin 1998. Detection of atomic deuterium in the upper atmosphere of Mars. *Science* **280**, 1576–80.

Kuiper, G. P. 1952. *The atmospheres of the Earth and planets.* Chicago: University of Chicago Press.

Kuzmin, R. 1988. The cryolithosphere of Mars. *Solar System Research* **22**, 195–203.

Le Bas, M. J., R. W. le Maitre, A. Streckheisen, B. Zanettin 1986. A chemical classification of volcanic rocks based on the total alkalies-silica diagram. *Journal of Petrology* **27**, 745–50.

Lee, S. W., P. C. Thomas, J. Veverka 1982. Wind streaks in Tharsis and Elysium: implications for sediment transport by slope winds. *Journal of Geophysical Research* **87**, 10025–10041.

Lee, S. W. & P. C. Thomas 1995. Longitudinal dunes on Mars: relation to current wind regimes. *Journal of Geophysical Research* **100**, 5381–95.

Leighton, R. B. & B. C. Murray 1966. Behaviour of carbon dioxide and other volatiles on Mars. *Science* **153**, 136–41.

Leighton, R. B. and 8 co-authors 1969. Mariner 6 and 7 television pictures: preliminary analysis. *Science* **166**, 49–67.

Leovy, C. B. & Y. Mintz 1969. Numerical simulation of the weather and climate of Mars. *Journal of Atmospheric Science* **26**, 1167–90.

Longhi, J. & V. Pan 1989. The parent magma of the SNC meteorites. *Proceedings of the Nineteenth Lunar and Planetary Science Conference*, G. Ryder (ed.), 451–64. Cambridge: Cambridge University Press (on behalf of the Lunar and Planetary Institute, Houston, Texas).

Lopes, R. M. C., J. E. Guest, C. J. Wilson 1980. Origin of the Olympus Mons aureole and perimeter scarp. *Moon and Planets* **22**, 221–34.
Lopes, R., J. Guest, K. Hiller, G. Neukeum 1982. Further evidence for a mass movement origin of the Olympus Mons aureole. *Journal of Geophysical Research* **87**, 9917–28.
Lorrel, J. and 9 co-authors 1972. Mariner 9 celestial mechanics experiment: gravity field and pole directions of Mars. *Science* **175**, 317–20.
Lowell, P. 1906. *Mars and its canals*. New York: Macmillan.
— 1909. *Mars as an abode for life*. New York: Macmillan.
— 1910. Schiaparelli. *Popular Astronomy* **18**, 466–98.
Lucchitta, B. K. 1978. A large landslide on Mars. *Bulletin of the Geological Society of America* **89**, 1601–609.
— 1979. Landslides in Valles Marineris, Mars. *Journal of Geophysical Research* **84**, 8097–8113.
— 1981. Mars and Earth: comparison of cold-climate features. *Icarus* **45**, 264–303.
— 1982. Ice sculpture in the Martian outflow channels. *Journal of Geophysical Research* **87**, 9951–73.
— 1987. Recent mafic volcanism on Mars. *Science* **235**, 565–7.
— 1993. Ice in the northern plain: relic of a frozen ocean? Pp. 72–4 in *Workshop on the Martian northern plains: sedimentological, periglacial and palaeoclimatic evolution*. Technical Report 93-04, Lunar and Planetary Institute, Houston, Texas.
— 1998. Pathfinder landing site: alternatives to catastrophic flooding and an Antarctic ice-flow analog for outflow channels on Mars [abstract]. *Twenty-ninth Lunar and Planetary Science Conference*, 1287–7. Lunar and Planetary Institute, Houston, Texas.
Lucchitta, B. K. & J. L. Klockenbrink 1981. Ridges and scarps in the equatorial belt of Mars. *Moon and Planets* **24**, 415–29.
Lucchitta, B. K., H. M. Ferguson, C. Summers 1986. Sedimentary deposits in the northern lowland plains. *Journal of Geophysical Research* **91**, 166-174.
Lucchitta, B. K. 7 co-authors 1989. Canyon systems on Mars [abstract]. Abstracts and program, Fourth International Conference on Mars, Tucson, 26–7.
Lucchitta, B. K. and 7 co-authors 1992. The canyon system on Mars. See Kieffer et al. (1992: 453–92).
Lucchitta, B. K., N. K. Isbell, A. Howington-Kraus 1994. Topography of Valles Marineris: implications for erosional and structural history. *Journal of Geophysical Research* **99**, 3783–98.

Malin, M. C. 1997. Comparison of volcanic features of Elysium (Mars) and Tibesti (Earth). *Bulletin of Geological Society of America* **84**, 908–919.
Malin, M. C. and 15 co-authors 1998. Early views of the Martian surface from the Mars orbiter camera of Mars Global Surveyor. *Science* **279**, 1681–5.
Malin, M. C. & K. S. Edgett 1999. Oceans or seas in the Martian northern lowlands: high-resolution imaging tests of proposed coastlines. *Geophysical Research Letters* **26**(19), 3049.
— 2000. Evidence for groundwater seepage and surface runoff on Mars. *Science* **288**, 2330–35.
Martin, T. 1995. Mass of dust in the Martian atmosphere. *Journal of Geophysical Research* **100**, 7509–513.
Masursky, H., J. M. Boyce, A. L. Dial, G. C. Schaber, M. E. Strobell 1977. Classification and time of formation of Martian channels based on Viking data. *Journal of Geophysical Research* **82**, 4016–4038.
Matyska, C., D. A. Yuen, D. Breuer, T. Spohn 1998. Symmetries of volcanic distribution on Mars and Earth, and their mantle plume dynamics. *Journal of Geophysical Research* **103**, 28587–98.
Maxwell, T. E. 1989. Structural mapping along the cratered terrain boundary, eastern hemisphere of Mars. See Frey (1989: 54–5).
McCauley, J. F. 1978. *Geological map of the Coprates Quadrangle of Mars*. Map I-897, USGS, Flagstaff, Arizona.
McCauley, J. F. and 7 co-authors 1972. Preliminary Mariner 9 report on the geology of Mars. *Icarus* **17**, 289–327.
McElroy, M. B., T. Y. Kong, Y. L. Yung 1977. Photochemistry and evolution of Mars' atmosphere: a Viking perspective. *Journal of Geophysical Research* **82**, 4379–88.
McGetchin, T. R., R. O. Pepin, R. J. Phillips (eds) 1981. *Basaltic volcanism on the terrestrial planets*. New York: Pergamon.
McGill, G. E. 1985a. Age and origin of large Martian polygons [abstract]. *Lunar and Planetary Science* **XVI**, 535–6.
— 1985b. Age of deposition and fracturing, Elysium/Utopia Region, northern Martian plains. *Geological Society of America Abstracts Programs* **17**, 659.
— 1989. The Martian crustal dichotomy. See Frey (1989: 59–61).
McGill, G. E. & L. Scott-Hills 1992. Origin of giant Martian polygons. *Journal of Geophysical Research* **97**, 2643–7.
McGovern, P. J. & S. C. Solomon 1993. State of stress, faulting and eruption characteristics of large volcanoes on Mars. *Journal of Geophysical Research* **98**, 23553–79.
McKay, D. S. and 8 co-authors 1996. Search for past life on Mars: possible relic biogenic activity in Martian meteorite ALH84001. *Science* **273**, 924–9.
McSween, H. Y. 1994. What we have learned about Mars from SNC meteorites. *Meteoritics* **29**, 757–9.
McSween, H. Y. & E. Jarosewich 1983. Petrogenesis of the Elephant Moraine A79001 meteorite; multiple magma pulses on the shergottite parent body. *Geochimica et Cosmochimica Acta* **47**, 1501–513.
McSween, P. J. and 19 co-authors 1999. Chemical, multispectral and textural constraints on the composition and origin of rocks at the Mars Pathfinder landing site. *Journal of Geophysical Research* **104**, 8679–715.
Mellon, M. T. & B. M. Jakosky 1993. The geographic variations in the thermal and diffusive stability of ground ice on Mars. *Journal of Geophysical Research* **98**, 3345–64.
Meyer, J. D. & M. J. Grollier 1977. *Geologic map of the Syrtis Major quadrangle of Mars*. Map I-995, USGS, Flagstaff, Arizona.
Michaux, C. M. & R. L. Newburn 1972. *Mars scientific model*. Publication 606-1, Jet Propulsion Laboratory, Pasadena, California.
Moore, J. M. and 7 co-authors 1995. The circum-Chryse region as a possible example of a hydrologic cycle on Mars: geologic observations and theoretical evaluation. *Journal of Geophysical Research* **100**, 5433–47.
Moroz, V. I. 1964. The infrared spectrum of Mars (1.1–4.1 | μm). *Astronomicheskii Zhurnal* **41**, 350–61.
Morris, E. C. 1979. A pyroclastic origin for the aureole deposits of Olympus Mons. *Reports of Planetary Geology Program 1981*, H. E. Holt & E. C. Koster (eds), 252–4. Technical Memorandum 82385, NASA, Washington DC.
— 1981. Structure of Olympus Mons and its basal scarp [abstract]. In *The Third International Colloquium on Mars – programme and abstracts*, 161–2. Lunar and Planetary Institute, Pasadena, California.
Mouginis-Mark, P. H. 1981a. Ejecta emplacement and modes of formation of Martian fluidized ejecta craters. *Icarus* **45**, 60–76.
— 1981b. Late-stage summit activity of Martian shield volcanoes. In *Proceedings of the Twelfth Lunar and Planetary Science Conference*, 1431–47. New York: Pergamon.
Mouginis-Mark, P. J., L. Wilson, J. W. Head 1982. Explosive volcanism on Hecates Tholus, Mars: investigation of eruption conditions. *Journal of Geophysical Research* **87**, 9890–904.
Mouginis-Mark, P. J., L. Wilson, J. W. Head, S. H. Brown, J. L. Hall, K. D. Sullivan 1984. Elysium Planitia Mars: regional geology, volcanology, and evidence for volcano–ground-ice interactions. *Earth, Moon and Planets* **30**, 149–73.
Mouginis-Mark, P. J., L. Wilson, J. R. Zimbelman 1988. Polygenetic eruptions on Alba Patera, Mars. *Bulletin Volcanologique* **50**, 361–79.
Mouginis-Mark, P. J. & M. S. Robinson 1992. Evolution of Olympus Mons caldera, Mars. *Bulletin Volcanologique* **54**, 347–60.
Mouginis-Mark, P. J., T. J. McCoy, C. J. Taylor, K. Keil 1997. Martian parent craters for SNC meteorites. *Journal of Geophysical Research* **97**, 10213–26.
Murchie, S. J., J. Mustard, J. Bishop, J. Head, C. Pieters, S. Erard 1993. Spatial variations in the spectral properties of bright regions of Mars. *Icarus* **105**, 454–68.
Murphy, J. R., R. M. Haberle, O. B. Toon, J. B. Pollack 1993. Martian global dust storms zonally symmetric numerical simulations including size dependent particle transport. *Journal of Geophysical*

Research **98**, 3197–220.
Mustard, J. F., S. S. Murchie, S. Erard, J. Sunshine 1997. In situ compositions of Martian volcanics: implications for the mantle. *Journal of Geophysical Research* **102**, 25605–616.
Mutch, P. & A. Woronow 1980. Martian rampart and pedestal crater's ejecta emplacement: Coprates quadrangle. *Icarus* **41**, 259–68.

Nedell, S. S., S. W. Squyres, D. W. Anderson 1987. Origin and evolution of the layered deposits in Valles Marineris. *Icarus* **70**, 409–441.
Nelson, D. M. & R. Greeley 1999. Geology of Xanthe Terra outflow channels and the Mars Pathfinder landing site. *Journal of Geophysical Research* **104**, 8653–70.
Neukum, D. & D. U. Wise 1976. A standard crater curve and possible new timescale. *Science* **194**, 1381–7.
Nier, A. O., M. B. McElroy, Y. L. Yung 1976. Isotopic composition of the Martian atmosphere. *Science* **194**, 68–70.
Nyquist, L. E. 1983. Do oblique impacts produce Martian meteorites? *Journal of Geophysical Research* **88**, 785–98.

Opik, E. J. 1965. Mariner IV and craters on Mars. *Irish Astronomical Journal* **7**, 92–104.
— 1966. The Martian surface. *Science* **153**, 255–65.
Owen, T. 1966. The composition and surface pressure of the Martian atmosphere: results from the 1965 opposition. *Astrophysical Journal* **146**, 257–70.
Owen, T., K. Biemann, D. R. Rushneck, J. E. Biller, D. W. Howarth, A. L. Lafleur 1977. The composition of the atmosphere at the surface of Mars. *Journal of Geophysical Research* **82**, 4635–9.
Owen, T., J. P. Maillard, C. de Bergh, B. L. Lutz 1988. Deuterium on Mars: the abundance of HDO and the value of D/H. *Science* **240**, 1767–70.

Palluconi, F. D. & H. H. Kieffer 1981. Thermal inertia mapping of Mars from 60°S to 60°N. *Icarus* **45**, 415–26.
Parker, T. J., R. S. Saunders, D. M. Schneeberger 1989. Transitional morphology in west Deuteronolus Mensae, Mars: implications for modification of the upland/lowland boundary. *Icarus* **82**, 111–45.
Parker, T. J., D. M. Schneeberger, D. C. Pieri, R. S. Saunders 1986. Geomorphic evidence for ancient seas in west Deuteronilus Mensae, Mars, 1: regional geomorphology. *Reports of Planetary Geology Program 1985*, 96–8. Technical Report 87-01, Lunary and Planetary Institute, Houston, Texas.
Parker, T. J., D. S. Gorsline, R. S. Saunders, D. C. Pieri, D. M. Schneeberger 1993. Coastal geomorphology of the Martian northern plains. *Journal of Geophysical Research* **98**, 11061–1078.
Pechmann, J. C. 1980. The origin of polygonal troughs on the northern plains of Mars. *Icarus* **42**, 185–210.
Peterson, J. E. 1977. *Geologic map of the Noachis quadrangle of Mars.* Map I-910, USGS, Flagstaff, Arizona.
Phillips, R. J. 1990. Geophysics at Mars: issues and answers. Lecture delivered at the Twenty-first Lunar and Planetary Science Conference, Houston, Texas.
Phillips, R. J., R. S. Saunders, J. E. Conel 1973. Mars: crustal structure as inferred from Bougeur gravity anomalies. *Journal of Geophysical Research* **78**, 4815–20.
Phillips, R. J. & R. S. Saunders 1975. The isostatic state of Martian topography. *Journal of Geophysical Research* **80**, 2893–8.
Phillips, R. J. & E. R. Ivins 1979. Geophysical observations pertaining to solid state convection in the terrestrial planets. *Physics of the Earth and Planetary Interiors* **19**, 107–148.
Phillips, R. J., N. H. Sleep, W. B. Banerdt 1990. Permanent uplift in magmatic systems with application to the Tharsis region, Mars. *Journal of Geophysical Research* **95**, 5089–5100.
Pieri, D. 1980. Geomorphology of Martian valleys. In *Advances in planetary geology*, A. Woronow (ed.), 353–6. Washington DC: NASA.
Pike, R. J. 1978. Volcanoes on the inner planets: some preliminary comparisons of gross topography. *Proceedings of the Ninth Lunar and Planetary Science Conference*, 3239–73. New York: Pergamon.
— 1979. Simple to complex craters: the transition on Mars. See Boyce & Collins (1979: 132–4).
— 1980a. Formation of complex impact craters: evidence from Mars and other planets. *Icarus* **43**, 1–19.
— 1980b. Terrain dependence of crater morphology on Mars: yes and no. *Lunar and Planetary Science* **XI**, 885–7.
Plescia, J. B. 1979. *Tectonism of the Tharsis region.* Technical Memorandum 80339 (pp. 47–9), NASA, Washington DC.
— 1980. *Cinder cones of Isidis and Elysium.* Technical Memorandum 82385 (pp. 263–5), NASA, Washington DC.
— 1981. The Tempe volcanic province of Mars and comparisons with the Snake River Plains of Idaho. *Icarus* **45**, 586–601.
— 1993. Wrinkle ridges of Arcadia Planitia, Mars. *Journal of Geophysical Research* **98**, 15049–60.
Plescia, J. B. & R. S. Saunders 1979. The chronology of Martian volcanoes. *Proceedings of the Tenth Lunar and Planetary Science Conference*, 2841–59. New York: Pergamon.
— 1982. Tectonic history of the Tharsis region, Mars. *Journal of Geophysical Research* **87**, 9775–91.
Plescia, J. B. & J. Crisp 1991. Recent Elysium volcanism: effects on the Martian atmosphere. In *Workshop of the Martian surface and atmosphere through time* [abstracts volume], pp. 102–3. Lunar and Planetary Institute, Houston, Texas.
Pollack, J. B., C. B. Leovy, Y. H. Mintz, W. Van Camp 1976. Winds on Mars during the Viking season: predictions based on a general circulation model with topography. *Geophysical Research Letters* **3**, 479–82.
Pollack, J. B. & D. C. Black 1979. Implications of the gas compositional measurements of Pioneer Venus for the origin of planetary atmospheres. *Science* **207**, 56–9.
Pollack, J. B, J. S. Colburn, F. M. Flasar, R. Kahn, C. E. Carlston, D. Pidek 1979. Properties and effects of dust particles suspended in the Martian atmosphere. *Journal of Geophysical Research* **84**, 2929–45.
Pollack, J. B., C. B. Leovy, P. W. Grieman, Y. Mintz 1981. A Martian general circulation experiment with large topography. *Journal of Atmospheric Science* **38**, 3–29.
Pollack, J. B., J. F. Kasting, S. M. Richardson, K. Poliakov 1987. The case for a wet, warm climate on early Mars. *Icarus* **71**, 203–224.
Pollack, J. B., R. M. Haberle, J. Scaeffer, H. Lee 1990. Simulations of the general circulation of the Martian atmosphere, I: polar processes. *Journal of Geophysical Research* **95**, 1447–73.
Postawko, S. E. & F. P. Fanale 1993. Changes in erosional style on early Mars: external versus internal influences. *Journal of Geophysical Research* **98**, 11017–24.
Potter, D. B. 1976. *Geologic map of the Hellas Quadrangle of Mars.* Map I-941, USGS, Flagstaff, Arizona.

Raitala, J. & K. Kauhanen 1989. Magma-chamber-related development of Alba Patera on Mars. *Earth, Moon and Planets* **45**, 187–204.
Raitala, J. 1990. Martian Tharsis bulge: insight into mantle-related processes. *Tectonophysics* **174**, 175–81.
Reasenberg, R. 1977. The moment of inertia and isostasy of Mars. *Journal of Geophysical Research* **82**, 369–75.
Reimers, P. E. & P. D. Komar 1979. Evidence for explosive volcanic density currents on certain Martian volcanoes. *Icarus* **39**, 88–110.
Rice, J. W. & K. S. Edgett 1997. Catastrophic flood sediments in Chryse Basin, Mars, and Quincy Basin, Washington: application of sandar facies models. *Journal of Geophysical Research* **102**, 4185–200.
Rieder, R. and 7 co-authors 1997a. The chemical composition of Martian soil and rocks returned by the mobile alpha-proton X-ray spectrometer: preliminary results from the X-ray mode. *Science* **278**, 1771–4.
Rieder, R., H. Wänke, T. Economou, A. Turkevich 1997b. Determination of the chemistry of Martian soil and rocks: the alpha-proton X-ray spectrometer. *Journal of Geophysical Research* **102**, 4027–44.
Riehle, J. R. 1973. Calculated compaction profiles of rhyolitic ash-flow tuffs. *Geological Society of America, Bulletin* **84**, 2193–216.
Ringwood, A. E. 1966. Chemical evolution of the terrestrial planets. *Geochimica et Cosmochimica Acta* **30**, 41–104.

Ringwood, A. E. & S. P. Clark 1971. Internal constitution of Mars. *Nature* **234**, 89–92.
Romanek, C. S., M. M. Grady, I. P. Wright, D. W. Mittlefehldt, R. A. Socki. C. T. Pillinger, E. K. Gibson 1994. Record of fluid–rock inclusions–interactions on Mars from the meteorite ALH84001. *Nature* **372**, 655–7.
Rossbacher, L. A. & S. Judson 1981. Ground ice on Mars: inventory, distribution, and resulting landforms. *Icarus* **45**, 39–59.
Roth, L. E., G. S. Downs, R. S. Saunders 1980. Radar altimetry of south Tharsis, Mars. *Icarus* **42**, 287–316.
Rotto, S. & K. L. Tanaka 1995. *Geologic/geomorphic map of the Chryse Planitia region of Mars.* Map I-2441, USGS, Flagstaff, Arizona.
Rover Team 1997. Characterization of the Martian surface deposits by the Mars Pathfinder rover, Sojourner. *Science* **278**, 1765–8.
Ryan, J. A., R. M. Henry, S. L. Hess, C. B. Leovy, J. E. Tillman, C. Walcek 1978. Mars meteorology: three seasons at the surface. *Geophysical Research Letters* **5**, 715–18.
Ryan, J. A. & R. D. Lucich 1983. Possible dust devils, vortices on Mars. *Journal of Geophysical Research* **88**, 11005–11011.

Sagan, C. and 12 co-authors 1973. Variable features on Mars, 2: Mariner 9 global results. *Journal of Geophysical Research* **78**, 4163–96.
Sagan, C. & C. Chyba 1983. The early faint Sun paradox: organic shielding of ultraviolet-labile greenhouse gases. *Science* **276**, 1217–21.
Sagdeef, R. Z. & A. V. Zakharov 1989. Brief history of the Phobos mission. *Nature* **342**, 581–5.
Sakimoto, S., J. Crisp, S. M. Baloga 1997. Eruption constraints on tube-fed planetary lava flows. *Journal of Geophysical Research* **102**, 6597–614.
Santee, M. L. & D. Crisp 1993. Thermal structure and dust loading of the Martian atmosphere during late southern summer: Mariner 9 revisited. *Journal of Geophysical Research* **98**, 3261–81.
— 1995. Diagnostic calculations of the circulation of the Martian atmosphere. *Journal of Geophysical Research* **100**, 5465–84.
Schaber, G. C., K. C. Horstman, A. L. Dial 1978. Lava flow materials in the Tharsis region of Mars. In *Proceedings of the Ninth Lunar and Planetary Science Conference*, 3433–58. New York: Pergamon.
Schofield, J. T. and 8 co-authors 1997. The Mars Pathfinder atmospheric structure investigation/meteorology (ASI/MET) experiment. *Science* **278**, 1752–8.
Schonfeld, E. 1979. Origin of Valles Marineris. In *Proceedings of the Tenth Lunar and Planetary Science Conference*, 3031–3038. New York: Pergamon.
Schopf, J. W. 1999. *Evolution! Facts and fallacies.* New York: Academic Press.
Schultz, P. H. 1977. Lunar and Martian floor-fractured craters. In *Basaltic Volcanism 2nd Inter-team meeting*, 53–5. Lunar and Planetary Institute, Houston, Texas.
— 1992. Atmospheric effects on ejecta emplacement. *Journal of Geophysical Research* **97**, 11623–62.
Schultz, P. H., R. A. Schultz, J. Rogers 1982. The structure and evolution of ancient impact basins on Mars. *Journal of Geophysical Research* **87**, 9803–820.
Schultz, P. H. & A. B. Lutz 1988. Polar wandering on Mars. *Icarus* **73**, 91.
Schultz, R. A. & H. V. Frey 1990. A new survey of large multi-ring impact basins on Mars. *Journal of Geophysical Research* **95**, 14175–89.
Schultz, R. A. & K. L. Tanaka 1994. Lithospheric-scale buckling and thrust structures on Mars: the Coprates Rise and South Tharsis ridge belt. *Journal of Geophysical Research* **99**, 8371–86.
Scott, D. H. 1969. *The geology of the southern Pancake Range and lunar crater volcanic field, Nye County, Nevada.* PhD thesis, University of California, Los Angeles.
— 1982. Volcanoes and volcanic provinces: western hemisphere of Mars. *Journal of Geophysical Research* **87**, 9839–51.
Scott, D. H. & M. H. Carr 1978. *Geologic map of Mars.* Map I-1083, USGS, Flagstaff, Arizona.
Scott, D. H. & K. L. Tanaka 1980. Mars Tharsis region: volcanotectonic events in the stratigraphic record. *Proceedings of the Eleventh Lunar and Planetary Science Conference*, 2403–421. New York: Pergamon.
— 1982. Ignimbrites of western Amazonis Planitia, Mars. *Journal of Geophysical Research* **87**, 1179–90.
— 1986. *Geologic map of the western hemisphere of Mars.* Map I-1802a, USGS, Flagstaff, Arizona.
Scott, D. H. & J. M. Dohm 1990. Chronology and global distribution of fault and ridge systems on Mars. *Proceedings of the Twentieth Lunar and Planetary Science Conference*, G. Ryder (ed.), 503–13. Cambridge: Cambridge University Press (on behalf of the Lunar and Planetary Institute, Houston, Texas).
Scott, D. H. & J. R. Underwood 1991. Mottled terrain: a continuing Martian enigma. *Proceedings of the Twenty-first Lunar and Planetary Science Conference*, G. Ryder & V. L. Sharpton (eds), 627–34. Cambridge: Cambridge University Press (on behalf of the Lunar and Planetary Institute, Houston, Texas).
Scott, D. H., J. D. Dohm, J. W. Rice 1995. *Map of Mars showing channels and possible palaeolake basins.* Map I-2461, USGS, Flagstaff, Arizona.
Seiff, A. and 10 co-authors 1997. The atmospheric structure and meteorology instrument on the Mars Pathfinder lander. *Journal of Geophysical Research* **102**, 4045–4056.
Sharp, R. P. 1973a. Mars: troughed terrain. *Journal of Geophysical Research* **78**, 4063–4072.
— 1973b. Mars: south polar pits and etched terrain. *Journal of Geophysical Research* **78**, 4222–30.
Sharp, R. P. & M. C. Malin 1975. Channels on Mars. *Bulletin of the Geological Society of America* **86**, 593–609.
Sheehan, W. 1998. *Planets and perception.* Tucson: University of Arizona Press.
Shelfer, T. D. & R. V. Morris 1998. Effect of a ferric weathering rind on the and Mossbauer spectra of a basaltic rock. In *Abstracts of Twenty-ninth Lunar and Planetary Science Conference*, 1327. Lunar and Planetary Institute, Houston, Texas.
Shreve, R. L. 1966. Sherman landslide, Alaska. *Science* **154**, 1639–43.
Singer, R. B., T. B. McCord, R. N. Clark 1979. Mars surface composition from reflectance spectra: a summary. *Journal of Geophysical Research* **84**, 8415–26.
Sinton, W. M. 1967. On the composition of Martian surface materials. *Icarus* **6**, 222–8.
Sjogren, W. L. 1979. Mars gravity: high-resolution results from Viking Orbiter 2. *Science* **203**, 1006–1010.
Sjogren, W. L., L. Wong, W. Downs 1975. Mars gravity field based on a short arc technique. *Journal of Geophysical Research* **80**, 2899–908.
Sleep, N. H. 1994. Martian plate tectonics. *Journal of Geophysical Research* **99**, 5639–56.
Sleep, N. H. & R. J. Phillips 1979. An isostatic model for the Tharsis province. *Geophysical Research Letters* **6**, 803–806.
— 1985. Gravity and lithospheric stress on the terrestrial planets with reference to the Tharsis region of Mars. *Journal of Geophysical Research* **90**, 4469–89.
Smith, D. E. and 6 co-authors 1993. An improved gravity model for Mars: Goddard Mars model 1. *Journal of Geophysical Research* **98**, 20871–90.
Smith, D. E. & M. T. Zuber 1996. The shape of Mars: the topographic signature of the hemispheric dichotomy. *Science* **271**, 184.
Smith, D. E. and 18 co-authors 1999a. The global topography of Mars and implications for surface evolution. *Science* **284**, 1495–503.
Smith, D. E., W. L. Sjogren, G. L. Tyler, G. Balmino, F. G. Lemoine, A. S. Konopliv 1999b. The gravity field of Mars: results from Mars Global Surveyor. *Science* **286**, 94–7.
Smith, P. H. and 28 co-authors 1997. Results from the Mars Pathfinder camera. *Science* **278**, 1758–65.
— and 12 co-authors 1998. Topography of the northern hemisphere of Mars from the Mars laser altimeter. *Science* **279**, 1686–92.
Soderblom, L. A. 1977. Historical variations in the density and distribution of impacting debris in the inner solar system: evi-

dence from planetary imaging. In *Impact and explosion cratering*, D. J. Roddy, R. D. Pepin, R. B. Merrill (eds), 240–41. New York: Pergamon.
Soderblom, L. A., C. D. Condit, R. A. West, B. M. Herman, T. J. Kriedler 1974. Martian planet-wide crater distributions: implications for geologic history and surface processes. *Icarus* **22**, 239–63.
Solomon, S. C. 1979. Formation, history and energetics of cores in terrestrial planets. *Physics of the Earth and Planetary Interiors* **19**, 168–82.
Solomon, S. C. & J. W. Head 1982. Evolution of the Tharsis province of Mars: the importance of heterogeneous lithosphere thickness and volcanic construction. *Journal of Geophysical Research* **87**, 9755–74.
Sparks, R. S. J. & L. Wilson 1976. A model for the formation of ignimbrite by gravitational column collapse. *Journal of the Geological Society of London* **132**, 441–51.
Sparks, R. S. J., L. Wilson, G. Hulme 1978. Theoretical modelling of the generation, movement and emplacement of pyroclastic flows by column collapse. *Journal of Geophysical Research* **83**, 1727–39.
Spudis P. D. & R. Greeley 1978. Volcanism in the cratered uplands of Mars. *Eos* **58**, 1182.
Squyres, S. W. 1978. Martian fretted terrain: flow of erosional debris. *Icarus* **34**, 600–613.
Squyres, S. W. & M. H. Carr 1986. Geomorphic evidence for the distribution of ground ice on Mars. *Science* **231**, 249–52.
Squyres, S. W., D. E. Wilhelms, A. C. Moosman 1987. Large-scale volcano–ground-ice interaction on Mars. *Icarus* **70**, 385–408.
Squyres, S. W. and 7 co-authors 1992. The channels of Mars. See Kieffer et al. (1992: 523–54).
Stolper E., H. Y. McSween, J. E. Hays 1979. A petrologic model for the relationships among achondrite meteorites. *Geochimica et Cosmochimica Acta* **43**, 589–602.

Takahashi, E. & C. M. Scarfe 1985. Melting of peridotite to 14 g and the genesis of komatiite. *Nature* **315**, 566–8.
Tanaka, K. L. 1986 The stratigraphy of Mars. *Journal of Geophysical Research* **91**, 139–58.
—— 1988. Chaotic materials and debris flows in the Simud–Tiu Valles outflow system of Mars. *Proceedings of the Nineteenth Lunar and Planetary Science Conference*, G. Ryder & V. L. Sharpton (eds), 1175–6. Cambridge: Cambridge University Press (on behalf of the Lunar and Planetary Institute, Houston, Texas).
—— 1997. Sedimentary history and mass flow structures of Chryse and Acidalia Planitia, Mars. *Journal of Geophysical Research* **102**, 4131–50.
—— 1999. Debris-flow origin for the Simud–Tiu deposit on Mars. *Journal of Geophysical Research* **104**, 8637–52.
—— 2000. Fountains of youth. *Science* **288**, 2325–6.
Tanaka, K. L. & D. H. Scott 1987. *Geologic maps of the polar regions of Mars.* Map I-1802c, USGS, Flagstaff, Arizona.
Tanaka, K. L. & M. P. Golombek 1989. Martian tension fractures and the formation of grabens and collapse features at Valles Marineris. *Proceedings of the Nineteenth Lunar and Planetary Science Conference*, G. Ryder & V. L. Sharpton (eds), 383–96. Cambridge: Cambridge University Press (on behalf of the Lunar and Planetary Institute, Houston, Texas).
Tanaka, K. L. & M. G. Chapman 1990. The relation of catastrophic flooding of Mangala Valles, Mars, to faulting of Mareotis Fossae and Tharsis volcanism. *Journal of Geophysical Research* **95**, 14315–23.
Tanaka, K. L. & G. J. Leonard 1995. Geology and landscape evolution of the Hellas region of Mars. *Journal of Geophysical Research* **100**, 5407–432.
Tanaka, K. L., J. M. Dohm, J. H. Lias, T. M. Hare 1998. Erosional valleys in the Thaumasia region of Mars: hydrothermal and seismic origins. *Journal of Geophysical Research* **103**, 31407–420.
Theilig, E. & R. Greeley 1979. Plains and channels in the Lunae Planum–Chryse Planitia region of Mars. *Journal of Geophysical Research* **84**, 7994–8010.
Thomas, P. 1982. Present wind activity on Mars: relation to large latitudinally zoned sediment deposits. *Journal of Geophysical Research* **87**, 9999–10008.
Thomas, P. & J. Veverka 1979. Seasonal and secular variations of wind streaks on Mars: an analysis of Mariner 9 and Viking data. *Journal of Geophysical Research* **84**, 8131–46.
Thomas, P. and 11 co-authors 1992. See Kieffer et al. (1992: 767–95).
Thomas, P. & P. J. Gierasch 1995. Polar margin dunes and winds on Mars. *Journal of Geophysical Research* **100**, 5397–406.
Thornhill, G. D., D. A. Rothery, J. B. Murray, A. C. Cook, T. Day, J. P. Muller, J. C. Iliffe 1993. Topography of Appolinaris Patera and Ma'adim Valles: automated extraction of digital elevation models. *Journal of Geophysical Research* **98**, 23581–8.
Toon, O. B., J. B. Pollack, W. Ward, J. A. Burns, K. Bilski 1980. The astronomical theory of climate change on Mars. *Icarus* **44**, 552–607.
Toulmin, P., A. K. Baird, B. C. Clark 1977. Geochemical and mineralogical interpretations of the Viking chemical results. *Journal of Geophysical Research* **82**, 4625–34.
Treiman A. H. 1995. A petrographic history of Martian meteorite ALH84001: two shocks and an ancient age. *Meteoritics* **30**, 294–302.
Tsoar, H., R. Greeley, A. R. Peterfreund 1979. Mars: the north polar sand sea and related wind patterns. *Journal of Geophysical Research* **84**, 8167–80.
Turtle, E. P. & H. J. Melosh 1997. Stress and flexural modelling of the Martian lithospheric response to Alba Patera. *Icarus* **126**, 197–211.

Underwood, J. R. & N. J. Trask 1978. *Geologic map of the Mare Acidalium region of Mars.* Map I-1048, USGS, Flagstaff, Arizona.
Urey, H. 1952. *The planets.* New Haven, Connecticut: Yale University Press.

Van Bemmelen, R. W. & M. G. Rutten 1955. *Table mountains of northern Iceland.* Leiden: E. J. Brill.
Veverka, J., P. Geirasch, P. Thomas 1981. Wind streaks on Mars: meteorological control of occurrences and mode of formation. *Icarus* **45**, 154–66.
Vickery, A. M. & J. H. Melosh 1987. The large crater origin of SNC meteorites. *Science* **237**, 738–43.
Viking Science Team 1977. Scientific results of the Viking Project. *Journal of Geophysical Research* **82**, 3959–4667.

Ward, A. W. 1974. Climatic variations on Mars, 1: astronomical theory of insolation. *Journal of Geophysical Research* **84**, 7934–9.
—— 1979. Yardangs on Mars: evidence of recent wind erosion. *Journal of Geophysical Research* **84**, 8147–66.
Ward, A. W. and 8 co-authors 1999. General geology and geomorphology of the Mars Pathfinder landing site. *Journal of Geophysical Research* **104**, 8555–71.
Watters, T. E. 1993. Compressional tectonism on Mars. *Journal of Geophysical Research* **98**, 17049–61.
Watters, T. R. & M. S. Robinson 1999. Lobate scarps and the Martian crustal dichotomy. *Journal of Geophysical Research* **104**, 25629–40.
Webster, P. J. 1977. The low-altitude circulation of Mars. *Icarus* **30**, 626–49.
Wentworth, S. J. & J. I. Gooding 1995. Carbonates in the Mars meteorite ALH84001: water-borne but not like SNCs [abstract]. *Proceedings of the Twenty-sixth Lunar and Planetary Science Conference* [abstracts]. Lunar and Planetary Institute, Houston, Texas.
Wilhelms, D. E. & S. W. Squyres 1984. The Martian hemispheric dichotomy may be due to a giant impact. *Nature* **309**, 138–40.
Wilhelms, D. E. & R. B. Baldwin 1989. The origin of igneous sills in shaping the Martian uplands. *Proceedings of the Nineteenth Lunar and Planetary Science Conference*, G. Ryder (ed.), 355–65. Cambridge: Cambridge University Press (on behalf of the Lunar and Planetary Institute, Houston, Texas).
Willemann, R. J. & D. L. Turcotte 1982. The role of lithospheric stress in the support of the Tharsis Rise. *Journal of Geophysical Research* **87**, 9793–801.
Williams, R. S. 1978. Geomorphic processes in Iceland and on Mars: a comparative appraisal from orbital images. *Geological Society of*

America 91st annual meeting, abstracts with programs, 517.

Wilson, L. & J. W. Head 1993. Mars: review and analysis of eruption theory and relationships to observed landforms. *Reviews of Geophysics* **32**, 221–64.

Wise, D. U. 1979. *Geologic map of the Acidalia quadrangle of Mars.* Map I-1154, USGS, Flagstaff, Arizona.

Wise, D. U., M. P. Golombek, G. E. McGill 1979. Tharsis province of Mars: geologic sequence, geometry, and a deformation mechanism. *Icarus* **38**, 456–72.

Witbeek, N. E., K. L. Tanaka, D. H. Scott 1991. *Geologic map of the Valles Marineris region, Mars.* Map I-2010, USGS, Flagstaff, Arizona.

Wolff, M. J., S. W. Lee, R. T. Clancy, L. J. Martin, J. F. Bell, P. B. James 1997. 1995 observations of Martian dust storms using the Hubble space telescope. *Journal of Geophysical Research* **102**, 1679–92.

Wood, C. A. 1980. New observations of Martian basins. *Abstracts of the Tenth Lunar and Planetary Science Conference*, 1271–2. Lunar and Planetary Institute, Houston, Texas.

Wood, C. A. & J. W. Head 1976. Comparison of impact basins on Mercury, Mars and the Moon. *Proceedings of the Seventh Lunar and Planetary Science Conference*, G. Black (ed.), 3629–51. New York: Pergamon.

Wood, C. A., J. W. Head, M. J. Cintala 1978. Interior morphology of fresh Martian craters: the effects of target characteristics. *Proceedings of the Ninth Lunar and Planetary Science Conference*, 3691–709. New York: Pergamon.

Woronow, A. 1981. Preflow stresses in Martian rampart ejecta blankets: a means of estimating the water content. *Icarus* **45**, 320–30.

Wu, S. S. C., P. A. Garcia and A. Howington-Kraus,. 1991. Volumetric determination of Valles Marineris, Mars [abstract]. *Twenty-first Lunar and Planetary Science Conference*, 1357–8, Lunar and Planetary Institute, Houston, Texas.

Wüllner, U. 1996. *Konventien unter lateralen, oberflächenahen Temperaturanomalien: bedeutung für der Oberfläschenheterogenitäten auf Mars und Mond.* PhD thesis, Westfälische-Wilhelms-Universität, Münster.

Yen, A. S., B. C. Murray, G. R. Rossman 1998. Water content of Martian soil: laboratory simulations of reflectance systems. *Journal of Geophysical Research* **103**, 11125–34.

Zimbelman J. R. 1985. Estimates of rheologic properties for flows on the Martian volcano Ascraeus Mons. In *Proceedings of the Sixteenth Lunar and Planetary Science Conference*, pp. D157–62, special issue of *Journal of Geophysical Research* **90**.

Zimbelman, J. R., S. C. Solomon, V. L. Sharpton (eds) 1988. *MEVTV workshop on the nature and composition of surface units on Mars.* Technical Report 88-05, Lunar and Planetary Institute, Houston, Texas.

Zimbelman, J. R., R. A. Craddock, R. Greeley, R. O. Kuzmin 1997. Volatile history of Mangala Valles, Mars. *Journal of Geophysical Research* **97**, 18309–318.

Zimbelman, J. R., S. M. Clifford, S. W. Williams 1998. Terrain softening revisited: photogeological considerations. *Abstracts of the Eighteenth Lunar and Planetary Science Conference*, 1321–2. Lunar and Planetary Institute, Houston, Texas.

Zisk, S. H., P. J. Mouginis-Mark, J. M. Goldspell, M. A. Slade, R. F. Jurgens 1992. Valley systems on Tyrrhena Patera, Mars: Earth-based radar measurements of slopes. *Icarus* **96**, 226–33.

Zuber, M. T. & L. L. Aist 1990. The shallow structure of the Martian lithosphere in the vicinity of the ridged plains. *Journal of Geophysical Research* **95**, 1215–30.

Zuber, M. T. & D. E. Smith 1997. Mars without Tharsis. *Journal of Geophysical Research* **102**, 28673–86.

Zuber, M. T. and 20 co-authors 1998. Observations of the north polar region of Mars from the Mars orbiter laser altimeter. *Science* **282**, 2053–2060.

Zurek, R. W. and 8 co-authors 1992. Dynamics of the atmosphere of Mars. See Kieffer et al. (1992: 835–933).

Index